Animal Spaces, Beastly Places

In a myriad of ways, animals help make up the societies in which we humans live. People eat animals, wear products made from them, watch them in zoos or on television, keep them in their houses and in factory farms, hunt them and experiment on them, and place them in mythology and stories.

Animal Spaces, Beastly Places examines how animals interact with people in different ways. Through a wide and comprehensive range of examples, which range from feral cats and wild wolves to domestic animals and intensively farmed cattle, the contributors explore the complex relations through which humans and non-human animals are mixed together. Our practices involving animals range from those of love and compassion to untold cruelty, force, violence and power. As humans, we have placed different animals into different categories according to notions of species, usefulness, domesticity or wildness. As a result of these varying and often contested orderings, animals are assigned to particular worldly spaces wherein they are supposed to remain.

Animal Spaces, Beastly Places shows us that there are many variations in the spatiality of human–animal orderings, within and across cultures, and over time. It develops new ways of thinking about human–animal interactions and encourages us to find better ways for humans and animals to live together.

Chris Philo is Professor of Geography at the University of Glasgow and **Chris Wilbert** is a Research Fellow at the School of Social and Human Sciences at City University, London.

Critical Geographies

Edited by **Tracey Skelton**, *Lecturer in International Studies, Nottingham Trent University* and **Gill Valentine**, *Professor of Geography, The University of Sheffield.*

This series offers cutting-edge research organised into four themes: concepts, scale, transformations and work. It is aimed at upper-level undergraduates, research students and academics, and will facilitate inter-disciplinary engagement between geography and other social sciences. It provides a forum for the innovative and vibrant debates which span the broad spectrum of this discipline.

Animal Spaces, Beastly Places

New geographies of human–animal relations

Edited by Chris Philo and Chris Wilbert

Lond

First published 2000
by Routledge
11 New Fetter Lane, London EC4P 4EE

Simultaneously published in the USA and Canada
by Routledge
29 West 35th Street, New York, NY 10001

Routledge is an imprint of the Taylor & Francis Group

© 2000 Selection and editorial material Chris Philo and Chris Wilbert; individual
chapters the contributors

Typeset in Perpetua by Taylor & Francis Books Ltd
Printed and bound in Great Britain by Biddles Ltd, Guildford and King's Lynn

British Library Cataloguing in Publication Data
A catalogue record for this book is available from the British Library

Library of Congress Cataloging in Publication Data
Animal spaces, beastly places: new geographies of human–animal relations / Edited
by Chris Philo and Chris Wilbert.
p. cm. – (Critical geographies)
Includes bibliographical references.
1. Human–animal relationships. I. Philo, Chris. II. Wilbert, Chris. III. Series.

QL85 .A52 2000
304.2'7–dc21

99-089955

ISBN 0–415–19846–1 (hbk)
ISBN 0–415–19847–X (pbk)

Contents

vi   *Contents*

Figures

Contributors

Alec Brownlow is a PhD student in Geography at Clark University's Graduate School of Geography, where he is investigating the social and cultural issues involved in landscape and ecological restoration. His wider interests include nature–society theory, theories of restoration, and social and cultural critiques of science. Wildlife ecologist turned human geographer, he lives with his dog, Maggie.

Gail Davies is a Lecturer in Geography at University College London (UCL), where she teaches on the cultural geographies of science and nature. She received her PhD from UCL in 1998 for her thesis entitled 'Networks of Nature: Stories of Natural History Filmmaking from the BBC'.

Nick Evans is a Senior Lecturer in Geography at University College, Worcester (UCW). He has teaching and research interests in agricultural geography. Recent research projects have included an investigation of environmentally friendly farming schemes in Britain, the impact of BSE on SSSIs, and the geography of rare breeds in the British Isles.

Huw Griffiths is a Senior Lecturer in Geography at the University of Hull. He is an animal biogeographer. Much of his work to date has been on conservation-related attitudes to predatory species, and he is presently involved in projects on the perception of wild space, on brown bears, and on Slovenian dormouse hunting clubs and landscape interactions.

Pyrs Gruffudd is a Lecturer in Geography at the University of Wales, Swansea. He has written on landscape and national identity in inter-war Wales, and is currently researching the relationships between modernist architecture and discourses of nature and health in the 1930s.

Philip Howell is a Lecturer in Geography at the University of Cambridge. He works in the field of nineteenth-century cultural and historical geography, and his principal current research concerns the role of sexuality in the construction of public space. He does not own a dog.

Owain Jones is a Research Fellow at the School of Geographical Sciences, University of Bristol, and Visiting Lecturer at the School of Geography, University of the West of England. He has published articles on childhood, gender, nature and space; discourses of rurality; childhood and rurality; governance and rurality (with Jo Little); and trees, nature, time, place and landscape (with Paul Cloke).

Unna Lassiter is a PhD student in Geography at the University of Southern California. She is working on a dissertation concerned with cultural diversity and attitudes towards marine wildlife.

David Matless is a Lecturer in Geography at the University of Nottingham. He is the author of *Landscape and Englishness* (1998: Reaktion) and co-editor of *The Place of Music* (1998: Guilford). His current research interests include the cultures of geography in the early twentieth century, geographies of nature and the body, and the cultural landscape of the Norfolk Broads since the late nineteenth century.

Chris Philo is a Professor of Geography at the University of Glasgow. He is interested in the geographies of mental health subjects, the social geography of 'outsiders', animal geographies, and other aspects of human geography's history and theory. He has recently co-edited both *Body Cultures: Essays on Sport, Space and Identity by Henning Eichberg* (1998: Routledge) and *Entanglements of Power: Geographies of Domination/Resistance* (1999: Routledge).

Ingrid Poulter graduated in psychology from the University of Hull in 1994. After a short 'career' in marketing, she returned to Hull as a Research Assistant in 1996 and worked on a number of inter-disciplinary projects. She is currently employed as a Community Health Promotions Worker in South Shields, Tyne and Wear.

James R. Ryan is a Lecturer in Human Geography at the Queen's University of Belfast. His most recent book is *Picturing Empire: Photography and the Visualisation of the British Empire* (1997: Reaktion).

David Sibley is a Reader in Geography at the University of Hull. He is trying to develop new perspectives on space which draw upon psychoanalytical arguments about aversions and desires. This obsessive compulsive interest came initially from research on Gypsy spaces, and later projects have considered childhood, the production of knowledge and, currently, bereavement and animal geographies. He has recently authored *Geographies of Exclusion: Society and Difference in the West* (1995: Routledge).

Paul Waley is a Lecturer in Geography at the University of Leeds. He is one of a small number of geographers in Britain specialising in Japan, and has spent over ten years in Tokyo at different times since 1977, working as an editor, translator and writer, as well as conducting research there. Current research includes work on the historical geography of urban areas in Japan and planning issues in contemporary Japanese cities.

Michael J. Watts is Director of the Institute of International Studies and Chancellor's Professor of Geography and Development Studies at the University of California, where he has taught for twenty years. He is currently working on, among other things, questions of globalisation and the politics of Islam.

Chris Wilbert is currently a Post-Doctoral Research Fellow in the School of Social and Human Sciences at City University, London. He is the author of various articles on political ecology, new technology and leisure, and problems surrounding 'community participation'. His current research examines relations between experts and lay people in natural history and ecology.

Jennifer Wolch is a Professor of Geography at the University of Southern California, where she studies human–animal interactions in cities and also human attitudes towards animals. Her co-edited book *Animal Geographies: Place, Politics and Identity in the Nature–Culture Borderlands* was published by Verso in 1998.

Michael Woods is a Lecturer in Human Geography in the Institute of Geography and Earth Sciences at the University of Wales, Aberystwyth, having previously studied at the University of Wales, Lampeter and the University of Bristol. His major research interests are in rural politics and cultural perspectives on political geography.

Richard Yarwood is a Lecturer in Geography at University College, Worcester (UCW), and Director of the Centre for Rural Research at UCW. He has teaching and research interests in rural and social geography. Recent research projects have dealt with neighbourhood watch and the fear of crime in rural places, affordable housing in rural areas, and the geography of rare breeds in the British Isles.

Preface and acknowledgements

This book originated in a conference session on 'Animals, Agency and Geography', organised by ourselves and sponsored by the Social and Cultural Geography Research Group of the Institute of British Geographers, which was held at the Annual Conference of the Royal Geographical Society and Institute of British Geographers, University of Exeter, in January 1997. Huge thanks are due to all those who gave papers and otherwise participated in this session. We decided that a volume of essays growing out of this session would aptly capture the sense of an emerging 'new' animal geography, and would complement the volume then being planned by Jennifer Wolch and Jody Emel (now published as the excellent *Animal Geographies: Place, Politics and Identity in the Nature–Culture Borderlands*, London: Verso, 1998). Many of the original speakers at the Exeter session were able to be involved, but in every case their chapters below are substantially revised, lengthened and updated versions of what they presented back in January 1997. Several additional people have come into the project subsequently, offering chapters which have widened the scope of the book still further, and we are also delighted to be able to include the Afterword by Michael Watts. We feel that the book as now published makes a significant and fresh contribution to research on the 'placement' of animals relative to the everyday spaces of human activity, as well as signalling the challenge of conceptualising the 'agency' of animals in evading, slipping out of or messing up these human placements. In addition, we feel that ethical and political questions to do with the possibilities for more equitably shared spaces, ones occupied conjointly by both humans and animals, are raised and even, on occasion, answered. We certainly hope that readers will enjoy reading the contributions here as much as we have.

We must therefore say an enormous thank you to all of our contributors: we have been overwhelmed by their enthusiasm for the project, by their readiness to take on board our editorial suggestions, however off-beam, and by their hitting of deadlines when commonly trying to meet many other commitments at the same time. Special thanks must go to Jennifer Wolch, though, for her advice at every stage of the project. We must also thank Tracey Skelton and Gill Valentine for their interest, and for inviting us to put the book in their 'Critical Geographies' series. Similarly, thanks are due to various people at Routledge for their assistance and patience, namely Allison Bell, Sally Carter, Sarah Carty, Andrew Mould and Ann Michael, and also to Justin Dyer for his careful copy-editing work. More generally, we would like to thank all of those friends and colleagues who have not thought it totally odd for us to be human geographers fascinated by animal worlds.

The editors and publishers would also like to thank the following for granting permission to reproduce images in this work:

Robert Thompson Artworks, Bath, for permission to reproduce the cartoon 'And this little piggy…'.

The Syndics of Cambridge University Library, Cambridge, for permission to reproduce the illustration 'No dog carts', and also in association with the Provost and Fellows of Eton College, Eton, for permission to reproduce the illustration Elizabeth Barrett Browning and Flush.

David Hancock, Director of the Shugborough, Staffordshire, for permission to reproduce Briton Riviere's Temptation (1879).

The Royal Society for the Prevention of Cruelty to Animals, Horsham, for permission to reproduce the advertisement 'It's hard to swim when you've been running for three hours', and the League Against Cruel Sports, London, for reproduction of the photograph used in the advertisement.

The International Fund for Animal Welfare, Crowborough, for permission to reproduce the advertisement 'The effects of a pack of twenty on an expectant mother', and the League Against Cruel Sports, London, for reproduction of the photograph used in the advertisement.

The Royal Geographical Society with the Institute of British Geographers, London, for permission to reproduce the illustration of Dallmeyer's telescopes and the photograph 'From the Settite & Royan R.s, N.E. Afr. 1876'.

The Architect's Journal, London, for permission to reproduce the photographs of the Penguin Pool at London Zoo and The Descent of Man.

The Royal Institute of British Architects (RIBA) for permission to reproduce the photograph of the home for apes at the London Zoo.

At-Bristol Images, Bristol, for permission to reproduce the computer-generated projection of Wildscreen@Bristol.

Every effort has been made to contact copyright holders for their permission to reprint material in this book. The publishers would be grateful to hear from any copyright holder who is not here acknowledged and will undertake to rectify any errors or omissions in future editions of this book.

Chris Philo and Chris Wilbert
Glasgow and London, January 2000

1 Animal spaces, beastly places

An introduction

Chris Philo and Chris Wilbert

Human–animal relations and the new animal geography

> Kahuzi Beiga [national park in eastern Congo] was the birthplace of gorilla tourism and the place where Dian Fossey, the anthropologist and subject of the film *Gorillas in the Mist*, first encountered the highland gorilla. ... But in the 1990s the park's once healthy population of gorillas has been severely reduced by poachers and Rwandan rebels, who use it as a hideout. ... Twenty gorillas have been killed since April, and the days when tourists rubbed shoulders and shook hands with these magnificent creatures are now long gone. ... Basengezi Katintima, the governor of South Kiva province, where the park is situated, said: 'In Rwanda they are talking about a human genocide, but here we are talking about an animal genocide.'
>
> (Gough 1999: 13)

> Scientists are secretly killing up to nine million laboratory animals a year because too many are being bred for research, an investigation has found. ... Last year more than 6.5 million mice and around 2,400,000 rats were culled. And more than 1,000 laboratory dogs, an estimated 25 percent, were killed because they were not needed. Most were bred for testing pharmaceuticals and spent their whole lives in kennels waiting to be experimented upon. [One anti-vivisectionist] said: 'This waste of life is totally outrageous. ... This is the hidden side of the research industry.'
>
> (Woolf 1999: 6)

> Scientists have discovered a startling new source of air pollution: pig and chicken farms. ... They have found that nitrogen emissions from units for intensively rearing animals are killing woods and forests at the same rate as the effects of industrial pollution. ... The emissions – most of them from animal farms' growing piles of manure – are causing serious damage to woodland in some areas. In Denmark and Holland, where large pig and chicken farms are a major industry, precious heathlands are being destroyed.
>
> (McKie 1999: 4)

These three extracts from recent newspaper stories show some of the myriad ways in which human relations with the world of non-human animals[1] can become the focus of attention, and all three point to the importance of the *spaces* and *places* bound up in the human–animal relations under scrutiny.[2] The first extract tells of a place which had initially witnessed one of the first sustained encounters between humans and gorillas, a remote and wild place, now undergoing serious changes: first as a designated national park encouraging a 'gorilla tourism' entertaining casual encounters between tourists and gorillas, and then as an environment for an 'animal genocide' occurring in the shadow of humans waging war. From being a 'natural' place where the gorillas remained relatively undisturbed by humans,[3] this place has become, variously, the scientist's site of fieldwork, the naturalist's site of biological conservation, the entrepreneur's site of capital accumulation, the poacher's site of prey, and the soldier's site of refuge. Different sets of humans possessing differing purposes and technologies have hence flowed into and out of this East African region, reflecting broader geographies of science, state intervention, capitalism, colonialism, politics and human struggle, and in the process they have shaped widely divergent kinds of human–animal relations (see also Haraway 1989: 263–268).

The second newspaper extract tells of a type of space, the scientific laboratory, which has seemingly seen the mass extermination of mice, rats, dogs and other mammals simply because many more of them have been bred than are actually 'necessary' for the conduct of pharmaceutical scientific experiments. The impression is of a whole hidden geography of such laboratories and breeding stations, tucked away in countryside complexes and university campuses, wherein many animals live and die as part of a highly unequal human–animal relation predicated on the utility, adaptability and expendability of the animals so incarcerated. Questions about science, state intervention (or lack of it) and capitalist industry are obviously again to the fore, as well as those of ethics, welfare and politics.

The third extract then tells of how animal products, their bodily wastes in this case, may have effects which can diffuse beyond the bounds of the spaces where they are immediately present, creating a spatial connection from the pigs and chickens on their farms to a range of environments beyond the farm boundaries (a 'spatial externality' neatly underlined in a cartoon accompanying the article: see Figure 1.1). In this instance, a complex human–animal relation is established which does not operate solely through the physical proximity of humans and animals, but rather entails a spread-out geography through which animals are able to have an effect on humans at-a-distance. Wider questions, for example about private property, the byproducts of economic activity, and the duty of the state to regulate agricultural activities in the interest of preventing pollution and preserving heathlands, are once more deeply implicated.

Humans are always, and have always been, enmeshed in social relations with animals to the extent that the latter, the animals, are undoubtedly constitutive of human societies in all sorts of ways. Humans are ecologically dependent on animals, principally as sources of food, clothing and many other materials which

Figure 1.1 'And this little piggy …'

Source: *Observer* (1999) 25 July, p. 4, © Robert Thompson Artworks

sustain 'our'[4] human existence, which means that animals, especially dead ones, enter centrally into what humans can themselves be and do in the world (Cockburn 1996; Cole and Ronning 1974). Rimbaud once declared that '[t]he man [*sic*] of the future will be filled with animals' (quoted by Deleuze in Rabinow 1992: 236); and, with the rise of global meat-eating, xenotransplantations of organs, and blood products grown in the bodies of animals, such an observation is gaining even more truth than Rimbaud could have anticipated.[5] In return, of course, animals are dramatically affected by the actions of humans, not least through farming and, increasingly, genetic engineering. Indeed, humans past and present have radically changed the life conditions of all manner of animals, whether pets, livestock or wild (Benton 1993: 68–69; see also Watts, this volume). Human–animal relations have hence been filled with power, commonly the wielding of an oppressive, dominating power by humans over animals, and only in relatively small measure have animals been able to evade this domination or to become themselves dominant over local humans. Examples can be adduced of the latter, such as plagues of locusts, rampaging elephants, or perhaps the ramifications of BSE or 'mad cow disease'.[6] Yet, usually animals have been the relatively powerless and marginalised 'other' partner in human–animal relations. What surely cannot be denied is the historical and global significance of such human–animal relations for both parties to the relationship – to be sure, they commonly entail matters of life and death for both parties, the animals in particular – and any social science which fails to pay at least some attention to these relations, to their differential constitutions and implications, is arguably deficient.

A 'new' animal geography has emerged to explore the dimensions of space and place which cannot but sit at the heart of these relations, and contributions here are now running alongside more established anthropological, sociological and psychological investigations into human–animal relations (see the journal *Society and Animals*). An older engagement between academic geography and the subject-matter of animals, often cast as 'zoogeography', had been preoccupied with mapping the distributions of animals – describing and sometimes striving to explain their spatial patterns and place associations – and in so doing it had tended to regard animals as 'natural' objects to be studied in isolation from their human neighbours. A branch of geographical inquiry expressly named as 'animal geography' secured a modest foothold for itself, notably when it was positioned as one of the 'systematic geographies' by Hartshorne in his famous diagram of the discipline's logical structure (Hartshorne 1939: 147, Figure 1). The importance to animal geography of the human influence on animal lives was noted in a few papers published in the 1950s and 1960s, particularly in Bennett's (1960) explicit call for a 'cultural animal geography' drawing upon the interest of the Berkeley School in themes such as animal domestication, but the connections to research in human geography (as opposed to physical geography) remained tenuous. In fact, Davies (1961: 412) complained that the concerns of animal geography remained 'too remote from the central problems of human geography'. It was therefore not until much more recently that geographers in any number began to recognise the possibilities for, and indeed socio-ecological importance of, a revived animal geography which would focus squarely on the complex entanglings of human–animal relations with space, place, location, environment and landscape. The publication of the 'Bringing the animals back in' theme issue of the journal *Society and Space* (Wolch and Emel 1995) was a landmark in this respect, to be followed by an edited book drawing upon contributions to this theme issue and adding others (Wolch and Emel 1998), and also by a theme issue of the journal *Society and Animals* (Wolch and Philo 1998) introducing a geographical perspective to a wider, inter-disciplinary audience.[7]

This history of animal geography and account of an emerging new animal geography have already been reviewed in some detail elsewhere (Philo 1995: 658–664; Philo and Wolch 1998: 104–108), and there is no need for us to cover this ground any further. Suffice to say that the emphasis now is indeed on excavating the kinds of networks of human–animal relations sketched out in the opening examples, tracing their 'topologies' (Whatmore and Thorne 1998), and showing how the spaces and places involved make a difference to the very constitution of the relations in play. The present volume should also be seen as a sustained attempt to contribute further to this small, but we would argue significant, corpus of research and scholarship.

One of the things a new animal geography seeks to do, moreover, is to follow how animals have been socially defined, used as food, labelled as pets or pests, as useful or not, classed as sentient, as fish, as insect, or as irrational 'others' which are evidently not human, by differing peoples in differing periods and worldly contexts. It thereby endeavours to discern the many ways in which animals are 'placed' by human societies in their local material spaces (settlements, fields, farms, factories, and so on), as well as in a host of imaginary, literary, psychological and even virtual spaces. It is thus not only the physical presence of animals which is of importance here, since animals also exist in our human imaginings – in the spoken and written spaces of folklore, nursery rhymes, novels and treatises; in the virtual spaces of television or cinema, in cartoons and animation – while they are also used as symbols to sell a huge variety of commodities and products (Rowland 1973; Wilbert 1993). All such imaginings of animals, as bound up with human uses made of them, must be seen as affected deeply by the form of 'animal–human mode of production' underlying the specific society in question, whether it be hunter-gatherer, feudal, industrial, capitalist, post-industrial or whatever (Tapper 1994).

Such an orientation, looking at how animals are imagined or represented in human societies, is only an element of a larger picture. If we concentrate solely on how animals are represented, the impression is that animals are merely passive surfaces on to which human groups inscribe imaginings and orderings of all kinds. In our view, it is also vital to give credence to the practices that are folded into the making of representations, and – at the core of the matter – to ask how animals themselves may figure in these practices. This question duly raises broader concerns about non-human agency, about the agency of animals, and the extent to which we can say that animals destabilise, transgress or even resist our human orderings, including spatial ones. Noske's (1989: 169) query, which she frames in terms of anthropology, can hence be paraphrased for geography: that is, can a 'real' geography of animals be developed, rather than an anthropocentric geography of *humans in relation to animals*? It is around precisely such themes that our collection of essays is composed, as we will now elaborate in the remainder of this introductory chapter.

'Proper places'; or specifying 'animal spaces'

> I do not like animals. Of any sort. I don't even like the idea of animals. Animals are no
> friends of mine. They are not welcome in my house. They occupy no space in my heart.
> Animals are off my list. … I might more accurately state that I do not like animals, with
> two exceptions. The first being in the past tense, at which point I like them just fine, in the
> form of nice crispy spareribs and Bass Weejun penny loafers. And the second being outside,
> by which I mean not merely outside, as in outside my house, but genuinely outside, as in
> outside in the woods, or preferably outside in the South American jungle. This is, after all,
> only fair. I don't go there; why should they come here?
>
> (Fran Lebowitz in Metcalf 1986: 16)

A key concern for the new animal geography, and particularly for this volume, is
the geographical conception of place in relation to animals (see also Philo and
Wolch 1998: 111–113). There are two main senses of place to be explored here,
between which there is a close, sometimes inseparable, connection. The first
refers to the 'place' which a particular animal, a given species of animal or even
non-human animals in general can be said to possess in human classifications or
orderings of the world. We can follow de Certeau (1984: 117), who argues as
follows:

> A place (*lieu*) is the order (of whatever kind) in accord with which elements
> are distributed in relationships of coexistence. It thus excludes the possibility
> of two things being in the same location (*place*). The law of the 'proper' rules
> in the place: the elements taken into consideration are *beside* one another,
> each situated in its own 'proper' and distinct location, a location it defines. A
> place is thus an instantaneous configuration of positions. It implies an indica-
> tion of stability.

The emphasis is on the setting up of classificatory schemes wherein each identified thing has its own 'proper place' relative to all other things, and can be neatly
identified, delimited and positioned in the relevant conceptual space so as to be
separate from, and not overlapping with, other things there identified, delimited
and positioned. Such a conceptual placing of animals in the wider 'scheme of
things' – such a specifying ('species-identifying') of animals – reflects an impulse
with deep roots and wide cultural diversity (Foucault 1970). These go back to
both pre-Neolithic totemic societies (Shepard 1993) and biblical classifications of
the different beasts (as in Leviticus: see Sibley 1995: 37), recur in ancient and
medieval 'great chain of being' thinking (Lovejoy 1936), and then reappear in
Linneaus's *Systema naturae* and countless more recent tabular representations of
the natural world (Frangsmyr 1983; Spary 1996). The result of such classifications, systems and tables is to fix animals in a series of abstract spaces, 'animal

spaces', which are cleaved apart from the messy time–space contexts, or concrete places, in which these animals actually live out their lives as beings in the world. (It might be added that most people do not necessarily live their lives through such abstract classifications and representations; such placings are indeed always unstable, relational productions, formed through diverse actions and interactions, as many of the chapters below clearly demonstrate.) Furthermore, as well as establishing the logical standings of different animals *vis-à-vis* one another – as in the biblical distinguishing of 'clean' animals from the 'unclean' – the conceptual work involved in these systems always runs up against the inevitable and fundamental question of just 'what is an animal?' Since this is also a question that cannot be avoided in our introductory chapter, we will now briefly examine what it entails – together with related points about the contesting of knowledges about animals – before returning to the theme of placing animals.

'What is an animal?' and related matters

The conceptual placing of animals is first about deciding what is or is not an animal, and such a question is pertinent to animal geography because how it is answered will in part determine the purview of the subfield, and also because new animal geographers must be aware of how different human societies have chosen to answer the question. So, where and how do the lines get drawn, particularly between humans and non-human animals? Do we (as animal geographers) follow systematics in constructing groupings based on multiple correlations of common bodily features, or certain molecular biologists in using chromosomal studies? Do we follow the philosophers who speculate on the differing capacities for agency between humans and animals (see below), or do we take more commonsense approaches such as those of animal rights campaigners who tend to limit the discussion of rights to 'sentient' beings which can 'sense' their environing worlds? Most people may be happy to include birds and reptiles as animals,[8] but not everyone includes fish or other non-mammalian sea creatures, and many people who regard themselves as vegetarians nonetheless eat fish because they do not regard them as 'proper' animals (Willetts 1997). For the record, we personally *would* include fish as animals, and we are pleased that in the present collection Waley considers fish in Japanese rivers while Jones offers a brief discussion of encountering fish and their aquatic spaces.[9] Fewer people still may be happy to count insects and other creatures with an external skeleton as animals, perhaps because they are so small and numerous that it is impossible to distinguish individuals among them in a fashion supposedly essential for the conferment of 'animalhood' (see also Jones, this volume). Again, we personally *would* include insects as animals, and would wish to see more work done on insect geographies which goes beyond merely a zoogeographical concern for their distributions.[10] Very few people may then wish to go so far as to include non-humans such as viruses in the category of animals, since this would amount to incorporating anything that is animate, thereby raising additional thorny questions about the dividing lines between animals, plants and organisms. We are attracted to the prospect of this extreme manoeuvre, reckoning that any animal geography must itself have no easy end to its compass, but at the same time we are aware that there are good reasons – conceptual, empirical, perhaps political and ethical – for needing a limit beyond which we are not prepared to extend the label of 'animal'. Whatever, we would certainly still acknowledge the need for geographers to take seriously plants and organisms, and especially viruses, if necessary in other subfields of geographical study.[11]

Answers to such questions can never really be fully decided, and so it seems sensible to allow the 'what is an animal?' question to remain partially open. A human-geographical approach to animals should also be alert to the differing ways in which this question is answered within different human societies, as tied into their particular economic, political, social, cultural and psychological horizons. Whether the humans involved be primatologists in Japan (Haraway 1989), fishers in Iceland (Einarsson 1993), Mende villagers of Sierra Leone (Richards 1993) or African-American women in Los Angeles (Wolch, Brownlow and Lassiter, this volume), understandings of what comprises an animal will be accepted tacitly, and sometimes even made explicit, but will be formulated more or less *differently* in each case. In other words, there will be a complex geography to these answers, and a new animal geography must be interested in these classificatory variations across the regions, localities and sites of the world.

To accept this relativity, however, is not to overlook the ways in which what come to be regarded as legitimate knowledges about animals are constructed and contested. For example, the natural sciences have for some time been regarded as *the* legitimate and primary form of knowledge in many societies, Western and non-Western (but see Waley, this volume). These sciences have shown themselves to be very effective in manipulating matter and in helping to develop productive relations, at least since the late nineteenth century (Jacobs 1997), and it has been argued that the centre of production has now become the scientific laboratory (Latour 1983). Such sciences, specifically the biological sciences, have often dovetailed with strains of philosophising about the agency of animals, creating in the process rigid orthodoxies regarding how distinctions should be drawn between humans and animals, and within the animal 'kingdom' itself. In comparison to these sciences and philosophies, lay and 'indigenous' knowledges (or ethnosciences), whether developed by the peoples of non-Western countries or by the 'lower classes' of Western countries, have long been derided by both dominant Western scientific cultures and the 'rational' institutions which adopt their logics as a guide and model. Yet in recent years there has been a growing interest in, and appreciation of, these 'other' forms of knowledge that have been for so long undervalued during phases of imperialism and (neo-)colonialism. This change has emerged partly from environmental movements and related anti-development organisations and politics (Leff 1995; Shiva 1989; Watts 1998), and one conse-quence has been a new basket of ideas for thinking about the status and character of animals (Baudrillard 1994: 133–134). Ingold (1994a), among others, argues that people in the developed world can learn much about different relations with animals from hunter-gatherer cultures, while a commentator like Shepard (1993) goes further to valorise hunter-gatherer relations with animals as being more 'mature' than those currently prevailing in the West. Elaborating, Ingold (1994a) suggests that hunter-gatherers generally view the distinction between humans and animals as permeable and easily crossed, unlike what is taken as read in much Western science and philosophy. Within this 'other' cosmology, humans and animals are supposed to coexist in a relation of trust, rather than one of enmity, which means that, if humans behave well towards animals, the latter can be trusted to provide for humans, to give their lives for human sustenance. Although those peoples who live a life of 'being-with' (as opposed to 'against') animals may genuinely have much to teach urban Western societies, it must also be acknowl-edged that those very peoples are themselves incredibly marginalised economically, politically, socially and culturally in global spaces (Shiva and Mies 1993; Tapper 1994: 56).

The domination of certain knowledges over others, as just discussed, also points to related divisions which have been constructed between experts, elites and what we might term lay, amateur or even popular knowledges, all of which have consequences for which understandings of animals, and of what is an animal, become sanctioned as 'proper'. All sorts of boundary-work are involved in social struggles over which group has authority, and hence over which form of knowing is taken as legitimate, and the participants in these struggles obviously all portray themselves – and seek to persuade others to portray them – as the relevant 'experts' in the field (Gieryn 1995). At one level of generalisation, it might be claimed that the legitimate spokespersons for animals have become the biological sciences, as just indicated, while the spokespersons of the human world have become those known as politicians, at least in the more developed areas of the world. This is a key division within what Latour (1993: Chap. 2) has termed 'the modern constitution'.[12]

Yet the divisions between expert and lay knowledges, or between elite and popular, are often not so clear-cut and opposed as such a binary implies (Chartier 1984). For instance, pro- and anti-hunting groups vie for recognition as the 'right' spokespeople for certain animals, as do anti- and pro-vivisectionists, and in both cases it is possible to find strange admixtures of expert, lay, political and moral discourses on both sides of the anti and pro schisms (see also Woods, this volume). Similarly, in Britain a distinction which arose in the later nineteenth century between 'amateur' naturalists undertaking field studies and 'professional' scientists undertaking laboratory-based experimental work would appear to reflect a rigorous separation of powers, of purposes, in at least some of the biological sciences. In practice, though, it is clear that the experts in their laboratories had to undertake fairly large amounts of fieldwork, especially in the new field and oceanographic research stations of the time. Moreover, professionals also often had to rely upon amateur naturalists for conducting surveys of populations and distributions of birds and other organisms. The British ecologist Charles Elton (1933) even remarked on the usefulness of the public in collecting raw data on animals for the scientist, just as the public could be useful for the police when seeking evidence to help solve a crime. This being said, the likes of Elton did not envisage a meeting of equals, since they still operated with a hierarchy of knowledge positioning the amateur's *local* knowledge or 'wisdom' as less valuable than the more *general* formulations of the biological sciences. (A more specific account of such a struggle over 'natural' knowledge, rooted in the English Broadlands, is rehearsed in Matless, this volume.) Notwithstanding this endemic hierarchy of which knowledges are to be valued, it can still be observed that lay peoples or popular cultures do generate their *own* knowledges, and that these can differ from, and even oppose, expert and elite science. Moreover, these knowledges may embrace understandings of animals which end up contesting powerful orthodoxies (Cooter and Pumfrey 1994: 248–250), while non-expert peoples may also use and interpret expert knowledges of animals in ways running counter to what the popularisers of expert knowledges might wish.[13]

Related to the conceptual placing of animals is also a strong human sense of the proper places which animals should occupy physically. An important link in this respect is the ecological concept of the 'niche', in which an organism is said to have its own distinctive place within both biological classifications and worldly environments, in which case the placement is envisaged as being not only conceptual but also material (Kingsland 1991).[14] It is not only a science such as ecology that embraces this double placement of animals, however, since many different forms of human discourse – economic, political, social and cultural – include a strong envisaging of both where animals are placed in the abstract 'scheme of things' and where they should be found in the non-discursive spaces and places of the world. Although it is now somewhat overused, we propose to press into service once again Said's (1978: 54–55, 71–72) famous term 'imaginative geography', and to suggest that many human discourses contain within them a definite imaginative geography serving to position 'them' (animals) relative to 'us' (humans) in a fashion that links a conceptual 'othering' (setting them apart from us in terms of character traits) to a geographical 'othering' (fixing them in worldly places and spaces different from those that we humans tend to occupy). In fact, we wish to propose that there are a number of overlapping imaginative geographies at work here, depending on exactly whose imaginative projections are under consideration, exactly where in the world is being talked about, exactly what kinds of animals are at stake, and so on.

For some people, as in the opening quote above from Lebowitz, the favoured situation is one facilitating as much distance as possible between them and animals of any sort. Unless said animals appear in the form of food, the suggestion is that animals should definitely remain 'outside' the perimeter of human existence, banished from the vicinity of everyday human life and work. It might be argued, though, that many people (in Western societies at least) internalise a slightly more complex baseline imaginative geography of animals – or at least of mammals – which effectively maps increasing distance from humans on to changes in the sorts of animals which should be present, implying that some species should properly be proximate to us while others should properly be more remote. Thus, zones of human settlement ('the city') are envisaged as the province of pets or 'companion animals' (such as cats and dogs), zones of agricultural activity ('the countryside') are envisaged as the province of livestock animals (such as sheep and cows), and zones of unoccupied lands beyond the margins of settlement and agriculture ('the wilderness'[15]) are envisaged as the province of wild animals (such as wolves and lions). It is this spatial structure which organises the progression of the chapters below, the spatial referents of which loosely shift from the city to the countryside and on to wilder-lands.

As these chapters nevertheless make clear, the imaginative geography described above is scrambled, and destabilised, in various ways which make the conjoint conceptual–physical placements involved both more interesting and more difficult to decode. Thinking first about the urban chapters, Howell stresses the equation of pets with the (gendered) domestic hearth and the well-to-do neighbourhoods of Victorian London, but not with the whole city, whereas Griffiths, Poulter and Sibley show that an urban area such as Hull is imagined not just as a city of pets but also as a city of feral cats (and, we may add, of many other wilder animals too). While animal welfare organisations such as the Cats Protection League tend to view feral cats as 'convicts on the loose' (Ingold 1994a: 3), needing to be re-domesticated and returned to the household, the suggestion is that some (if not all) Hull residents are prepared to view feral cats as legitimate residents of certain (but not all) city locations. Wolch, Brownlow and Lassiter reveal differing senses of which animals should be where in a city like Los Angeles, as particularly linked with the rural pasts of some African-American women which lead them to expect the proximity of livestock animals (alive or dead) in the city.[16] For other of their interviewees, though, it was easier to accept wild animals coming into the city, perhaps trying to visit old habitats now subject to building, than to accept the proximity of animals being reared (as it were) for their table. In the one strictly countryside chapter, Yarwood and Evans show how the issue is not merely that livestock animals are regarded as the correct occupants of rolling pasture, but also that much more detailed imaginings are coming into play which associate specific breeds of livestock (such as Aberdeen Angus cattle) with particular regional agricultural landscapes. Moving into the chapters dealing with different varieties of wilder-lands, Matless discusses the English Broadlands, a region of attractive landscape that is inscribed by a contested 'moral geography' (see also Matless 1994) comprising different visions of which wild animals should properly be present and which are 'alien' invaders. The issue in this regard is not merely that Broadland should be home to wild animals, but precisely which wild animals are to be deemed the rightful occupants of the region, the best suited to the local environment and the most aesthetic relative to the local landscape. Brownlow covers parallel ground when surveying debates about whether or not the gray wolf is an appropriate wild animal for restoration to the Adirondack mountainscape, while Waley reconstructs debates about the ecological restoration of fish to Japanese rivers which reveal the interleavings of science and spirituality in the imaginative placings of fish and other animals across Japan's spaces (urban, rural and wild).

As most of the chapters relating to wilder-lands also indicate, a common assumption is that wild lands beyond the circle of normal human activity are, and should be, stocked with all manner of large and/or dangerous animals with whom humans would not normally wish to have encounters. Many people, as in the quote from Lebowitz above, are doubtless happy for a multitude of animals, from ferocious lions to snappy crocodiles and poisonous spiders, to reside 'genuinely outside ... in the ... jungle'. A host of symbolic connotations are nonetheless attached to Western depictions of wilderness (Benton 1993: 66; Wilbert 1999), as a place of freedom, anarchy, redemption, contemplation and the like, and there is a common idealisation that humans are really the strangers here (they are the beings 'out of place') who should not allow their own search for extreme experiences in the wild to interfere with the animals occupying these remote spaces. Yet, in reality the capitalist appropriation of lands and seas as private property, or as objects of contractual agreement for a multitude of extractive rights by nation-states, sees the entirety of the global biosphere, any wilder-lands included, coming within the orbit of human control and intervention (Benton 1993: 66–67). More narrowly, wilder-lands are sometimes regarded as repositories of wild creatures available for sports such as hunting and shooting, and the hunting debate currently raging in Britain is full of contested imaginings of the differing wild animals present in the country districts of counties like Somerset and Devon, and of their rights to be 'left in peace' (or not) in these districts (see Woods, this volume; and also Woods 1998a, 1998b). 'Big-game' hunting in the 'distant' lands of Africa and India was prompted in the nineteenth century by stereotypical visions of forests and plains stuffed full of lions, tigers, elephants and rhinoceroses,[17] even if to a limited extent there was also conflict over whether or not such animals should be left to their own devices (see Chapter 10).

A crucial moment in the history of human–animal relations was the emergence of zoos in Europe and North America, the main objectives of which have been precisely to translate wild animals *from* 'the wilderness' *to* the special, enclosed and policed enclaves nearer to our human homes in 'the city' (reversing the baseline imaginative geography noted earlier). A staple ingredient in Western imaginings relating to animals has thus become the zoo as a space (or set of spaces) specifically put aside for wild animals no longer 'in the wild', thereby leading many people to 'naturalise' the zoo in the sense of accepting it unproblematically as an appropriate location for many animals (see both Gruffudd and Davies, this volume). Zoos are often highly contested places as well, however, where what has developed into a tripartite relation between science, entertainment and education has always been uneasy. Indeed, apart from during the early days of some zoos such as London Zoo, there seems to have been little scientific interest in them until perhaps recently. Although they may have had some importance during the nineteenth century in refining classifications, there appears to be little need or desire nowadays to have a representative of most wild species placed on display behind bars in a zoo compound (Secord 1996). In recent years zoos have reinvented themselves as 'arks' seeking to conserve animal species, of course, utilising a new conservation biological science and even returning animals to the wild, but once again controversy surrounds zoos as being the most effective places to undertake such work. For all this, the zoo does nonetheless stand as a highly tangible expression of the dual conceptual and material placements of animals under discussion here, and offers a good point at which to flip to the next stage in our argument.

'Other spaces'; or making 'beastly places'

> The longest spell of liberty ever enjoyed by an inhabitant of the Zoo was over two years. It was a very rare legless newt, which looks like a shiny black worm, banded at intervals with startling rings of white. One morning it simply wasn't there. ... Over two years passed and it had been forgotten. Then a gardener was turning out a dead plant in the Reptile House. There, in its roots, was the missing newt, who had been flourishing on casual cockroaches.
>
> (Mainland 1927: 84)

Echoing what we said earlier about the dangers of concentrating solely on how human societies imagine or represent animals, we now wish to consider ways in which animals, as embodied, 'meaty'[18] beings, often end up evading the places to which humans seek to allot them, whether the basket in the kitchen, the garden, the paddock, the field, the cage or whatever. Such evasions can occur at an individual level, when a pet wanders from the house into the surrounding streets, or at a societal level, when relatively large numbers of animals, say coypus, escape from fur farms (see Matless, this volume). Another example comes to mind: that of Cholmondeley the chimpanzee, who, in 1951, reputedly escaped from London Zoo, boarded a Number 53 bus and jumped upon a woman passenger (Vevers 1976: 97). In Cresswell's (1996) terms, Cholmondeley became 'out of place', transgressing the taken-for-granted (human) notions (or *doxa*) about what sort of beings should properly be present in a Number 53 bus driving through the streets of London. Indeed, Cholmondeley transgressed a number of spatial boundaries widely accepted by the human occupants of London: not just that which separates animals in cages from the public, but also those which separate the zoo from the immediate locality, which designate a bus as something for travelling on with a purpose (we wonder whether it was popularly entertained whether Cholmondeley had purposefully entered the bus to effect his get-away), and which frown upon buses as sites for jumping on people (whether or not with some sexual motive).[19] It is possible to identify many appealing examples of such trangressions – Philo (1995) lists a few associated with livestock animals in nineteenth-century London – but the key lesson is that in such cases, as in many more mundane ones, it is animals themselves who inject what might be termed their own agency into the scene, thereby transgressing, perhaps even resisting, the human placements of them. It might be said that in so doing the animals begin to forge their own 'other spaces', countering the proper places stipulated for them by humans, thus creating their own 'beastly places' reflective of their own 'beastly' ways, ends, doings, joys and sufferings. Many commentators will doubtless see problems with such wordings, spotting the dangers of both anthropomorphism and hylozoism, and accusing us of misconstruing what should be meant by 'agency' in serious academic research. Since such issues are so relevant to how we are framing our collection as a whole, and to the 'animal-centred' geography surfacing in many chapters, let us now insert a second subsection that, while interrupting the flow of our narrative, is currently very pertinent.

'What is the agency of animals?' and related matters

It can be argued that many animals have their own 'territories' (see below), that they live in certain environments to which they have adapted, and which they have helped to mould in some way over time. It can also be argued that many animals do have the capacity to transgress the imagined and materially constructed spatial orderings of human societies, but can we then talk about some animals having the capacity to *resist* these orderings? To ask such a question does indeed raise the thorny issue of animal agency, given that the concept of 'resistance' is generally taken to entail the presence of conscious intentionality, seemingly only a property of *human* agency in that only humans are widely recognised to possess self-consciousness and the facility for acting on intentions with a view to converting plans into outcomes.[20] Such a claim returns us to the broader theme, as aired above, of distinguishing humans from animals. It also forces us to meet the claims of philosophers about the supposedly special status of the human being as a self-aware entity, the only such entity, possessing the mental 'wiring' allowing inner thought and the related development of language (e.g. Hegel 1975: 48–51; Rousseau 1973).

At the same time, notions of this kind clearly link to a long-standing human belief in a basic distinction between what is often termed the 'civilised' or 'rational' being who can think and act in the world (the human), and what are often identified as the base passions and instincts which allegedly obliterate a being's potential for agency (the primal basis of the animal, present within humans, but most obviously displayed in non-human animals) (Horigan 1988; Ingold 1994b).[21] From this distinction arises the widespread use of terms such as 'animal' and 'bestial' to describe groups of people perceived to engage in anti-social, possibly inhumane, activities. Indeed, all kinds of cultural cross-codings have historically been constructed between some humans and some animals. Sometimes human groups may be regarded as lesser or marginal to other more dominant groups, and may thereby be associated with animals (or the bestial in general) also viewed as lesser or marginal or as inhabiting marginal spaces. For example, white colonists sometimes associated native American Indians with wild animals such as the wolf, ones which were seen as pests or even 'vermin', and this is a point made by Brownlow (this volume; and also Ryan, this volume). Many groups may erroneously be coded by a dominant culture with non-human pests and vermin, and it is obvious that all kinds of metaphors of threat, contagion and pollution come to surround such associations and cross-codings of peoples and animals. Occasionally, of course, the identification of people with certain animals can be more 'positive': consider their uses in totemic societies, and to a lesser extent in such things as heraldry or their adoption as symbols by nation-states.

Assumptions and ramifications which render the human strictly separable from the animal may appear at first sight entirely 'reasonable', but on closer inspection it transpires that they flow from a particular set of discourses which began to gain currency in Europe from the sixteenth and seventeenth centuries onwards. Such discourses formalised a specific notion of agency rooted in a particular conception of the human and its supposedly distinctive qualities, attributes and capacities

(Daston 1995). The implication is that this knowledge, as a bulwark of Western science and philosophy, stems from a quite circumscribed historical–geographical context, and does not necessarily have the universal provenance commonly assumed (Latour 1993: 120). With the taking seriously of 'other' knowledges – notably non-Western 'indigenous' knowledges or ethnosciences – which provide a less dualistic account of the differences between humans and animals, many people (outside the West, but in it too) have started to deconstruct seemingly obvious claims about the privileged status of the human, in contradistinction to the animal, as *the* source of agency in the world. Many other societies and cultural worldviews have been prepared to see capacities for agency distributed much more widely across the many different things of creation – humans, animals, spirits and the elements all included – thereby disrupting what Westerners have normally taken to constitute the properties of consciousness, self-awareness, intentions, thought and language. This readiness to suppose that such properties of being are also possessed – to some extent, in some form – by many non-human animals *has* now been transmitted into the scholarship of a few Western academics, however, resulting in sustained research and writing on animal consciousness, self-awareness, decision-making, emotions, and the like (Dawkins 1993; DeGrazia 1996: Chap. 7).

In certain respects a sophisticated intellectual innovation such as 'actor-network theory' (ANT) also takes us in this direction, since ANT questions any neat attribution of specific capacities to specific things in the world (e.g. Callon 1986; Callon and Law 1995; Latour 1993).[22] ANT prefers to conceive of such capacities being distributed much more widely, perhaps unpredictably, across many different kinds of things associated with (what are conventionally taken as) rather different orders of reality (the natural, the cultural, the discursive, the economic, the psychological). Such orders of reality (e.g. nature and society) are hence no longer seen as readily delimitable realms and causes, and instead become viewed as more or less stable and durable outcomes, effects or products of struggles, enrolments, translations of interests and processes of purification involving often vast networks of 'micro-actors' both human and non-human. This is not to suggest that ANT ignores differences in the world (cf. Laurier and Philo 1999: 1064–1069), and writers such as Serres or Latour are prepared to identify differences between human societies and those of, say, baboons.[23] To be more precise, they propose that the key difference here lies in the historical uses and mediations of objects, in that for humans, unlike for baboons, objects are employed to solidify social bonds. Objects in use thus serve to stabilise present human relationships or to maintain the effects of past actions, which then allows human accomplishments from the past, or perhaps human actors from far away, to bear upon the localised present and perhaps to circumscribe actions in the local situation (Murdoch 1997b: 329; Serres 1995: 87).[24] Murdoch (1997b: 328–329) outlines Latour's further argument that human societies differ from simian ones through the increased mixing of humans and non-humans in the former, such that, beginning with the delegation of basic activities to bone and stone tools, human societies gradually attain a level of durability which in simian societies does not exist. With the human domestication of animals and plants, the number of non-humans existing alongside people proliferates exponentially, making it impossible to recognise a pure 'human' society. Industry and science subsequently increase the proliferation of non-humans mixed up with humans, all layering on top of older mixings.[25] Resources, technologies, animals, and so on, all actively participate in, refine and frame such processes of interaction, and this is an observation which spirals back to the basic point about ANT *not* accepting the standard divisions of conventional thought. Within ANT, agency is conceived of not as some innate or static thing which an organism always possesses, but rather in a relational sense which sees agency emerging as an effect generated and performed in configurations of different materials (Callon and Law 1995: 502). This means that anything can potentially have the power to act, whether human or non-human, and the semiotic term 'actant' is used to refer to this symmetry of powers.

One upshot for Latour's own approach is thus to rescue the non-human things of the world for social science, to recover the 'missing masses', and to give them a 'place', a role, an ability to prompt changes, a capacity for agency. In one of his most recent works, he is led in all seriousness to ascribe thought-and-action to non-human entities such as a prototype 'rapid personal transit' (RPT) system called Aramis (Latour 1996). While there are reservations to be voiced about this ascription (Laurier and Philo 1999; see also below), it is evident that Latour's pioneering experiments open up a 'space' for contemplating the agency of non-humans, animals included, and for speculating about them taking their seats in his envisaged 'parliament of things' (Latour 1993: 142–145). In fact, a thought-experiment along these lines has already been conducted by Callon (1986), when considering what scallops, exercising a measure of agency, brought to one particular situation of scientific inquiry. This means that it is not just non-Western or lay knowledges pushing along this road, nor just a few eccentric Western animal psychologists; rather, it is also arising in the challenge to social theory occasioned by ANT and its derivations. These latter approaches are providing an impetus to the statements of a new animal geography ready to debate, for example, the agency of animals (see esp. Wilbert 1999), implying that it is indeed not foolish to be talking about animals possessing a measure of agency and being capable of resisting – not just transgressing – human norms such as the spatial orderings of the city, the farm or the zoo.

What is more, Latour explicitly debates his anthropomorphism, his readiness to relocate qualities, attributes and capacities normally only associated with

humans to other things in the world. Laurier and Philo (1999) discuss the stance
on anthropomorphism adopted in his book *Aramis* (Latour 1996), noting the host
of 'x-morphisms' which Latour allows to run between all manner of phenomena
when aspects of one order (say, machines) are deployed to illuminate the characteristics of another (say, humans), and vice versa. In so doing, Latour admits to
being anthropomorphic in how he describes machines, and is unrepentant for
doing so because his view is that it illuminates shadowy dimensions of the
machinic thing-world all too easily neglected by conventional social science (see
Laurier and Philo 1999: esp. 1056–1060). It is helpful to chart this aspect of
Latour's work because the charge of anthropomorphism is often laid at the door
of many who toy with the notion of agency in animals, and it is refreshing to see
how he responds to this kind of charge in an academic setting.

There is nothing new about anthropomorphism in animal studies, of course,
and one lovely example that we came across recently is a book entitled *Tiny Toilers
and Their Works* (Glenwood Clark n.d.), which systematically discusses many
different insects through the lens of different human labourers and how they
organise their activities, products and spaces. For example:

> Surrounding the cities of these ants are their gardens, large circular terraces
> from which the industrious farmers have carefully removed all vegetation
> except their favourite rice-plant. … The land-workers are fond of open air
> and sunlight, and they use their gardens as exercise grounds in which they
> may walk and enjoy the air. … From the edges of the circular clearings about
> the nest wide highways run off in every direction into the surrounding
> country, like the spokes radiating from the hub of a wheel. These roads are
> often sixty feet long, and from two to five inches wide, real monuments to
> the engineering skills of their builders. … In the open garden areas about the
> ant-city usually grow the plants whose seeds are used as food by the horticul-
> turalists. … The ant farmer … is well aware of the importance of keeping
> her farm-land free from weeds and useless plants.
>
> (Glenwood Clark n.d.: 9–10)

Wasps as paper-makers, spiders as aviators, cicadas as miners, termites as architects: all of these anthropomorphisms and many more feature in this book, which
in many regards, as the above quote hints, could definitely be termed a study of
'insect geography'. Its anthropomorphic cast is acutely obvious, and it is certainly
not an example which we are advocating be emulated by animal geographers. To a
great extent we do agree with criticising work, as here, which goes overboard in
attributing human intentions, goals, mental states and material practices to non-
human animals, thus seemingly erasing the 'otherness' of these animals.[26] Such
works can be seen to engage in hylozoism, the imputation of intentions to non-

humans, and this is also something of which Latour (1988) has been accused (we would probably argue rightly) in his work on Pasteur's microbes (see Schaffer 1991). We would suggest that there is no need to go so far as imputing conscious intentions to non-humans, even though it may sometimes seem desirable as a narrative strategy (and Latour [1996] evidently feels this to be a valuable tactic with respect to Aramis). Callon's (1986) celebrated study refrains from going so far as imputing intentions to scallops; instead, he argues that the scientists' goals are to enrol the scallops within a network, but that ultimately the scallops impede these moves. He thereby adopts a similar line to Pickering (1995: 17), who argues that in studying scientific practice, organised around specific plans and goals, there *is* a need to have knowledge of the scientists' intentions but *not* necessarily to have insight into the intentions of things.

Yet, we would also caution that the criticism of anthropomorphism is not as self-evidently correct as it might first appear. First, we could go some way with Latour's semiotic method and his assertions that, rather than talking of anthropomorphisms, we should be talking of many different possible 'morphisms', whether these be technomorphisms, zoomorphisms or whatever. The claim is that the term 'anthropomorphism' actually 'underestimates our humanity,' in that the '*anthropos*' and the '*morphos*' together mean both that which has human shape and that which gives shape to humans (Latour 1992: 235; 1993: 137). People thus give form to non-humans, but are themselves acted upon and given form by non-humans, in part through the effects of other 'morphisms' which enlarge the imaginative resources available to humans for conceptualising their own humanity. Second, we may argue against the critique of anthropomorphism from a more *Verstehende* approach. The basic logic to the anthropomorphism critique is that a category mistake is occurring because humans are radically different from animals (Fisher 1996: 4), and that to portray the latter in terms of the former is to misrepresent their quite different 'true nature', and thus to foster woeful misunderstandings of what they, the animals, are really about. Yet, if the possibility is entertained that humans and animals may *not* be so completely different after all, as has been proposed on the basis of alternative cultural worldviews, scientific findings and even ANT, then the logical grounding for the charge of anthropomorphism becomes much more rickety. The option is instead raised of a measured, hesitant and reflected-upon form of anthropomorphism, one whose self-critical proclivities would prevent the excesses of a Glenwood Clark, but one which would allow the possibility of insights to be produced from considering *some* non-humans in *some* situations *as if* they could perceive, feel, emote, make decisions and perhaps even 'reason' something like a human being. In addition, this position rebounds back on the hard anthropocentrism of the standard anthropomorphism critique, given that, from its determinedly human-referenced starting-point, it implies that there are no continuities between humans and non-humans (animals included) and thereby no basis for us (humans) sharing anything with other beings in the world. Humans are effectively sealed off from the rest of creation by such a vision, and in the process prioritised as the only beings genuinely able both to enjoy and to suffer their worldly lot. Such an aggressive anthropocentrism undermines any basis for a more inclusive 'politics', an imagining that Elder, Wolch and Emel (1998a, 1998b) term the '*pratique sauvage*', which might allow humans to make common cause with animals in creating genuinely shared spaces able to sustain them simultaneously if differently. For these and other reasons, there does seem to be merit in stalling on the anthropomorphism critique, so as to explore the theoretical, political and ethical gains which may result from permitting a guarded anthropomorphism into our studies.[27]

Let us return to our basic argument by considering wild animals seeking to stay wild in their wild places, getting on with their own lives and worlds without anything to do with us humans, performing their specific forms of agency to one another, creating their own worlds, their own beastly places, without reference to us.[28] It is important that animal geographers exercise their imaginations in trying to glimpse something of these beastly places as lived by such animals themselves, in part to gain a better sense of the implications that follow for wild animals when humans turn up and start altering the configurations of their worlds, removing food sources, blocking off migration tracks, bringing in noise and other pollutants, and so on. Gullo, Lassiter and Wolch (1998: 141–142) briefly attempt something along these lines in their cougars paper, pondering 'animal ideas about humans', and in discussion at the Exeter conference Wolch (1997) speculated further on how cougars might apprehend the humans, houses, roads and tree-felling suddenly arriving in their hunting grounds. The whole political economy, and basic geography, of humans seizing, occupying and transforming wilder-lands must also be grasped in such research, and Gullo *et al.* (1998) detail the expansion of suburban housing schemes into the territories of cougars in northern California. Sometimes humans seek to destroy wild animals because they cause problems, in their wildness and violence, when intruding into settlement and agricultural spaces, as in the case of cougars menacing pets in gardens or foxes killing chickens in their runs (see also both Wolch, Brownlow and Lassiter, and Woods, this volume). As discussed above, humans also try to catch wild animals on occasion, and hence strive to remove them from the wild to other sites such as zoos, circuses and sports arenas, an enforced transgression of their proper place. Ryan (this volume) documents the transplanting of captured beasts, or more typically trophies, stuffed parts, drawings or photographs of animals, from the wilds of Africa to the zoos, museums and drawing rooms of nineteenth-century Europe. Davies (this volume), meanwhile, compares such processes with the appropriation of animal images from the wild and their subsequent circulation through the 'electronic zoo' of late twentieth-century nature film-making. What we would personally celebrate are many of those instances of what surely should be termed 'resistance', occurring whenever wild animals seek to evade being trapped or killed by human beings. From their vantage point, they are struggling to remain defiantly 'in place', and in the process they are resisting acts of 'replacing' or 'displacing' which many human visitors to their wild homes might wish to effect.

Numerous animals inhabit localities which do not warrant the designation 'wild', but rather comprise the marginal spaces in and around the towns, cities and countrysides where most humans live, work and rest. These tend to be the spaces that humans avoid, ugly to their eyes and surplus to their requirements, although it should be admitted that some human 'outsiders' such as homeless people, travellers and law-breakers can also seek out these marginal spaces away from the normal public round.[29] These may include the sorts of spaces inhabited by rodents in cities, such as sewers, or the neglected tracts of land often situated next to urban railway lines where a diversity of creatures may be found (see also Wolch *et al.* 1995 on the 'habitats' which animals can find in the city; and also Wolch 1998). Such spaces and their occupants are commonly regarded as transgressive of settled human society, divorced from the civilised goings-on of towns, cities and villages, but still sufficiently close-by to prompt enough distaste, fear and loathing to be coded as 'out of place' in proximity to everyday houses, businesses and streets. Sometimes the very presence of certain animals seeking to live their lives in these spaces — think particularly of rats in sewers — can help to render such spaces marginal in the minds of many humans, as ones to be shunned by all 'decent' people. Moreover, marginal spaces such as sewers, as well as becoming associated with both animals such as rats and the more 'animalistic' aspects of human behaviour (urine, faeces and other dirt), also constitute some of the symbolic recesses of urban societies in the developed world (Gandy 1999: esp. 24, 36; Stallybrass and White 1986: Chap. 3). They are the underground places where 'monsters' may lurk, 'uncanny' places where other creatures, other hauntings, may be found at the heart of culture.[30] Mention should also be made here of wild animals moving into built-up areas and surviving in marginal spaces, given that they may sometimes be viewed as unwanted aliens by their new human neighbours. Similarly, mention should be made of feral cats and dogs setting themselves up in the marginal spaces of settlements such as Hull (see Griffiths, Poulter and Sibley, this volume), particularly because feral creatures are in themselves curiously transgressive beings, neither purely wild nor purely tame, existing as 'in-between' animals finding themselves, appropriately enough, utilising 'in-between' spaces.[31]

Many other examples of animals disturbing human spatial orderings could be adduced here, but let us return once again to zoos. As already stated, zoos have emerged as spaces wherein wild animals can be removed from the wild to cages, enclosures and pools set within overall zoo complexes, often located adjacent to centres of human population, ready to be viewed by the general (human) public (for histories of zoos, see Hoage and Deiss 1996; Jordan and Ormrod 1978). In this context, the animals are 'in place' (although see endnote 19), since, in their wildness and essential *lack* of adaptation to human ways, they must be kept confined and not allowed to escape from their proper place within the zoo grounds. Furthermore, they must be disciplined to display behaviours appropriate to their place of confinement and display in the zoo, a claim made with remarkable directness by the author of a 1920s text on *Secrets of the Zoo*:[32]

> For their own sakes, you *have* to make the Zoo's animals behave themselves as far as their limited minds will permit of training and discipline. … For example, they must take their food decently and allow their cages to be cleaned, and if they have any little tricks or peculiarities they are encouraged to show them off as object-lessons to visitors. Of course there are a certain number of 'hard cases' – creatures such as the British wild cat, the Tasmanian devil, or the rhinoceros – who are permanent mutineers against any form of restraint. Broadly speaking, however, discipline is enforced just as much as in any other place where the inmates are kept against their will, from a school for small boys, to Dartmoor Prison.
>
> (Mainland 1927: 31)

Intriguingly, Mainland also includes in his text detailed notes on how animals were moved around the zoo when changing cages or enclosures, or simply when needing to be moved out for a period to allow these spaces to be cleaned and repaired. The emphasis was very much on how to retain control over the animals in the process, notably in the discussion of an elaborate system of trenches, tunnels and dens through which polar bears could be coaxed at such times, although some potential for the bears to resist the wishes of their keepers – if only by 'taking their time' – evidently remained (see Figure 1.2: see also the discussion of zoo design in Gruffudd, this volume). What we begin to see here are complex spatial expectations being imposed upon animals such as these zoo bears, albeit expectations which the animals themselves do not know or may only know to a limited extent. Although some animals can be taught which places are 'out of bounds' and which are not, most animals will wander in and out of the relevant human spatial orderings without necessarily knowing that they are doing this.[33] Any animals absconding from zoos, or indeed getting themselves into the wrong parts of the zoo (as in the case of the legless newt), are hence transgressing the zoo's spatial regulations – maybe even doing so as what might be construed a wilful act – and thereby effecting a particularly dramatic act of animal 'out of placeness'. The case of Cholmondeley the chimpanzee has already been mentioned, and other instances could be recorded such as the frenzy surrounding the escape of an eagle from London Zoo in 1965, with huge crowds and media activity in and around Regents Park hoping to catch sight of the bird (Vevers 1976: 97). The 'moral panics' which are produced by zoo escapees, as occasioned by the news of the hunt for Barbara the great female polar bear ('a thoroughly annoyed bear in a fog within three miles of Charing Cross': Mainland 1927: 81), exemplify the deep unease often spurred by animal transgressions of human spatial orderings; and they also cannot but suggest a measure of (resistant) agency on the part of animals (the polar bear, and maybe even the legless newt) in getting out of their allotted spaces in the zoo order.[34]

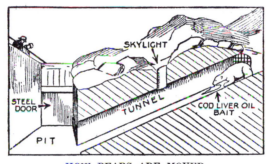

HOW BEARS ARE MOVED.

The drawing on this page shows a section of the bears dens on the lower slopes of the Mappin Terraces.

There are no bars between the animals and the visitors. There *is*, however, an unjumpable trench which is a perfect protection. The only way in which a bear can get into the trench (unless he falls over the precipice) is down a tunnel. The upper end of this opens in the animal's enclosure, and the other end at the bottom of the slope allows the beast to walk out into the trench.

Each den has one of these tunnels, and by controlling the various gates leading to them the retreat of the animals is cut off at each stage of their progress from one den to another. Every gate they pass through takes them one more step towards their new quarters.

You can see in the sketch the place where cod-liver oil was "slopped" in the hope of luring "Sam" to face the journey down the tunnel.

This well-thought-out arrangement enables the animals to be moved about without being boxed up in travelling cages—but they take their own time about moving.

Figure 1.2   'How bears are moved.'

Source: Mainland, L.G. (1927) *Secrets of the Zoo*, London: Partridge, p. 63

A territory, space and place for animals

> Nothing [is] more errant, more nomadic in appearance than animals, and yet their law is that of territory. But one must push aside all the countermeanings on this notion of territory. It is not at all the enlarged relation of a subject or of a group to its own space, a sort of organic right to private property of the individual, of the clan or of the species — such is the phantasm of psychology and of sociology extended to all of ecology. ... The territory is the site of a completed cycle of parentage and exchanges. ... Neither the beasts nor the savages [*sic*] know 'nature' in our way: they only know *territories*, limited, marked, which are the spaces of insurmountable reciprocity.
>
> (Baudrillard 1994: 139, 141, note 4)

The chief themes of this collection have now begun to emerge, and it should be evident that the prime aim of the following essays is to explore the conjoint conceptual and material placements of animals, as decided upon by humans in a variety of situations, and also to probe the disruptions of these placements as achieved on occasion by the animals themselves. It is to record many of the animal spaces specified in one way or another by humans, but also to spy on something of the beastly places made by animals themselves, whether wholly independent of humans or when transgressing, even resisting, human spatial orderings. At the same time, it is to weave these themes through inquiries tackling a whole range of human–animal relations associated with different worldly contexts. More specifically, it involves documenting all manner of encounters between humans and animals, partly as shaped by the former's economic, political, social and cultural requirements, but also encounters in which animals affect humans. It might be added that there are various modes of human–animal relation and encounter which are *not* tackled here as much as they might be, the most obvious perhaps being the sexual, and as yet the new animal geographers have neglected the spaces and places of bestiality (thus failing to link across to recent research on sexuality and space: see Bell and Valentine 1995).[35] The further ambition of the collection proceeds from the familiar claim (after Lévi-Strauss 1968) that animals are 'good to think with', and that 'stories' of animals are especially valuable in helping their human tellers and hearers to develop their own moral identities and psychological interiorities (Shepard 1993; Tambiah 1969; Tapper 1994). Hence, by telling numerous different 'stories' about the roles of space and place as integral to human–animal relations, the essays below will provide much substance for reflection, particularly in the sense of suggesting the ethical charge that is carried by *all* situated human–animal encounters (a claim elaborated by Jones, this volume; but see also Watts, this volume, for cautions about *too* 'individualising' a take on such ethics as they translate into debates over animal 'rights').

Moreover, there is a sense in which we would like readers to catch the drift of the above quote from Baudrillard,[36] wherein he proposes that an inescapable feature of animals is their activity within territories – their beingness within these territories, which are the vital supports of their own animal lives, their 'cycle[s] of parentage and exchange' – and emphasises this quality of animal territoriality in the face of all 'human' understandings of both territory (as something conditioned by capitalist constructs of private property) and animals (which should not be sentimentalised, given voice or supposed to have a 'psychic life').[37] In this respect Baudrillard appears to oppose elements of the anthropomorphism advocated above, but it might be countered that we are actually agreed in refuting anthropocentrism while allowing a hesitant anthropomorphism, speculating that some animals may have some qualities akin to humans *alongside* much that will be different, other and unavailable to human ken. Whatever, the implication is that humans should respect this 'law' of territory bound into animal lives, and thereby strive to cultivate the *'pratique sauvage'* (Elder *et al.* 1998a, 1998b) which demands a new mode of human–animal geographical coexistence. This will call upon us (humans) to desist fixing animals rigidly into our spatial orderings, and instead will ask us to allow animals more space: to grant them more room (even, say, for animals travelling to the slaughterhouse in trucks),[38] and to furnish them with more ecological resources within human settlements (as in creating and leaving corridors of appropriate habitat threading through cities, towns and villages). The ideal will be to open up spaces wherein they can indeed exist and ignore us for most of the time, and which they can occupy and convert into their own beastly places, as many animals are continually seeking to do (even in the bustling city). Us humans will be around as well, facilitating where appropriate, maybe watching (if we are interested and concerned), but always effacing ourselves and not doing harm. We should look to extend human 'courtesies' to animals, almost a sense of allowing them the decencies of life, space and place that we (humans) would expect and want for ourselves and others, in a manner that maybe does stem from a certain anthropomorphism (reflecting the possibility that in certain respects animals are *not* so different from humans) but which also objects to a crass anthropocentrism (one that only thinks about the world in terms of what we humans see, want and take to be important). It is to imagine a new animal geography on the sharp end of such issues, not 'ducking'.

Notes

1 We make a choice in advance to refer to non-human animals simply as 'animals', dropping the 'non-human', while realising that such categories are in many senses relational (and that on many counts humans *are* animals). We develop the notion of what exactly is an animal later in this introductory essay.

2 'Space' and 'place' are basic concepts deployed by geographers, and considerable philosophical and social-theoretical complexity now attaches to their usage. Rather than review this complexity at the outset, however, we will allow our takes on these two concepts to emerge as the essay progresses. In Philo and Wolch (1998: 108–113), an attempt *is* made to be more

explicit about the meanings of 'space' and 'place' in relation to animal geography both old and new.

3   Nothing is said in the report, however, about how 'indigenous' peoples in the region had previously interacted with, or thought of, the gorillas.

4   Throughout this chapter terms such as 'we', 'us' and 'our' will be deployed to refer to humans and 'our' human actions, possessions, and so on. Such terms should really be placed in 'scare quotes' throughout, because one message of our essay is that the implied baseline distinction between 'us' (humans) and 'them' (animals) should *not* be taken as so self-evidently obvious. We have refrained from doing this, however, so as to avoid making the text look more laden down with such quote marks than it is already.

5   In a related piece, Deleuze (1988: 132) writes that '[t]he superman, in accordance with Rimbaud's formula, is the man who is even in charge of the animals (a code that can capture the fragments from other codes …).' He also adds in an endnote, with reference to the same letter from Rimbaud, that 'the future man is in charge not only of the new language, but also of animals and whatever is unformed' (Deleuze 1998: 153, endnote 18).

6   If organisms such as viruses are seen as animals, then many of them *do* exert a power of life and death over many humans. This being said, it is unclear whether the prion blamed by scientists for the disease of BSE is more than a form of protein. The BSE crisis in the UK may not really be thought of as an assertion of dominance by animals over local humans, moreover, since some three million cattle have been killed since 1996 in order to try to limit the spread of the disease. (Actually, an attempt to underwrite confidence in the export meat market is arguably a more likely reason for these substantial animal sacrifices.) See also Hinchliffe and Naylor (1998).

7   Special reference should also be made to Anderson's contributions in this connection. Her initial piece on the Adelaide Zoo (Anderson 1995) was particularly influential, but so too have been her subsequent papers on the long-term historical geography of domestication, wherein she considers both the material processes involved and the ideological codings written into these processes (Anderson 1997, 1998).

8   Examples of new animal-geographical work which tackles birds and reptiles include Proctor (1998; Proctor and Pincetl 1996) on owls, and Whatmore and Thorne (1998) on crocodiles.

9   Little geographical work on fish has departed from the zoogeographical mapping tradition (a recent example is Kracker 1999), although a few geographical studies of the fishing industry have in effect been inquiries into human–fish relations. Of particular note has been Coull's detailed research on Scotland's sea fisheries, as brought together in Coull (1996; and also Cruickshank 1984).

10 Very little geographical work has been conducted on insects, although recent Meetings of the Association of American Geographers have included occasional zoogeographical papers on ants (Davis 1998; DeMers 1994) and papers touching on the human dimensions of 'plagues of locusts' (Richardson 1994, 1998).

11 A reformulated plant geography, perhaps along the lines being advocated for an animal geography, may be interesting, although questions to do with plants, gardens, forests and other plant habitats (such as heathlands) have already been well covered in the study of landscape. We certainly support ongoing research into the geographies of viruses, as linked into medical and health care geography. See, for instance, Gould (1993). While not being entirely happy with aspects of this book, the attention paid to the micro-spaces of HIV (see Gould 1993: Chap. 1), as connected to inquiries into different spatial scales of human–virus interactions from the city to the continent to the globe, does strike us as imaginative and important. See also Hinchliffe and Naylor (1998).

12 'The modern constitution' is what Latour, following Shapin and Schaffer (1985), terms the separation of powers between representatives of non-humans, the scientists, and those of humans, the politicians. This historical separation between epistemology and sociology, or between nature and society, is seen to be 'invented' in the dispute between Robert Boyle and Thomas Hobbes in the seventeenth century. The constitution thus 'defines humans and non-humans, their properties and their relations, their abilities and their groupings' (Latour 1993:

15). Latour (1999) makes some further elaborations upon this constitution through a reading of Plato's *Gorgias*.

13 Cooter and Pumfrey (1994: 250) make the further point that when experts appeal to a public for support, they seek to appeal to its interests. If a lay public accepts such an appeal, it allows itself to become part of a network of alliances sustaining that enterprise. But just as the public have been enrolled by experts, so too have the public enrolled the experts, and according to its influence in this network it may alter the kinds of science or politics pursued in the future

14 The niche is classically regarded as the organism's proper place, exclusive to it. Indeed, Darwin, among others, regarded the niche as a space pre-existing the organism (Kingsland 1991). Only when an organism is introduced from another ecological system may another organism be seen to be competing in the same niche, and then that introduced organism is often referred to as 'alien', as 'out of place'.

15 We resist the urge to use the term 'wilderness' too freely, however, as the term is problematic, implying areas somehow free of human impacts. As Marx himself argued in the nineteenth century, there may be very few, if any, areas left that can be thought of in such terms as wilderness. Even areas commonly thought of as being wilderness have often been shown to have been managed by peoples, especially by now-excluded 'indigenous' peoples, at times in the past. See also Cronon (1995). We prefer to use the term 'wilder-lands'.

16 This position would not, of course, be unusual to many people around the world. Philo (1995) shows the history of exclusions of such animals from British cities, but in recent years various environmental campaigners have argued for livestock animals to be reintroduced to cities in countries such as Britain, and for more food growing to be undertaken in cities.

17 Images of wild animals in abundance in wild lands can sanction indiscriminate hunting, as Thorne (1998) shows most persuasively in a discussion of the slaughter of kangaroos in the Australian outback.

18 This term is used by Nast and Pile (1998a: e.g. 3; 1998b: e.g. 412). The term becomes even more evocative, perhaps disturbing, when deployed in relation to non-human creatures which often end up as 'meat' for human tables, or indeed for other animals. More generally, it might be remarked that the recent turn to considering 'the bodily' in human geography, as well as in the social sciences and cultural studies, has yet to say much about animal bodies (but see some suggestions in Jones, this volume).

19 It must be noted that from another perspective the chimpanzee in the zoo is also 'out of place'. For many critics, it should not be there, but rather in the wild, where it would be 'in place', developing its own social ways, or, as Sorensen (1995: 10) puts it, 'striving to maintain the power of existence and development that is the *telos* of animals'.

20 Indeed, this is why both authors of this editorial have worked with and against Cresswell's (1996: esp. 22–23) well-known distinction between transgression and resistance. Transgression, for Cresswell, entails actions whose consequences overstep certain limits defined by humans, but which are not necessarily intended by anybody (or anything) to do so. Resistance, for him, entails actions deliberately, wilfully, purposefully, knowingly, pursued with conscious intention on the part of the doer to overstep (known) limits. See Wilbert (1999: 245, 249) and Philo (1995: 656).

21 Paralleling this distinction are long-running fears in many societies regarding the risk of certain humans reverting to the animal, or of some humans not reaching the civilised heights of culture (the 'truly' human). Commonly, this is thought of as a modern 'racist' view used to justify slavery, colonialism, genocide or other horrors, but such fears have also not infrequently been expressed in connection with people identified as 'idiot' or 'lunatic' (to use older parlances) or 'mentally subnormal' or 'mentally ill' (to use more recent, if also unsatisfactory, terms). Conversely, Richards (1993: 151) – discussing the desires of conservationists to increase chimpanzee numbers in Sierra Leone – points to how Mende people recognise chimpanzees as their ancestors who still live in the woods, while also fearing that increasing numbers of chimpanzees will encourage some humans to revert to, or be tempted to copy, these excessively strong, violent and cannibalistic beings.

22 Outline accounts of the potential for deploying ANT in human geography have been fairly extensive in recent years: see Bingham (1996), Braun and Castree (1998), Demeritt (1994), Hinchliffe (1996), Murdoch (1997a, 1997b) and Thrift (1996).

23 We wish to emphasise that Latour and other ANT writers *do* see important differences between humans and animals, here in the specific sense of distinguishing human societies from simian societies, so as to underline that the broader conceptual manoeuvres attempted by ANT – ones which *do* scramble conventional notions of agency – are designed to collapse the distinctions between humans and non-humans, animals included, in certain respects *but not necessarily* in others. These are extremely complicated and controversial questions, however, and we recognise the diversity of claims and counterclaims currently entering the relevant literature (as hinted at in outline by Laurier and Philo 1999: see also Hinchliffe 1999; Whatmore 1997, 1999). We also acknowledge our own ambivalence about many of these claims and counterclaims.

24 Conventionally, these effects also take the form of *attributions* which localise agency as singularity, most often as humans with intentions and languages, thus endowing one part of a configuration as a prime mover or strategic speaker. As Callon and Law (1995: 503) argue, such attributions 'efface the other entities and relations in the *collectif*, or consign these to a supporting and infrastructural role'. The focus on strategic speakers, or spokespersons, has also been a criticism of ANT.

25 Latour's outline of enduring sets of heterogeneous relations runs into Marxism in some ways, for these proliferating objects are not simply neutral; rather, they embody delegations, and 'they contain and reproduce the "congealed labour"' (Murdoch 1997b: 329), the knowledges, skills and capacities of other humans and non-humans. It is increasingly through such objects that social order, power, scale or hierarchy are consolidated and preserved. This begins to sound much like Marx's description of objectified (or dead) labour, indeed objectified social relations, and its appropriation of living labour (see Marx 1973: 693ff). Latour, of course, will not accept the kinds of dualisms or explanatory powers that Marx developed.

26 We could think here of differing forms of anthropomorphism along the same lines as Fisher (1996), where the above example of Glenwood Clark would probably be classed as 'imaginative' anthropomorphism. This is the same kind of anthropomorphism that we see when animals, or representatives of them, are used to 'people' situations in such things as cartoons or adverts. But there are other kinds according to Fisher, forms which we engage in all of the time, where we try to make sense of others, both human and non-human others. See next endnote.

27 In effect we are splitting apart anthropomorphism and anthropocentrism as two different, even opposed, movements. Anthropomorphism, on our understanding, allows the human to 'explode' into many different registers of the real, so that fragments of what we normally deem 'human' can be traced throughout the animal and 'thing' world. Anthropocentrism, meanwhile, wishes to close its gates around the existing figure of the human, referencing everything back to the human, and not paying much attention to anything that does not directly pertain to the conventional human and its usual scale and scope of operation. For a more sustained defence of adopting a measured anthropomorphism in our studies of animals, see Midgley (1983: Chap. 11).

28 The new animal geography should here acknowledge a link back to older approaches which did focus on animals 'making' their own distributions relative to underlying variations in environmental conditions. Many (all?) animal distributions in the world today are surely influenced in one way or another by human interventions, however, and to ignore this human dimension would be mistaken.

29 Numerous works on the social geographies of specific 'outsider' groups will verify this claim: see many passages in the texts of Sibley (1981, 1995). It is perhaps unsurprising, then, that Sibley has developed an interest in feral cats in the city to complement his writing elsewhere on human 'outsiders in urban societies'.

30 A whole host of urban myths surround places such as sewers and other marginal spaces of the city, and often these have to do with animals, such as the well-known stories of crocodiles in

the New York sewers. Indeed, sewers and other underground or marginal spaces have often been depicted in literature and the cinema as places where vagabonds or monsters hide.

31 See also work on geographies of 'feral livestock', a further instance of certain animals escaping from their allotted human placements in the world: see McKnight (1999).

32 Mainland, a journalist with the *Daily Mail*, wrote this book to explain the secrets of how London Zoo worked, particularly emphasising how the animals were persuaded to fit in with the requirements of the zoo. The quote here comes from the beginning of a chapter entitled 'Discipline at the zoo'. Other chapters have such titles as: 'When the big chaps get angry'; 'When zoo animals fight back'; 'When elephants go on strike'; and 'Escapes from the zoo'.

33 To say that animals may know which spaces are in and out of bounds does not necessarily imply the same conscious reflexivity as humans may possess. It could be argued that a system of punishments and rewards functions to prompt animals into learning such human spatial rules, but it may be that still more complex processes could be occurring here in the 'minds' of certain animals. Obviously, such reflections circle back to the whole question of animal agency, as already debated.

34 However, there are other transgressions of spaces which may be tolerated, or even welcomed: for example, some of the other animals that turn up at feeding time at the zoo, like the herons that turn up at the penguin enclosure (see Gruffudd, this volume).

35 There are many intriguing hints in Dekkers' (1994) somewhat controversial text, and we might also speculate about connections between rural-agricultural locales where 'peasant' communities endure a greater closeness to livestock and the prevalence of bestial practices (although this is probably too simplistic and mythic an association nowadays). Bagemihl (1999, as reviewed by Mars-Jones 1999) raises the issue of animal sexuality, in that he claims to reveal the extent to which much animal sexual behaviour is not 'heterosexual' but 'homosexual'.

36 We must thank Marcus Doel for pointing us to this Baudrillard reference.

37 Baudrillard (1994: 133) discusses the conceptual 'abyss' between humans and animals which has opened up in Western societies – 'animals were only demoted to the status of inhumanity as reason and humanism progressed' – itself bound up with the gradual acceleration of practices which have seen humans (ab)using animals:

> the abyss which separates them [humans and animals] today ... [is] the one that permits us to send beasts, in our place, to respond to the terrifying universes of space [monkeys in rockets] and laboratories, the one that permits the liquidation of species even as they are archived as specimens in the African reserves or in the hell of zoos.

He suggests that this process has recently been accompanied by humans seeking to get animals to disclose their experiences of this negation:

> Animals ... have followed this uninterrupted process of annexation through extermination, which consists of liquidation, then of making the extinct species speak, of making them present the confession of their disappearance. Making animals speak, as one has made the insane, children, sex (Foucault) speak.
>
> (Baudrillard 1994: 136)

All sorts of human tactics, from forms of animal psychology to various sentimental accounts, hence try to make animals 'speak' to us, in which case 'the psychic life of the animal' (Baudrillard 1994: 132) emerges as a source of great fascination (paralleling the invention of 'the unconscious' as a device for effectively forcing 'the mad' to reveal their truths). Baudrillard strenuously objects to this development, reiterating over and again that animals cannot speak and that they do not have an unconscious, and instead his point is that we (humans) should confront the fundamental questions prompted by the silence of animals: about their quite 'other' ways of being, but also about what we (humans) are really attempting to be and to do in a world where so much, after all, *does* 'escape the empire of meaning, the sharing of meaning' (Baudrillard 1994: 137). His claims comprise a possible criticism of some elements present in our own arguments, although we too are seeking strategies which will enable humans and animals to coexist (to share territories, spaces and places) in a fashion somehow better than is often currently the case, and which does not do this by forcing animals to talk to us or, indeed, in any simple way to be like us.

38   Such a statement might be unexpected here, since it appears to signal our acceptance of animals being killed by humans, particularly (but certainly not exclusively) for meat. This is not quite our position, but we do want to make positive suggestions about how all humans (most of whom are meat-eaters) might rethink their relations to animals (even ones that they end up eating). Hence, insofar that countless animals will continue to go to their deaths at the hands of humans in slaughterhouses, laboratories, zoos and elsewhere, there is a warrant for ensuring that our final claims here are more sophisticated than the simple plea 'humans should not kill animals' (much as we might wish to end on such a note).

References

Anderson, K. (1995) 'Culture and nature at the Adelaide Zoo: at the frontiers of "human" geography', *Transactions of the Institute of British Geographers* 20: 275–294.

Anderson, K. (1997) 'A walk on the wild side: a critical geography of domestication', *Progress in Human Geography* 21: 463–485.

Anderson, K. (1998) 'Animal domestication in geographic perspective', *Society and Animals* 6: 119–136.

Bagemihl, B. (1999) *Biological Exuberance: Animal Homosexuality and Natural Diversity*, London: Profile.

Baudrillard, J. (1994) 'The animals: territory and metamorphoses', in J. Baudrillard, *Simulacra and Simulation*, Ann Arbor: University of Michigan Press.

Bell, D. and Valentine, G. (eds) (1995) *Mapping Desire: Geographies of Sexualities*, London: Routledge.

Bennett, C.F. (1960) 'Cultural animal geography: an inviting field of research', *Professional Geographer* 12(5): 12–14

Benton, T. (1993) *Natural Relations: Ecology, Animal Rights and Social Justice*, London: Verso.

Bingham, N. (1996) 'Object-ions: from technological determinism towards geographies of relations', *Environment and Planning D: Society and Space* 14: 635–658.

Braun, B. and Castree, N. (1998) 'The construction of nature and the nature of construction: analytical and political tools for building survivable futures', in B. Braun and N. Castree (eds) *Remaking Reality: Nature at the Millennium*, London: Routledge.

Callon, M. (1986) 'Some elements of a sociology of translation: the domestication of the scallops and the fishermen of St. Brieuc Bay', in J. Law (ed.) *Power, Action and Belief: A New Sociology of Knowledge*, London: Routledge and Kegan Paul.

Callon, M. and Law, J (1995) 'Agency and the hybrid *Collectif*', *South Atlantic Quarterly* 94: 481–508.

Chartier, R. (1984) 'Culture as appropriation: popular cultural uses in early modern France', in S.L. Kaplan, *Understanding Popular Culture: Europe from the Middle Ages to the Nineteenth Century*, Berlin: Mouton Publishers.

Cockburn, C. (1996) 'A short meat-oriented history of the world', *New Left Review* 215: 16–42.

Cole, H.H. and Ronning, M. (eds) (1974) *Animal Agriculture: The Biology of Domestic Animals and Their Use by Man*, San Francisco: W.H Freeman & Co.

Cooter, R. and Pumfrey, S. (1994) 'Separate spheres and public places: reflections on the history of science popularisation and science in popular culture', *History of Science* 32: 237–267.

Coull, J.R. (1996) *The Sea Fisheries of Scotland: An Historical Geography*, Edinburgh: John Donald.

Cresswell, T. (1996) *In Place/Out of Place: Geography, Ideology and Transgression*, Minneapolis: University of Minnesota Press.

Cronon, W. (1995) 'The trouble with wilderness; or getting back to the wrong nature', in W. Cronon (ed.) *Uncommon Ground: Toward Reinventing Nature*, New York: W.W. Norton.

Cruickshank, A.B. (1984) 'The Lofoten spawning cod fishery: remote area resource development – a model under threat', University of Glasgow, Department of Geography, Occasional Paper No. 14.

Daston, L. (1995) 'How nature became the other: anthropocentrism and anthropomorphism in early natural philosophy', in S. Maasen. E. Mendelsson and P. Weingart (eds) *Biology as Society, Society as Biology: Metaphors*, Dordrecht: Kluwer.

Davies, J.L. (1961) 'Aim and method in zoogeography', *Geographical Review* 51: 412–417.

Davis, J.M. (1998) 'Remnant ant biodiversity along a fragmented arid riparian corridor', Paper given at the Annual Meeting of the *Annals of the Association of American Geographers*, Boston, March.

Dawkins, M.S. (1993) *Through Our Eyes Only? The Search for Animal Consciousness*, Oxford: Freeman.

de Certeau, M. (1984) *The Practice of Everyday Life*, Berkeley: University of California Press.

DeGrazia, D. (1996) *Taking Animals Seriously*, Cambridge: Cambridge University Press.

Dekkers, M. (1994) *Dearest Pet: On Bestiality*, London: Verso.

Deleuze, G. (1988) *Foucault*, Minneapolis: University of Minnesota Press.

Demeritt, D. 1994 'The nature of metaphors in cultural geography and environmental history', *Progress in Human Geography* 18: 163–185.

DeMers, M.N. (1994) 'Range expansion of the Western Harvester Ant', Paper given at the Annual Meeting of the *Annals of the Association of American Geographers*, San Francisco, March–April.

Einarsson, N. (1993) 'All animals are equal but some are cetaceans: conservation and culture conflict', in K. Milton (ed.) *Environmentalism: The View from Anthropology*, London: Routledge.

Elder, G., Wolch, J. and Emel, J. (1998a) 'Race place and the human–animal divide', *Society and Animals* 6: 183–202.

Elder, G., Wolch, J. and Emel, J. (1998b) '*Le pratique sauvage*: race, place and the human–animal divide', in J. Wolch and J. Emel (eds) *Animal Geographies: Place, Politics and Identity in the Nature–Culture Borderlands*, London: Verso.

Elton, C. (1933) *Exploring the Animal World*, London: Allen & Unwin.

Fisher, J.A. (1996) 'The myth of anthropomorphism', in M. Beckoff and D. Jamieson (eds) *Readings in Animal Cognition*, Cambridge, MA: MIT Press.

Foucault, M. (1970) *The Order of Things: An Archaeology of the Human Sciences*, London: Tavistock Publications.

Frangsmyr, T. (ed.) (1988) *Linnaeus: The Man and His Work*, Berkeley: University of California Press.

Gandy, M. (1999) 'The Paris sewers and the rationalisation of urban space', *Transactions of the Institute of British Geographers* 24: 23–44.

Gieryn, T.F. (1995) 'Boundaries of science', in S. Jasanoff, G. Markle, J. Petersen and T. Pinch (eds) *Handbook of Science and Technology Studies*, London: Sage.

Glenwood Clark, G. (n.d.) *Tiny Toilers and Their Works*, London: J.Coker & Co.

Gough, D. (1999) 'Gorillas in the midst of war', *Guardian*, 28 July: 13.

Gould, P. (1993) *The Slow Plague: A Geography of the AIDS Pandemic*, Oxford: Blackwell.

Gullo, A., Lassiter, U. and Wolch, J. (1998) 'The cougar's tale', in J. Wolch and J. Emel (eds) *Animal Geographies: Place Politics and Identity in the Nature–Culture Borderlands*, London: Verso.

Haraway, D. (1989) *Primate Visions: Gender, Race and Nature in the World of Modern Sciences*, London: Verso.

Hartshorne, R. (1939) *The Nature of Geography: A Critical Survey of Current Thought in the Light of the Past*, Lancaster, PA: Association of American Geographers.

Hegel, G.W.F. (1975) *Lecture on the Philosophy of World History*, Cambridge: Cambridge University Press.

Hinchliffe, S. (1996) 'Technology, power and space – the means and the end of geographies of technology', *Environment and Planning D: Society and Space* 14: 659–682.

Hinchliffe, S. (1999) 'Entangled humans: specifying powers and their spatialities', in J. Sharp, P. Routledge, C. Philo and R. Paddison (eds) *Entanglements of Power: Geographies of Domination/Resistance*, London: Routledge.

Hinchliffe, S. and Naylor, S. (1998) 'Spaces, agencies and ontologies – the whereabouts of unclean cows', Paper given at the Annual Conference of the Royal Geographical Society/Institute of British Geographhers, University of Surrey, January.

Hoage, R.J.and Deiss, W.A. (eds) (1996) *New Worlds, New Animals: From Menageries to Zoological Park in the Nineteenth Century*, Cambridge: Cambridge University Press.

Horigan, S. (1988) *Nature and Culture in Western Discourses*, London: Routledge.

Ingold, T. (1994a) 'From trust to domination: an alternative history of human–animal relations', in A. Manning and J. Serpell (eds) *Animals and Human Society*, London: Routledge.

Ingold, T. (1994b) 'Humanity and animality', in T. Ingold (ed.) *Companion Encyclopaedia of Anthropology*, London: Routledge.

Jacobs, M.C. (1997) *Scientific Culture and the Making of the Industrial West*, Oxford: Oxford University Press.

Jordan, B. and Ormrod, S. (1978) *The Last Great Wild Beast Show: A Discussion of the Failure of British Animal Collections*, London: Constable.

Kingsland, S. (1991) 'Defining ecology as a science', in A. Real and J.H. Brown (eds) *Foundations of Ecology*, Chicago: University of Chicago Press.

Kracker, L.M. (1999) 'The geography of fish: the use of remote sensing and spatial analysis tools in fisheries research', *Professional Geographer* 51: 440–450.

Latour, B. (1983) 'Give me a laboratory and I will raise the world', in. K. Knorr-Cetina and M. Mulkay (eds) *Science Observed*, London: Sage.

Latour, B. (1988) *The Pasteurisation of France and Irreductions*, Cambridge, MA: Harvard University Press.

Latour, B. (1992) 'Where are the missing masses? The sociology of a few mundane artifacts', in W.E. Bijker and J.Law (eds) *Shaping Technology/Building Society: Studies in Sociotechnical Change*, Cambridge, MA: MIT Press.

Latour, B. (1993) *We Have Never Been Modern*, Hemel Hempstead: Harvester Wheatsheaf.

Latour, B. (1996) *Aramis or the Love of Technology*, Cambridge, MA: Harvard University Press.

Latour, B. (1999) *Pandora's Hope: Essays on the Reality of Science Studies*, Cambridge, MA: Harvard University Press.

Laurier, E. and Philo, C. (1999) 'X-morphising: review essay of Bruno Latour's *Aramis or the Love of Technology*', *Environment and Planning A* 31: 1047–1071.

Leff, E. (1995) *Green Production: Towards an Environmental Rationality*, New York: Guilford Press.

Lévi-Strauss, C. (1968) *Totemism*, Harmondsworth: Penguin.

Lovejoy, A.O. (1936) *The Great Chain of Being: A Study of the History of an Idea*, Cambridge, MA: Harvard University Press.

McKie, R. (1999) 'Q: What causes as much air pollution as power station chimneys? A: Pig farms', *The Observer*, 25 July: 4.

McKnight, T. (1999) *Friendly Vermin: A Survey of Feral Livestock in Australia*, Berkeley: University of California Press.

Mainland, L.G. (1927) *Secrets of the Zoo*, London: Partridge.

Mars-Jones, A. (1999) 'Nowt so queer as animals: review of Bruce Bagemihl's *Biological Exuberance: Animal Homosexuality and Natural Diversity*', *The Observer Review*, 25 July: 11.

Marx, K. (1973) *Grundrisse*, London: Penguin.

Matless, D. (1994) 'Moral geography in Broadland', *Ecumene* 2: 127–155.

Metcalf, F. (ed.) (1986) *The Penguin Dictionary of Modern Humorous Quotations*, London: Penguin.

Midgley, M. (1983) *Animals and Why They Matter: A Journey Around the Species Barrier*, Harmondsworth: Penguin.

Murdoch, J. (1997a) 'Inhuman/nonhuman/human: actor-network theory and the prospects for a nondualistic and symmetrical perspective on nature and society', *Environment and Planning D: Society and Space* 15: 731–756.

Murdoch, J. (1997b) 'Towards a geography of heterogeneous associations', *Progress in Human Geography* 21: 321–337.

Nast, H.J. and Pile, S. (1998a) 'Introduction: MakingPlacesBodies', in H.J. Nast and S. Pile (eds) *Places Through the Body*, London: Routledge.

Nast, H.J. and Pile, S. (1998b) 'EverydayPlacesBodies', in H.J. Nast and S. Pile (eds) *Places Through the Body*, London: Routledge.

Noske, B. (1989) *Humans and Other Bodies: Beyond the Boundaries of Anthropology*, London: Pluto Press.

Philo, C. (1995) 'Animals, geography and the city: notes on inclusions and exclusions', *Environment and Planning D: Society and Space* 13: 655–681.

Philo, C. and Wolch, J. (1998) 'Through the geographical looking glass: space, place and human–animal relations', *Society and Animals* 6: 103–118.

Pickering, A. 1995 *The Mangle of Practice: Time, Agency and Science*, Chicago: Unversity of Chicago Press.

Proctor, J.D. (1998) 'The spotted owl and the contested moral landscape of the Pacific Northwest', in J. Wolch and J. Emel (eds) *Animal Geographies: Place, Politics and Identity in the Nature–Culture Borderlands*, London: Verso.

Proctor, J.D. and Pincetl, S. (1996) 'Nature and the reproduction of endangered space: the spotted owl in the Pacific Northwest and southern California', *Environment and Planning D: Society and Space* 14: 683–708.

Rabinow, P. (1992) 'Artificiality and enlightenment: from sociology to biosociality', in J.Crary and S. Kwinter (eds) *Incorporations*, New York: Zone 6: 234–252.

Richards, P. (1993) 'Natural symbols and natural history: chimpanzees, elephants and experiments in Mende thought', in K. Milton (ed.) *Environmentalism: The View from Anthropology*, London: Routledge.

Richardson, C.H. (1994) 'The desert locust, *Schistocerca gregaria*, and its pattern of invasions during 1993', Paper given at the Annual Meeting of the *Annals of the Association of American Geographers*, San Francisco, March–April.

Richardson, C.H. (1998) 'Desert locust plagues demonstrate urgency of true preventive control', Paper given at the Annual Meeting of the *Annals of the Association of American Geographers*, Boston, March.

Rousseau, J.J (1973) 'A discourse on the origin of inequality', in J.J. Rousseau, *The Social Contract and Discourses*, London: Everyman.

Rowland, B. (1973) *Animals with Human Faces: A Guide to Animal Symbolism*, Knoxville: University of Tennessee Press.

Said, E. (1978) *Orientalism*, London: Routledge and Kegan Paul.

Schaffer S. (1991) 'The Eighteenth Brumaire of Bruno Latour', *Studies in the History and Philosophy of Sciences* 22: 174–192.

Secord, J.A. (1996) 'The crisis of nature', in N. Jardine, J.A. Secord and E. Spary (eds) *Cultures of Natural History*, Cambridge: Cambridge University Press.

Serres, M. (1995) *Genesis*, Ann Arbor: University of Michigan Press.

Shapin, S. and Schaffer, S. (1985) *Leviathan and the Air-Pump: Hobbes, Boyle and the Experimental Life*, Princeton: Princeton University Press.

Shepard, P. (1993) 'On animal friends', in S.R. Kellert and E.O. Wilson (eds) *The Biophilia Hypothesis*, Washington, DC: Island Press.

Shiva, V. (1989) *Staying Alive*, London: Zed Books.

Shiva, V. and Mies, M. (1993) *Ecofeminism* , London: Pluto Press.

Sibley, D. (1981) *Outsiders in Urban Societies*, Oxford: Blackwell.

Sibley, D. (1995) *Geographies of Exclusion: Society and Difference in the West*, London: Routledge.

Sorensen M. (1995) *Transgenic Animals: Promising Ethical Perspectives or Hybris?*, Denmark: Odense University Humanities Research Centre, 'Man and Nature', Working Paper No. 69.

Spary, E. (1996) 'Political, natural and bodily economies', in N. Jardine, J.A Secord and E. Spary (eds) *Cultures of Natural History*, Cambridge: Cambridge University Press.

Stallybrass, P. and White, A. (1986) *The Politics and Poetics of Transgression*, London: Methuen.

Tambiah, S.J. (1969) 'Animals are good to think and good to prohibit', *Ethnology* 8: 424–459.

Tapper, R.L. (1994) 'Animality, humanity, morality, society', in T. Ingold (ed.) *What is an Animal?*, London: Routledge.

Thorne, L. (1998) 'Kangaroos: the non-issue', *Society and Animals* 6: 167–182.

Thrift, N. (1996) *Spatial Formations*, London: Sage.

Vevers, G. (1976) *London Zoo*, London: Bodley Head.

Watts, M. (1998) 'Nature as artifice and artifact', in B. Braun and N. Castree (eds) *Remaking Reality: Nature at the Millennium*, London: Routledge.

Whatmore, S. (1997) 'Dissecting the autonomous self: hybrid cartographies for a relational ethics', *Environment and Planning D: Society and Space* 15: 37–53.

Whatmore, S. (1999) 'Rethinking the "human" in human geography', in D. Massey, J. Allen and P. Sarre (eds) *Human Geography Today*, Oxford: Polity.

Whatmore, S. and Thorne, L. (1998) 'Wild(er)ness: reconfiguring the geographies of wildlife', *Transactions of the Institute of British Geographers* 23: 435–454.

Wilbert, C. (1993) 'The apple falls from grace', in M. Bookchin, G. Purchase, B. Morris, R. Aitchley, R. Hart and C. Wilbert, *Deep Ecology and Anarchism*, London: Freedom Press.

Wilbert, C. (1999) 'Anti-this – against-that: resistances along a human–non-human axis', in J. Sharp P. Routledge C. Philo and R. Paddison (eds) *Entanglements of Power: Geographies of Domination/Resistance*, London: Routledge.

Willetts, A. (1997) 'Bacon sandwiches got the better of me: meat eating and vegetarianism in S.E London', in P. Kaplan (ed.) *Food Health and Identity*, London: Routledge.

Wolch, J. (1997) 'The cougar's tale', Paper given at the Annual Conference of the Royal Geographical Society/Institute of British Geographers, University of Exeter, January.

Wolch, J. (1998) 'Zopolis', in J. Wolch and J. Emel (eds) *Animal Geographies: Place, Politics and Identity in the Nature–Culture Borderlands*, London: Verso.

Wolch, J. and Emel, J. (eds) (1995) Guest edited issue: 'Bringing the animals back in', *Environment and Planning D: Society and Space* 13: 631–730.

Wolch, J. and Emel, J. (eds) (1998) *Animal Geographies: Place, Politics and Identity in the Nature–Culture Borderlands*, London: Verso.

Wolch, J. and Philo, C. (eds) (1998) Guest edited issue: 'Animals and geography', *Society and Animals* 6: 103–202.

Wolch, J., West, K. and Gaines, T.E. (1995) 'Transspecies urban theory', *Environment and Planning D: Society and Space* 13: 735–760.

Woods, M. (1998a) 'Mad cows and hounded deer: political representations of animals in the British countryside', *Environment and Planning A* 30: 1219–1234.

Woods, M. (1998b) 'Researching rural conflicts: hunting local politics and actor-networks', *Journal of Rural Studies* 14: 321–340.

Woolf, M. (1999) 'Revealed: secret slaughter of 9 million "useless" lab animals', *Independent on Sunday*, 15 August: 6.

2   Flush and the *banditti*

Dog-stealing in Victorian London

Philip Howell

The Victorians and other animals

On the first day of September 1846, the poet Elizabeth Barrett, who was at the time making plans for her secret marriage to Robert Browning, stepped out of a shop in Vere Street in London, and into her carriage, only to find that her beloved cocker spaniel, Flush (see Figure 2.1), had been caught up from under the wheels and spirited away by thieves. For the third and last time, Flush had fallen victim to the dog-stealers who made their living from abducting pet dogs and holding them for substantial ransoms. Elizabeth Barrett's reaction, as twice before, was a mixture of alarm and resignation. There was nothing to be done except to pay the ransom the dog-stealers were sure to demand; she would be informed that an 'interme-diary' might be able to locate her dog, and, for a price, his recovery might be effected. And if not, she should expect the worst, as the dreadful tradition went in her neighbourhood of one lady who had refused the dog-stealers' demands, only to find her dog's head sent to her through the post. Elizabeth Barrett, who would countenance neither her father's nor Robert Browning's advice to face down the extortionists, insisted that all must be done to procure Flush's swift release. And in the end it was she who made the extraordinary journey to the criminal under-world of Whitechapel, to negotiate with a Mr Taylor, the head of the dog-stealing gang, or the *banditti*, as she called them. It would be five long days before Flush was returned to the Barretts' home in Wimpole Street, the 'archfiend' Taylor (Karlin 1989: 306) having extracted six guineas for his release. In all, Flush had cost his mistress some twenty pounds' ransom. But safe he eventually was, and within a week of his return, Elizabeth Barrett had married Robert Browning. Within another week she had left Wimpole Street for ever, making her way with him to the continent and finally to Italy, and to a new life, a world away from London and its dog-stealing fraternity.[1]

What do we know of her spaniel in his captivity? Fortunately, we have Virginia Woolf's *Flush: A Biography*, and her reconstruction of his traumatic experience in the dog-stealers' lair:

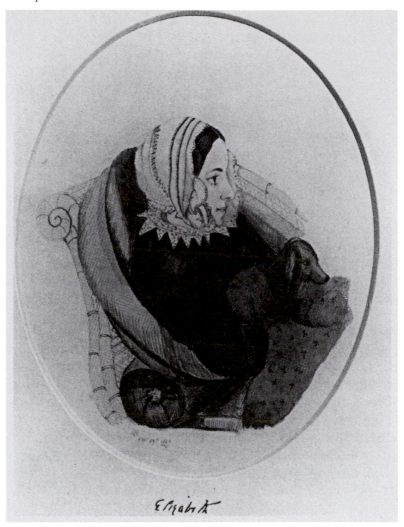

Figure 2.1   Elizabeth Barrett Browning and Flush.

Source: Dennis, B. (1996) *Elizabeth Barrett Browning: The Hope End Years*, Bridgend: Seren, Plate 7; reproduced by permission of the Syndics of Cambridge University Library and the Provost and Fellows of Eton College

Flush was going through the most terrible experience of his life. He was bewildered in the extreme. One moment he was in Vere Street, among ribbons and laces; the next he was tumbled head over heels into a bag; jolted rapidly across streets, and at length was tumbled out – here. He found himself in complete darkness. He found himself in chillness and dampness. As his giddiness left him he made out a few shapes in a low dark room – broken chairs, a tumbled mattress. Then he was seized and tied tightly by the leg to some obstacle. Something sprawled on the floor – whether beast or human being, he could not tell. Great boots and draggled skirts kept stumbling in and out. Flies buzzed on scraps of old meat that were decaying on the floor. Children crawled out from dark corners and pinched his ears. He whined, and a heavy hand beat him over the head. ...

Whenever the door was kicked open he looked up. Miss Barrett – was it Miss Barrett? Had she come at last? But it was only a hairy ruffian, who kicked them all aside and stumbled to a broken chair upon which he flung himself. Then gradually the darkness thickened. He could scarcely make out what shapes those were, on the floor, on the mattress, on the broken chairs. ... A stump of candle was stuck on the ledge over the fireplace. A flare burnt in the gutter outside. By its flickering, coarse light Flush could see terrible faces passing outside, leering at the window.[2] Then in they came, until the small crowded room became so crowded that he had to shrink back and lie even closer against the wall. These horrible monsters – some were ragged, others were flaring with paint and feathers – squatted on the floor; hunched themselves over the table. They began to drink; they cursed and struck each other. ...

The room was dark. It grew steadily hotter and hotter; the smell, the heat were unbearable, Flush's nose burnt; his coat twitched. And still Miss Barrett did not come.

(Woolf 1933: 78–80)

Virginia Woolf's purpose in writing a biography of a Victorian pet is satiric, of course (King 1994: 473–474; Lee 1996: 620–621); it is designed, at least in part, as a distancing commentary on a Victorian age which was still too close for comfort. Still, Woolf faithfully captures both something of the Victorian temper, and something of her own time and experiences.[3] *Flush: A Biography* should be read as a fable, a way of commenting on human society from the perspective of the animal world, and it has an honourable place among animal satires.[4] It is a way of reimagining the Victorians and their world, their values, and it is also a commentary on the way in which that world could be appropriated and written, one that is necessarily revealing of the assumptions and prejudices of its own age. In the passage quoted above, for instance, Flush's status as hero and biographical subject is counterpointed by the anonymous beastliness of his abductors, the brutalised poor of Whitechapel. In this story, the 'aristocratic' Flush is rudely thrust among thieves; a dog whose mistress had tried to teach him to count and to play dominoes here relegated once more to the animal kingdom. Like all good fables, then, beasts and humans have swapped places, turning the world upside down. This kind of inversion is no idle or ironic anthropomorphism, but a serious strategy for reimagining the shared history of people and beasts in the Victorian age. For this reason, if no other, Flush's experiences, and the world of the dog-stealing *banditti*, lay a claim on our attention.

We can borrow Flush's story (and Woolf's strategy) to take up the challenge of the 'animal turn' (Wolch and Emel 1995) in geography, 'giving precedence to the animal perspective' (Munich 1996: 127) in order to tell a fable about a world in which neither animals nor humans know their *place*. This strategy allows us to reimagine the entwined *geography* of the human and non-human world, for the places of Flush's nightmare ordeal among the dog-stealers are central to this story and its significance. Virginia Woolf is keen on her part to emphasise that this narrative of pet abduction is properly a city story, a story about the ways in which, again turning the world upside down, London's poor might be understood to exploit the rich, with Whitechapel or St Giles's preying on the comfortable bourgeoisie of Wimpole Street. Squier's excellent analysis of the sexual politics of Woolf's London points us in the right direction here. In particular, she emphasises that the spatial proximity of Whitechapel and Wimpole Street alerts us to the symbiotic relationship that existed between neighbourhoods only apparently worlds apart (Squier 1985: 127). 'The terms upon which Wimpole Street lived cheek by jowl with St. Giles's were well known', as Virginia Woolf (1933: 76) wrote: 'St. Giles's stole what St. Giles's could; Wimpole Street paid what Wimpole Street must.' For Flush and for bourgeois pet-owners in general, the city was a place of danger and anxiety, rather than simple, confident mastery. London's others, the thieves and the *banditti* who shared public space with the bourgeoisie and their pets, were never in fact very far away. The phenomenon of dog-stealing gives us a glimpse therefore not just of human and animal relations, but also of the complex geographies in which and by which relations between different sets of people in the Victorian city were played out.

Trading in affection: dog-stealing as a profession

While dog-stealing was hardly a new phenomenon, it was the spread of pet ownership among the urban bourgeoisie that called up the opportunities for a crime that directly attacked bourgeois sensibilities. Animal-related discourse had certainly always extended to both sentimental and economic relationships (Ritvo 1987), but pet-stealing encompassed and linked the worlds of affection and commerce. This was, quite literally, trading in affection. It is easy enough to see why, economically, such a crime should arise. By the Victorian era astounding prices, up to 150*l.*, were paid for breed dogs, particularly fancy dogs like spaniels, constituting an alluring temptation to men living in a world where that amount might easily represent several years' wages. The emotional value of such pets, in any case, might far exceed their sale value: a terrier worth at most 5*s.* might fetch, so it was said, as much as 14*l.* when ransomed (*Report from the Select Committee on Dog Stealing* (Metropolis), hereafter *Report*: iv). Dog-stealing could thus be a highly profitable business, a criminal profession or a lucrative sideline for men who might earn a considerable sum by snatching and successfully ransoming a favourite pet.

The dog-stealers' *modus operandi* was a simple one, as Henry Mayhew informs us, a matter of prowling the metropolis for a likely dog out in public with its master or mistress, in order to follow it home, and to wait for an opportunity to lure it away with a piece of coarse liver – fried and pulverised, mixed up with tincture of myrrh or opium – or perhaps with a bitch in heat (Mayhew 1950: 232). Mayhew describes the practice of a typical 'lurker' by the name of 'Chelsea George', who was said to clear 150*l*. yearly through his dog-stealing activities:

> Chelsea George caresses every animal who seems 'a likely spec,' and when his fingers have been rubbed over the dogs' noses they become easy and perhaps willing captives. A bag carried for the purpose, receives the victim, and away goes George, bag and all, to his printer's in Seven Dials. Two bills and no less – two and no more, for such is George's style of work – are issued to describe the animal that has thus been *found*, and which will be 'restored to its owner on payment of expenses.' One of these George puts in his pocket, the other he pastes up at a public-house whose landlord is 'fly' to its meaning, and poor 'bow-wow' is sold to a 'dealer in dogs,' not very far from Sharp's alley. In course of time the dog is discovered; the possessor refers to the 'establishment' where he bought it; the 'dealer makes himself *square*,' by giving the address of 'the chap he bought 'un of,' and Chelsea George shows a copy of the advertisement, calls in the publican as a witness, and leaves the place 'without the slightest imputation on his character'.
>
> (Mayhew 1968: 52, emphases in original)

Now the dog-stealer was always hidden in this way behind the figure of the 'restorer' or go-between who offered to help return the stolen dog. In this connection, Elizabeth Barrett's Mr Taylor, of Manning Street, Edgware Road, or of Shoreditch, or perhaps of Paddington, crops up time and again in police and parliamentary reports on the dog-stealing trade. An owner of a Scotch terrier eventually restored for 4*l*. was referred to this same Mr Taylor, who is said to have replied, 'with surprising familiarity': 'Ah, and you have got a valuable greyhound; I heard a man say he should have that too before long.' Nominally a shoemaker, this Mr Taylor was only the most notorious associate of the numerous thieves said to exist in places like Bayswater, Bethnal Green, Spitalfields, Paddington and Hammersmith.[5] The Metropolitan Police Report for 1837 identified as many as 141 dog-stealers in the capital, of whom 45 lived wholly by the trade, 48 augmenting their ostensible occupations, and 48 being associates (Mayhew 1968: 41–42). The corresponding figures for stolen dogs and prosecutions for 1841–3 were given in the same report, shown in Table 2.1.

Table 2.1   Metropolitan police: recorded incidents relating to dog-stealing

	Dogs stolen	Dogs lost	Persons charged	Convicted	Discharged
1841	43	521	51	19	32
1842	54	561	45	17	28
1843	60	606	38	18	20

Source: Mayhew, H. (1968) *London Labour and the London Poor*, Volume II, New York: Dover, p. 50

According to estimates produced by William Bishop, the dog fancier and champion of a society for prosecuting dog-stealers and their confederates, up to July 1844 there were 151 known victims of the dog-stealers, 62 of whom had paid out ransoms in 1843 and the first half of 1844. These persons all together had paid the dog-stealers nearly a thousand pounds, at an average of 6*l*.10*s*. each. From these same figures Mayhew listed 36 specially prominent individuals who had paid 438*l*.5*s*.6*d*., an average of upwards of 12*l*. each, and he suggested that over 2000*l*. a year was pocketed by the fifty or sixty professional dog-stealers in the city. These dog-stealers' victims included such illustrious names as the Honourable William Ashley, Sir Francis Burdett, the Duke of Cambridge, Sir Richard Peel and the French Ambassador. These men were clearly no respecters of titles or status:

> One of the coolest instances of the organization and boldness of the dog-stealers was in the case of Mr. Fitzroy Kelly's 'favourite Scotch terrier.' The 'parties,' possessing it through theft, asked 12*l*. for it, and urged that it was a reasonable offer, considering the trouble they were obliged to take. 'The dog-stealers were obliged to watch every night,' they contended … 'and very diligently; Mr. Kelly kept them out very late from their homes, before they could get the dog; he used to go out to dinner or down to the Temple, and take the dog with him; they had a deal of trouble before they could get it.' So Mr. Kelly was expected not only to pay more than the value of his dog, but an extra amount on account of the care he had taken of his terrier, and for the trouble his vigilance had given to the thieves!
>
> (Mayhew 1968: 50)

There were therefore two things that were specially striking about dog-stealing as a criminal activity. The first was its apparent professionalism and organisation, and the sense that this was a systematic attack on bourgeois property and propriety, as befitting Elizabeth Barrett's *banditti* or 'Confederacy' of dog-stealers (Karlin 1990: 303). This professionalism extended to the surveillance of bourgeois homes with valuable pets, a neat reversal of the usual relation between the bourgeoisie and the city's poor. As in the case of Fitzroy Kelly, Elizabeth Barrett's home was said to have been watched for some time before Flush was first snatched. Furthermore, the role of the go-between, and 'the burlesque dignities of mediation' that went with it, to use Robert Browning's words (Karlin 1990: 308), meant that men like Taylor were virtually immune from prosecution. Few observers failed to comment on the ingenious and alarming system of extortion that was the result:

> Obviously there is great art in dog stealing. The skulking vagabonds whom one meets in the streets, carrying puppies under their arms, are too clever for the general public. They belong to a society. If they sell a dog, they know how to regain it in less than a month. They suffer a sufficient interval to elapse for the owner to become attached to his purchase, and then the dog is safely lodged at the office of some 'society.' In due time the inevitable adver-tisement appears, the reward is secured, and the dog is returned. .... [T]hus we see how it is that people gain a livelihood by dogs. They are perpetually on the lookout for favourable specimens of that interesting tribe. They know exactly when a handsome dog may be picked up at any moment.
>
> (*London Review*, quoted in *The Times*, 7 October 1861: 9)

The second notable element was the viciousness, implied or actual, of the dog-stealers' trade. As well as Elizabeth Barrett's story 'of a lady ... having her dog's head sent to her in a parcel' (Karlin 1990: 304), we might note the experience of a Miss Mildmay whose dog was threatened with ill treatment if she did not pay a ransom of 6*l*., or the Miss Brown of Bolton Street, who came 'in great terror' to William Bishop saying that 'to have the poor dumb animal's throat cut was a very frightful affair' and who was supposed to have subsequently quitted the country as a result (*Report*: 26). Such casual cruelty was visited in cases like these on women made vulnerable precisely by their capacity for emotion and tenderness, and on their dogs, themselves noted for their loyalty and nobility.[6] It is not hard to see why the dog-stealers were so reviled.

The class and even racial significance of dog-stealing ought then to be immedi-ately evident. The 1840s were revolutionary times, of course, and, as markers of class distinction, pets were functional to the symbolic economy of the bourgeois city. The dog-stealers in their turn figured equally as a class enemy and as an alien threat – thus note Barrett's use of the words '*banditti*' or 'philistines' when refer-ring to these representatives of London's criminal underworld. Dog-stealing may be taken to raise the spectre of class conflict, something that the proximity of Wimpole Street and Whitechapel also suggested most forcefully. We might also see, in the whole dreadful business of dog-stealing, the threat of regression as well as revolution, for, if Turner (1980: 78) is right to argue that compassion for animals 'quelled the fears of man's bestial past' and 'served as an emblem of the heart and an example to the human race', so that kindness to animals was the *sine qua non* of civilisation itself, then dog-stealing threatened to take this civility away and bring to the surface the tensions induced by human kinship with animals. Dog-stealing ultimately suggested the inversion of social hierarchies, putting human dependence on animals at the centre of the problem; people were dominated and exploited through their dependence on animals and their own affections and sentiments, which could not be excluded wholly from their lives. Dog-stealing was thus a phenomenon that called into question conventional assumptions about human relations with animals: it recognised that, as in Dickens's *Great Expectations*, 'the uncivilized may be banished but may return' (Munich 1996: 140).

A peculiar species of property: on utility and pricelessness

Dog-stealing, then, was seen as systematic, professional, cruel, heartless and virtually beyond the reach of the law. *The Times* (2 August 1844: 4) concluded that, 'in the matter of dog-lifting the "great fact" which meets the public eye is, that nobody can keep a dog that is worth anything, or for which they can be ascertained to feel any affection'; 'On this phenomenon of the day', it went on, 'policemen, dog-dealers, and the losing public are wholly agreed, and Parliament bestirs itself to probe the evil.' As *The Times* noted, the problem was acute enough for Parliament to sanction a Select Committee to consider changing the law to make it more effective in combating the dog-stealers. One of the major obstacles here, however, was that it was not at all clear what *kind* of crime dog-stealing actually was, and the ineffectiveness of the existing law resulted precisely from this legal ambiguity. The whole question turned on the difficult question, in English law, as to the value of dogs as *property*. Given a dog's liability to stray, for instance, in contrast to most other types of property, it was difficult to draw sharp legal distinctions between stealing, receiving and simply 'finding' 'lost' dogs. As the Metropolitan Police Commissioner Richard Mayne reported to Parliament:

> A dog is that sort of animal, allowed such liberty, and that runs about in that kind of way, that he may be enticed away in a manner that no other property can be. It is not possible for a horse or a sheep, or any such animal, to be enticed away like a dog. You do not find them straying about as dogs do.
>
> (*Report*: 3)

The dog was, as the Radical MP Joseph Hume informed the House of Commons, a very peculiar and ill-defined 'species of property' (*The Times*, 12 June 1845: 3).

For one thing, it was difficult to assess the *value* of a pet dog. While long-standing tradition insisted that animals were to be understood in terms of their *utility* for human beings, the notion of 'aesthetic utility' – pets as sources of pleasure pure and simple – was difficult to reconcile with traditions of utility derived from a more bluffly pragmatic rural society.[7] In this vein, a dog such as a toy spaniel was particularly difficult to justify on grounds of utility alone, all the more so in that it was stereotypically a woman's pet.[8] So while Ritvo (1987: 2) can assert that '[n]ineteenth-century English law viewed animals simply as the property of human owners, only trivially different from less mobile goods', it is clear that there was nothing simple about this.

Indeed, whether one could even have property in a dog was a legal problem of some complexity (Seipp 1994: 85). From Tudor times, certain animals such as dogs, cats and animals kept for sport, as well as certain types of landed property, were specifically excluded from the larceny statutes. Judges predisposed to leniency could thus propose that certain types of stolen goods, if they could not be undervalued in the defendant's favour, might be taken as formally without value in law (Milsom 1981: 426). This 'strange doctrine of pricelessness', as the legal historian Baker (1978: 318) puts it, was designed to mitigate the rigours of a vengeful penal code. Scruples over the theft of objects of luxury, things of 'pleasure', that is, rather than things of 'profit', might be extended even to such items as jewels and banknotes, but the classic examples were birds, hawks and – especially – hounds.[9] Now though this was a legal fiction, and though this was strictly a question of value rather than of property, by the nineteenth century the argument based on lack of utility remained central. Unlike working animals or animals consumed as food, because of their lack of obvious utility dogs were undervalued as property. According to Blackstone, it was a felony to receive any money for the restoration of stolen property; but only animals that served for food were recognised as property, not creatures kept for whim and pleasure. Civil actions for loss were then acceptable, but not those for larceny. For Richard Mayne, it was as a result absurdly inconsistent to make dog-stealing only a misdemeanour while the taking of rewards for recovery would be a felony subject to the maximum penalty of life transportation (*Report*: 3).

In the middle of the nineteenth century, however, there were still many who felt that these distinctions made by the law with respect to property remained wise and beneficial, and that to alter them would produce bad effects on criminal jurisprudence. They felt, for instance, that to add to the felony statutes was inappropriate given the nature and worth of the animals involved, and also because such a move was contrary to the spirit of the age. In the debate over the proposed dog-stealing bill, the Liberal MP for Kinsale objected to a sentence of up to eighteen months 'even if a person stole all the dogs in England' (*The Times*, 26 June 1845: 3). Another pointedly asked why dogs should be singled out for special treatment: 'A person did not commit felony by stealing a ferret or any such animal', he noted, 'and he would wish to know what distinction could be drawn between a favourite cat and a favourite dog, that would justify them in making the stealing of one a larceny, while the stealing of the other was not larceny' (*Hansard*, LXXXI, 1845: 1185–1186).[10]

The bill's supporters did their best to assert the honour of the canine tribe. First of all, they argued, the distinction based on utility was misleading, because dogs may be said to be equally as useful as those animals who provide food for human beings. The promoter of the bill, H.T. Liddell, pointed to the example of drovers' dogs, watchdogs and sheepdogs. Secondly, the definition based on utility was argued to be clearly unsatisfactory. The bill's antagonist Joseph Hume, taking a strict stand on the principle of utility, would, as *The Times* (12 June 1845: 3) noted, 'by the same rule, license a thief to take a lady's necklace, whilst he protected her watch because it told her the time of day'. Finally, it was argued, dogs ought to be recognised, irrespective of these legal niceties, as the valuable property they clearly were for the benefit of the Exchequer, and thus deserving of the protection of the law: one letter-writer thus insisted that '[t]hose who indulge their fancy by keeping dogs pay a heavy tax to Government for that indulgence, and I cannot see why they are not to be equally protected as horses, or any other property which we may keep for our pleasure or amusement' (*The Times*, 25 March 1845: 7). The *Report* similarly based its arguments on the intrinsic worth of dogs that might, in the case of a beautiful spaniel, fetch a hundred guineas on the open market:

> However extravagant these prices may appear, Your Committee conceive that the value of any description of property must be measured by the price which such property will fetch in the market.
>
> (*Report*: iv)

The irony here, in that this mirrors the free trade in affections so complained of in the practice of the dog-stealers, is nowhere apparent to the bill's authors: the argument is simply deployed to the effect that any doubts as to dogs being *property*, in the legal sense, should be removed. Crucially, the Home Secretary, Sir James Graham, was won over by these arguments, straightforwardly insisting that '[t]he possession of a dog was a possession recognised in law' (*Hansard*, LXXXI, 1845: 385). Graham likewise noted the absurd legal anomaly by which an offender might be transported for stealing a collar, as a felony, but not for stealing the dog that wore it:

> A very short time ago, the penalty of death was attached to the larceny of a sheep; and it was now transportation for life. The same state of the law

applied, he believed, to the stealing of a jackass: and he would wish to know why they were to transport a person for life for stealing a jackass, or for seven years for stealing a dog-collar worth 7s. 6d., while no indictment could be preferred for stealing a dog worth 20l. and upwards?

(*Hansard* LXXXI, 1845: 1186)

The dog-stealing bill duly passed into law in 1845 (*8 and 9 Vict., c.47*). Dogs were now clearly identifiable as property, and offenders could be visited with the full rigour of the law, with the hitherto untouchable receivers now liable to punishment for misdemeanour.[11] This seemed to be effective enough in curbing their depredations. Though *The Times* continued to report dog-stealing cases throughout the nineteenth century, Mayhew could report at mid-century that while 'there might still be a little doing in the "restoring" way … it was a mere nothing to what it was formerly' (Mayhew 1968: 57), with only some half-a-dozen remaining in the professional dog-stealing line.

Domesticity and domestication: geographies of sentiment and economy

While the essentials of the problem are easy enough to sketch in, dog-stealing cannot be fully understood without appreciating the wider cultural geography in which the problem was embedded. We should remember that the principal *spaces* on which the crime of dog-stealing impacted were the private spaces of the home, from whose protection pet dogs were abducted, and the public spaces in which they were most vulnerable. These were not, of course, the rigidly separated spheres of Victorian bourgeois ideology. However much they tried, the Victorian bourgeoisie could never prevent these spaces from mixing and merging into one another. And it was not simply that the domestic sphere allowed the pressures of the market economy to be temporarily suspended and resolved (Armstrong 1987; Poovey 1988). As Nunokawa (1994) has emphasised, the zones of circulation associated with the public world of the marketplace were intimately connected to the zones of possession associated with the home. Property, being ephemeral and vulnerable, flew in the face of the protocols of propriety that the Victorians taught as the virtue of the home. The domestic sphere, in other words, was not a refuge from the market, as the Victorians liked to imagine. It was both a prop to the workings of the public world of men and the marketplace, and constantly, necessarily, infused with the values of the market. The concept of domestic security, presided over by the female 'angel of the house', was nothing more therefore than an elaborate ideological fiction which attempted to shield the Victorian bourgeois imagination from the forces of commodification associated with the world of property. Domesticity was thus a decisively *gendered* discourse that marked out a particular bourgeois world which stood for values *beyond* price at the same time as it was constantly shot through and threatened by a world in which everything was shown to have a price. Dog-stealing brought both this threat to domestic virtues, and the exposure of the market value of domesticity, uncomfortably and cruelly to the surface: it testified to the vulnerability of domestic property and the siege mentality which characterised the Victorian discourse of domesticity. As Nunokawa (1994: 4) puts it, resonantly, 'everywhere the shades of the countinghouse fall upon the home.'

The gendered nature of the parliamentary debate about dog-stealing is revealing of these ambiguities surrounding the issue of domesticity, utility and property. While supporters of the dog-stealing bill did their best to pose the problem as an outrage against *men*'s property, with Liddell insisting that 'no man's dog was safe a moment – and the more valuable it was the more certain it was to be stolen' (*Hansard*, LXXVI, 1844: 555), the bill's opponents were quick to express their reservations about legislating in the realm of womanly sentimentality.[12] In a general chorus of resentment at the ridiculousness of extending protection 'to the poodles and lap-dogs of the metropolis' (*Hansard*, LXCI, 1844: 691), unsympathetic MPs insisted that dog-stealing was already punishable by fine, imprisonment and whipping, 'and surely that was enough to protect the pug-dogs of the old ladies of England' (*Hansard*, LXXXI, 1845: 385). Dog-stealing was difficult to combat, then, precisely because the question of property, founded on the principle of utility, was preeminently a gendered issue.[13] This was not simply because lapdogs, the most 'useless' and vulnerable of pets, typically belonged to women. More than this, the association of non-useful animals with women confined to the house as domestic essence and ornament was particularly telling.[14]

The bourgeois ideal of domesticity, so crucial to constructions of femininity, was at the centre of these concerns. Indeed, the link between the ideal of domesticity and the highly charged social-symbolic process of animal domestication (Anderson 1997) deserves to be underlined. The keeping of dogs as pets was a powerful emblem of this domestic ideology, of course, and 'domestication' must be taken quite literally in the case of dogs: for 'the locus of the "house" came to lie within the parameters of a more *inclusively domesticated nature*' (Anderson 1997: 476, emphasis in original). Pets came to express the ideal of the Victorian family, from Queen Victoria's household menagerie downwards: 'Victorian pets … expressed themselves within domestic spaces resembling ideal Victorian homes' (Munich 1996: 129). The dog was regarded as instinctively understanding the concept of 'family', and the pet dog became part of a household whose domestic geography was central to its sense of security. What is more, in this period we may note the removal of animals like dogs from the public spaces of the city to the domestic confines of the bourgeois home. Pet dogs became more important in the city as an object of non-utilitarian 'fancy', far removed from the place of the utility dogs of the countryside, but just as important is the fact that the working dogs of the city increasingly disappeared from view. London's dog carts, for instance, were banned in 1840 (McMullan 1998) after considerable lobbying arising from anti-cruelty sensibilities and from a concern with urban nuisances and sanitation (see Figure 2.2). Dogs came off the public, working, streets, in other words, and into the bosom of the bourgeois home in the explicitly 'useless' form of 'pets'. This process closely parallels the removal of other animals from the city streets to specialised and marginalised spaces – meat markets and slaughter-houses, for instance – as Philo's (1995) discussion of animals, geography and the city has demonstrated; except in this case the business of animal removal is a quite literal process of 'domestication'.

We can trace in the 'bourgeoisification' of dogs a specific geography of exclusion and inclusion to set against the ancient histories of animal domestication. Dog-stealing clearly threatened the domestic security of the household and its property. But it did much more than this: by actively trading on the sentiments of the household, and particularly those of its female members, dog-stealing recon-

Figure 2.2   'No dog carts': London's dogs rejoice at their freedom from labour in the public streets.
Source: Ash, E.C. (*c.* 1934) *This Doggie Business*, London: Hutchinson & Co., opposite p. 74; reproduced by
permission of the Syndics of Cambridge University Library

nected the geographies of sentiment and economy that Victorian bourgeois ideology attempted, through its public/private and male/female distinctions, to hold apart. Dog-stealing was an innately vicious response to the effects of the practical and ideological separation of these geographies. For one thing, the reversal of bourgeois polarities – the servility and subordination of the owner forced to pay a ransom for his/her dog's restoration, the passivity and helplessness of an owner held to ransom by his/her own sentimental attachments to a 'useless' pet – was desperately cutting. It left the owner, whether male or female, in the 'feminine' role associated with emotion, attachment and tender-heartedness. That is why both Robert Browning and the men of the Barrett household advised Elizabeth Barrett to defy the demands of the dog-stealers, this representing a 'manly' refusal to parlay which would put an end to the endless cycle of ransom and abduction. Dog-stealing, moreover, put a price on things – sentiments, attachments, domestic companionship – which should have no price. As a leading article in *The Times* (26 March 1845: 4) put it in reference to a stolen dog which might be valuable either in monetary or emotional terms: such a dog 'may be the companion for years, which are not to be calculated by the rule of "Just so much as it will bring in market overt"'. The *legal* fiction of pricelessness, which shielded the dog-stealers and their trade, was matched by an equivalent *domestic* fiction which both made the trade possible and prompted bourgeois outrage. A correspondent to *The Times* (28 March 1845: 5) summed up the problem succinctly:

> Of all the combinations of rascals which infest this mighty city, there is not one, in my opinion, more hateful than the dog-stealer. Other thieves take our property, – these rob us of our friends. ... They make trade of our affections. They take from us one whose good qualities we should be happy to recognize in many of our human friends; and they compel us to a course of sordid bargaining with a knife at our favourite's throat.

And that this is a gendered question, shot through with the bourgeois ideal of domestic womanhood, is made plain by the Parliamentary Report: 'A worse plan could not be laid, even to set fire to premises, than is laid to get possession of dogs, such as pet dogs belonging to ladies' (*Report*: 30).

What made dog-stealing such a dreadful activity, we might thus conclude, was that it mixed up in a peculiarly outrageous way, for the connected norms of gender and class, the worlds of sentiment and the market, the private and the public worlds. The central problem was putting a price on love and affection, forcing pet owners to assess the monetary value of a pet to them and to their families. In Flush's case, Elizabeth Barrett fiercely rebuked Robert Browning for advising defiance of the dog-stealers and their demands, arguing to him that '[a] man may love justice intensely; but the love of an abstract principle is not the strongest love – now is it?' (Karlin 1990: 306).[15] Stealing dogs put that 'feminine', emotional argument dramatically to the test, revealing pets to their bourgeois owners as a special form of *property* rather than just innocent domestic companions. The phenomenon of dog-stealing thus revealed nothing less than the boundaries of an affective or sentimental economy in which capitalism and culture were complexly interwined. Things which were not supposed to be valued in monetary terms could be revealed as intimately penetrated by market values, so that the simple domestication of animals acted as a marker of the terrible vulnerability of the domestic realm to the public world of cash, commerce and calculation.

The most critical aspect of this penetration of the values of the market into the very world which was supposed to be shielded from it is arguably the position of women within a patriarchal society. Dog-stealing revealed not just the absence of domestic security, but also the ambivalent status of women within that household, a status which women shared with their pets. Squier (1985: 125) in this regard has amply demonstrated the elision in Victorian society of women and their pets, both of whom were 'routinely relegated to a marginal position in modern urban society'.[16] For Squier, in direct reference to Woolf's *Flush: A Biography*, Elizabeth Barrett and her dog were alike victims of a patriarchal society which confined them to a suffocating domestic captivity. The removal from Wimpole Street, in Flush's case involuntarily, and in Barrett's case firstly in the daring expedition to the dog-stealers' Whitechapel and secondly in the escape from the Barrett household altogether, takes on signal power accordingly.[17] Flush's capture by the dog-stealers revealed not only the vulnerability of the Barretts' domestic refuge in Wimpole Street, but also the symbiotic male domination of both Mr Taylor's den of *banditti* and the Barrett household itself. Indeed, for Elizabeth Barrett, the abduction of Flush may be taken to speak to her own personal domestic subjection quite directly:

> The Whitechapel episode is a temptation scene; forced to choose between winning the approval of her male counterparts and saving Flush, Barrett is also being asked, symbolically, to choose between two systems of morality – one masculine and impersonal, the other feminine and personal.
>
> (Squier 1985: 128)

That Barrett chooses to ransom her pet dog, whatever the cost, is thus of tremendous personal and political significance. Since 'Flush's value to patriarchal society is analogous to woman's value' (Squier 1985: 132), the message of the dog-stealing episode was that women, like pets, are worth only what their masters will assign to them; and it is this realisation that connects the serio-comic *Flush* with the high seriousness of a feminist epic like Barrett Browning's *Aurora Leigh*, much preoccupied as it is with the victimisation of women within patriarchal society.[18]

To those who might cavil at regarding dog-stealing as a feminist issue, we can point to a painting by the genre and animal painter Briton Riviere – in his day regarded as second only to Landseer – entitled *Temptation* (1879), in which the dog-stealers' challenge to domesticity, and the concession of the vulnerability of property, domestic security and even woman's virtue, are neatly illustrated (see Figure 2.3). The painting captures the key moment before the abduction of a prized spaniel. In the words of one critic:

> An untrustworthy-looking individual is trying to win the confidence of a King Charles Spaniel with a seductive bit of liver. The drama is unfinished. Whether fidelity or *gourmandise* will turn out the stonger passion we can only guess.
>
> (Armstrong 1891: 18)

Figure 2.3   Briton Riviere's *Temptation* (1879).

Source: Hancock, D. (1998) 'In cavalier style', *Dogs Monthly* 16(9): 11

In this picture, the open doorway to an otherwise solidly prepossessing domestic fortress represents the ideological fiction of domestic security. The ragged 'lurker' preys upon the dog's trusting affections, one hand offering temptation, the other poised to grab his quarry and thrust it into his coat pocket. And the dog with its bow is clearly feminised, both in reference to its position as a lapdog and in its possession by a doting mistress. There are few better illustrations of the vulnerability of property and the siege mentality of the Victorian bourgeoisie. It is not difficult, moreover, to read the painting as not merely denoting dog-stealing but really connoting sexual temptation and the vulnerability of domestic female virtue. A picture like *Temptation*, with its resonant title, and in its emphasis on the fragility of domestic security and virtue, points in fact to two forms of patriarchal tyranny. We have here a fable of woman and pet caught between two worlds – the private space of the domestic hearth, and the public space of the streets and the criminal underworld.[19] There is a political geography to dog-stealing, therefore, not just in the exploitation of the rich by the poor, of Wimpole Street by Whitechapel, but also because the geography of bourgeois domesticity cannot be divorced from the domination inherent in the political geography of 'domestication' itself.

Conclusions: to the doghouse

Few would in fact disagree that domestication is closely related to several strands of *domination*.[20] For Ritvo (1987: 2), the Victorian age marked the working out of 'a fundamental shift in the relationship between humans and their fellow creatures, as a result of which people systematically appropriated power they had previously attributed to animals, and animals became significant primarily as the objects of human manipulation'. But 'domestication' is hardly confined to human–animal relations alone: relations of domination are also evident in the 'domestication' of human beings. Munich (1996) comments in this vein that women and servants (including non-Europeans) were similarly patronised and domesticated, victims of a sadoerotic kind of power which reduced humans to the status of 'pets'. And Tuan (1984) makes this same point most insistently, by extending the definition of 'pets' from animals to people and even landscapes.[21] But if affection and domination are linked, in this *geography* of domesticated nature, then dog-stealing cannot be seen simply in terms of criminal opportunism and bourgeois vulnerability. Flush's horrible experiences point, as several critics have noted, to victimisation at the hands of the dog-stealers *and* in the protective space of the bourgeois home from which he was abducted. The equation of dog-stealers and domestic tyrants is indeed central to Woolf's biography of Flush, and it should make us pause when considering the ultimate significance of dog-stealing. The invalid Elizabeth Barrett in Wimpole Street was as much a captive as poor Flush in Whitechapel. If dogs like Flush are vulnerable in public spaces when off the leash, their chains forgotten in a moment of woman's weakness, as Woolf points out, so are women let out into the public space of the streets only under the restrictions of propriety and fear. Woolf's *Flush* is abundantly illustrative of how an instinctive love of freedom can be turned in patriarchal society to domestic anxiety and confinement: these are the lessons of Victorian bourgeois morality (DeSalvo 1989: 286–287). Just as Whitechapel and Wimpole Street are spatially proximate, just as the geographies of affection and commerce are inseparable, just as property and propriety are linked in zones of circulation and possession, so too in a terrible way is the dog-stealer revealed as an associate rather than an antagonist of the bourgeois patriarch. Dog-stealing is not (just) an attack on the home, it might be taken actively to bolster domestic ideology, acting to confine women, like pets, to the safety and security of the household. Flush's brutal transportation from domestic hearth to criminal rookery is a challenge to the moral geography of the bourgeoisie specifically in the ideal of domesticity for which pets were such potent symbols. The phenomenon of dog-stealing tested the limits to this domestication, revealing the vulnerability and ambiguity of the bourgeois domestic world by putting a price on its interior world of affection and sentiment, and by revealing the patriarchal domination at the heart of both the domestication of animals and the ideal of domesticity itself.

Acknowledgements

For various references, many thanks to Iain Black, Laura Cameron and Liz and Paul Crisp. Thanks also to Chris Philo for his encouragement. Above all, though, Elizabeth Mozzillo collaborated from the very beginning on the research and discussion that went into this chapter, and this is for her, 'for love's sake only'.

Notes

1   This episode is related in Elizabeth Barrett's correspondence, and from this source in numerous biographies (see, for instance, Markus 1995: 58–64).
2   Compare Elizabeth Barrett's reaction to the denizens of Whitechapel – 'The faces of those men!' (Karlin 1990: 311).
3   The best short introduction to the book and its themes is by Flint (1998) in the recent World's Classics edition, which came to my attention after this article had been written. Flint is especially strong and perceptive on the personal and societal contexts for Woolf's sustained concern for the connections between humans and animals. As Flint notes, *Flush* has not received much critical attention, but see Szladits (1970), DeSalvo (1989: 285–288) and, especially, Squier (1985: esp. 122), whose judgement of the book, that 'it ultimately framed a serious critique of the values organizing London's social and political life', ought to be enthusiastically endorsed.
4   An early reviewer, Burra in *The Nineteenth Century*, noted that Woolf used Flush 'almost as Swift used the Houyhnhnms, but without anger, rather with the gentlest irony, weaving together the lives of beast and man' (see Majumdar and McLaurin 1997: 321). If its satire is not quite Swiftian, however, *Flush* is certainly no mere anthropomorphic whimsy.

5   The *Report from the Select Committee on Dog Stealing* gives the addresses of the dog-stealers as follows: Whetstone Park, Lincoln's Inn Fields; Chapel Mews, Foley Place; Keppel Mews South and Woburn Mews, Russell Square; Kensal New Town, Paddington; New Road, St Pancras; Hammersmith; Shepherdess Walk City Road; Bateman Street, Spitalfields; and Cottage Place, Goswell Road.

6   In this regard note the reaction of *The Times* (2 August 1844: 4) to the evidence of William Bishop at the Select Committee, acknowledging that '[t]he mode in which the feelings of the fair sex are played upon forms one of the most instructive points of his evidence.'

7   Anderson (1997) has argued that domestication was the preeminent marker of animal virtue, so that animals were routinely arranged in terms of the nature of their domestication conceived in terms of utility for 'man'. Dogs, of course, have a privileged place in this story, as arguably the first domesticated animals; their gift of aesthetic satisfaction, their assumption of the role of pets, is an important element of this argument. But discussion of domestication must allow for the multiple and competing meanings of the word, its insertion into different discourses, such as, here, *legal* discourse, in which various antagonistic meanings of 'domestication' are obvious.

8   Ironically, while such a spaniel might derive its value from its distinguishing marks, or '*properties*', it did not necessarily qualify as *legal* property.

9   These distinctions were all reassessed by the Criminal Code Commissioners of 1879, who simplified the law on theft by making all tame living creatures (except pigeons, if not in a dove-cote or on their owner's land) equally capable of being stolen (see Stephen 1883: 163).

10 Other opponents of the dog-stealing bill insisted, rather less scrupulously, that dogs were simply a great nuisance, 'generally speaking, of the most deteriorated species, and, like Indian idols, they were prized on account of their ugliness. … [T]here were more instances of human beings dying through hydrophobia from the bite of dogs of the pet class than from the bite of any other' (*The Times*, 26 June 1845: 3).

11 Liddell wanted it to be considered in the first offence a misdemeanour, and only in the second offence a matter of seven years' transportation. Graham advised against holding out for transportation, which was a barrier to parliamentary acceptance of the new legislation.

12 *The Times* (11 November 1844: 5), before it changed its mind on the seriousness of the question, lamented that 'nothing is too trivial or too ludicrously absurd to be made the subject of Parliamentary discussion or inquiry'. When Sir James Graham regretted the inconvenient summer timing of the session, one wit piped up from the back benches, to laughter, that he was referring to the 'dog days' (*Hansard*, LXXVI, 1844: 556). *Punch* noted with rather less wit that the report on dog-stealing was presented to the House in a terribly dog-eared state (see *The Times*, 8 August 1844: 5).

13 Thomas (1984: 102) notes that the 'passion for unnecessary dogs' was a marker of class distinction, but it was also crucially a gendered issue. It is better to think of class and gender in this respect as mutually articulated categories.

14 Note the importance for domestication discourse of the virtue attached to submitting to 'man's' dominion, with the dog as the very emblem of servility and subordination, physically malleable and genetically manipulable. This is clearly matched by the gender hierarchies and proprieties of the Victorian age.

15 Lorenz, in his book on human–canine relations, *Man Meets Dog* (1954: 142), quotes Elizabeth Barrett Browning to this effect: 'If thou must love me, let it be for nought/ Except for love's sake only.'

16 A character in Wilson's *Gentlemen in England* (1986: 318, emphasis in original) notes that Elizabeth Barrett Browning 'couldn't write poetry any more than her … husband but she was a *pet*'.

17 Of course this is why the escape to Italy, for Flush and Elizabeth Barrett Browning, is so significant. Italy represented, for both, an escape from the confining class, gender and racial oppressions of Victorian London. Flush, in Italy, is divested of his 'aristocratic' pedigree along with his domestic captivity. Freed from dog-stealers and from patriarchal tyranny alike, he, like Elizabeth Barrett Browning, and like her creation Aurora Leigh, is given access to a new type of city life, democratic, affective, 'marked by equality between genders and classes' (Squier 1985: 134). Ironically, in the home of the real *banditti*, in a land of mongrels, there was nothing to be

feared from dog-stealers. Without domestic oppression, the menace of the dog-stealer melts
away. As Flush realises: 'Where were chains now? … Gone, with the dog-stealers. … He was
the friend of all the world now. All dogs were his brothers. He had no need of a chain in this
new world; he had no need of protection' (Woolf 1933: 110).

18 Thus Flint (1998: xxi–xxii) comments: '[W]e are faced with the fact that if for Elizabeth
Barrett, Flush was a love object, a person in his own right … for another set of people he was
an object of economic exchange: valuable for his breeding, his pedigree, and because emotional
worth could be turned into a site of blackmail.'

19 Drawn from life, Briton Riviere's painting of a dog-stealer in action comes layered in irony. An
animal painter like Riviere needed a dealer to provide canine models, but, as the critic
Armstrong (1891) notes, such dealers often dabbled in the illicit trade of stealing and
ransoming dogs. It is entirely possible, therefore, that a producer of sentimental animal pictures
like Riviere, engravings of whose pictures might be found on the walls of many bourgeois
homes, would employ dog-stealers to provide the very subjects of his pictures; another form of
trading in affection, perhaps.

20 The view that animals were in the Victorian age granted autonomous significance and freed
from moral subordination to human beings (Turner 1980) has largely been superseded by the
argument that sees in the ethic of care and affection another form of domination and conquest.
In her recent critical cultural geography of domestication, Anderson (1997: 478) demurs somewhat, insisting on a mix of moralities of control and care, so that there is 'no simple imposition
of mastery,' but nevertheless she reinforces the point that the social-symbolic process of
'domestication' remains enduring and problematic.

21 Elizabeth Barrett knew this, for sure – in poems like *Aurora Leigh* (1996: 24) she regrets an
England whose nature is 'tamed/ And grown domestic' (Book I: 634–635). In her poetry, Italy,
the refuge from the dog-stealers, stood in for a world of wildness and passion; there is a vital
critique of the domestic subjection of women here, wholly as powerful and compelling as
Virginia Woolf's plea for a room of one's own.

References

Anderson, K. (1997) 'A walk on the wild side: a critical geography of domestication', *Progress in
Human Geography* 21: 463–485.

Armstrong, N. (1987) *Desire and Domestic Fiction: A Political History of the Novel*, Oxford: Oxford
University Press.

Armstrong, W. (1891) 'Briton Riviere R.A.', *Art Annual*: 1–32.

Baker, J.H. (ed.) (1978) *The Reports of Sir John Spelman, Volume II*, London: Selden Society.

Browning, E.B. (1996) *Aurora Leigh*, New York: Norton.

DeSalvo, L. (1989) *Virginia Woolf: The Impact of Childhood Sexual Abuse on Her Life and Work*, London: The
Women's Press.

Flint, K. (1998) 'Introduction' to V. Woolf, *Flush*, Oxford: Oxford University Press.

Karlin, D. (ed.) (1989) *Robert Browning and Elizabeth Barrett: The Courtship Correspondence*, Oxford:
Oxford University Press.

King, J. (1994) *Virginia Woolf*, Harmondsworth: Penguin.

Lee, H. (1996) *Virginia Woolf*, London: Chatto & Windus.

Lorenz, K. (1954) *Man Meets Dog*, London: Methuen.

McMullan, M. (1998) 'The day the dogs died in London', *London Journal* 23: 32–40

Majumdar, R. and McLaurin, A. (eds) (1997) *Virginia Woolf: The Critical Heritage*, London: Routledge.

Markus, J. (1997) *Dared and Done: The Marriage of Elizabeth Barrett and Robert Browning*, London:
Bloomsbury.

Mayhew, H. (1950) *London's Underworld*, London: Spring Books.

Mayhew, H. (1968) *London Labour and the London Poor*, New York: Dover.

Milsom, S.F.C. (1981) *Historical Foundations of the Common Law* (Second Edition), London: Butterworths.

Munich, A. (1996) 'Domesticity; or her life as a dog', in *Queen Victoria's Secrets*, New York: Columbia University Press.

Nunokawa, J. (1994) *The Afterlife of Property: Domestic Security and the Victorian Novel*, Princeton: Princeton University Press.

Philo, C. (1995) 'Animals, geography and the city: notes on inclusions and exclusions', *Environment and Planning D: Society and Space* 13: 655–681.

Poovey, M. (1988) *Uneven Developments: The Ideological Work of Gender in Mid-Victorian England*, Chicago: University of Chicago Press.

Report from the Select Committee on Dog Stealing (Metropolis); together with the minutes of evidence taken before them, British Parliamentary Papers 1844 (549) XIV. 291.

Ritvo, H. (1987) *The Animal Estate: The English and Other Creatures in the Victorian Age*, Harmondsworth: Penguin.

Seipp, D.J. (1994) 'The concept of property in the early common law', *Law and History Review* 12: 29–91.

Squier, S.M. (1985) *Virginia Woolf and London: The Sexual Politics of the City*, Chapel Hill: The University of North Carolina Press.

Stephen, J.F. (1883) *A History of the Criminal Law of England, Volume III*, London: Macmillan & Co.

Szladits, L.L. (1970) 'The life, character and opinions of Flush the spaniel', *Bulletin of the New York Public Library* 74: 211–218.

Thomas, K. (1984) *Man and the Natural World: Changing Attitudes in England, 1500–1800*, Harmondsworth: Penguin.

Turner, J. (1980) *Reckoning with the Beast: Animals, Pain and Humanity in the Victorian Mind*, Baltimore: Johns Hopkins University Press.

Tuan, Y.-F. (1984) *Dominance and Affection: The Making of Pets*, New Haven: Yale University Press.

Wilson, A.N. (1986) *Gentlemen in England*, Harmondsworth: Penguin.

Wolch, J. and Emel, J. (1995) 'Bringing the animals back in', *Environment and Planning D: Society and Space* 13: 632–636.

Woolf, V. (1933) *Flush: A Biography,* London: Hogarth Press.

3   Feral cats in the city

Huw Griffiths, Ingrid Poulter and David Sibley

Introduction

In a recent advertisement for the Cat Action Trust, cute 'little Darcy' is shown stepping from an existence in the wild towards a 'normal' domestic life as a pet kitten. The text of this advertisement signals a particularly negative view of feral animals in urban environments. It maintains that:

> This little kitten was born in the wild and was facing a life of hunger, disease and terror. … Within a few months, little Darcy would have been on his own. Feral cats are *the inhabitants of dereliction*. They gather wherever they can find food and shelter. The colonies MUST be controlled or the neighbours complain and the cats are killed by the authorities. Please help us to save cats and kittens by sending a donation to help our work. We have Groups, nation-wide, saving and caring for feral cats who need love and so much attention if they ever hope to have a happy life.
>
> (our emphasis)

The clear message of the Cat Action Trust is that these cats need to be saved from their feral misery.

This representation of feral cats conveys nothing of the ambiguity of people's attitudes towards cats, and yet, whether domestic or feral, they evoke various responses which demonstrate a mixing of affection, fear and distaste. In the case of feral colonies, we suggest that their material environment also influences views of cats, whether benign or malign. Our study of feral cats in Hull suggests a number of possible relationships between people, animals and place, and a complexity which is at variance with the singular, negative view projected by the Cat Action Trust. More generally, representations of feral cats in their relationship with the built environment and urban 'wilderness' provide a commentary on atti-tudes to nature and civilisation.

The domestic and the feral: human–animal relationships in the home and in the wild

Human responses to animals can be captured in a dialectic of desire and disgust, domination and affection. Perin (1988: 117) remarks on '[a] remarkable ambivalence ... built into the dog–human bond'. On one hand, dogs serve to complete a person or a family, and are treated with affection similar to that given to people. On the other hand, as Perin (1988: 118) points out, in common English and American usage, ' "a dog's life" is an unhappy, slavish existence; and a "dog's death" is a miserable, shameful end.' Tuan (1984) argues similarly that the treatment of pet animals by people manifests a mix of affection and cruelty, and that affection itself cannot be separated from a desire to dominate. A pet may be expected to fit into the home, to become, as Tuan (1984: 107) puts it, 'as unobtrusive as a piece of furniture'. This may be achieved by de-naturing them, by spaying females and castrating males so that they become 'less smelly and dirty'. The spayed or castrated and regularly shampooed animal is thus rendered suitable for home life. Also, de-clawing, although illegal in the UK, is still practiced, as it is in other Western European countries. Cats' and dogs' defiling traits are removed, and the animal becomes a suitable object for affection. This suggests a common aversion to *untamed* nature if it appears as such in a domestic setting. Baker (1993: 104) suggests that

> animals – and cultural constructions of 'the animal' – will invariably figure as the negative term when used in binary oppositions. This is perhaps why, in the post-Cartesian West with its continuing appetite for the dualistic and oppositional, animals seem to figure so overwhelmingly negatively in our imaginative and our visual rhetoric.

In some cultures, it is the 'naturalness' of familiar animals which renders them unambiguously defiled. Thus, for Gypsies, cats are defiled because they lick their genitals and thus fail to maintain a distinction between the 'pure' outer body and the 'impure' inner body. For this reason, they are totally excluded from Gypsy space. Similarly, Gypsy dogs, which have the same defiling traits as cats, are only kept because they work – as hunters or as guard dogs – and they are never allowed in the living space (except, occasionally, if small breeds like Yorkshire terriers – so much for pollution taboos). Agee and Evans in *Let Us Now Praise Famous Men* (1965) made a similar observation about dogs in poor rural communities in the American South in the 1940s. They might have been useful, like farm cats kept as ratters, but they were not objects of affection. According to modern Western standards of animal welfare, they were neglected.

There are a number of themes here: domination, domestication and, through

the elision of the animal and the human, racism and a more general dehumanisa-
tion. These relationships between people and animals in the recent past, and in
societies other than the dominant ones in the West, serve to highlight their
cultural specificity. Historical changes in the treatment and perception of animals
underline this point. Thomas's (1996) history of animals in British society demonstrates that a number of species have moved from inside to outside of the
domestic economy, and have been variously treated with affection and disgust. In
seventeenth-century England, for example, bees were allegedly treated as if they
were a part of the human community. 'Bees would hate you, said an authority, if
you did not love them' (Thomas 1996: 96), but what Thomas refers to as the
'Eastern' view of dogs as filthy scavengers was still current in the sixteenth and
seventeenth centuries. 'In 1662, the preacher Thomas Brooks classified dogs with
vermin,' and, jumping a few centuries, Freud also saw dogs as reprehensible
because they had no horror of excrement and no shame about their sexual functions (Thomas 1996: 105–106). Thomas's historical survey and cross-cultural
comparisons demonstrate that there is nothing necessary or immutable in people's
responses to the wildness of animals. Their inclusion or exclusion, whether or not
they are treated with affection or disgust, depends on their placing within a
particular cosmology.

Clearly, animal categorisations do have particular cultural and historical
contexts, but they can also be rather slippery. Animals may resist categorisation.
Pet cats in modern Western societies are interesting in this regard because it is
generally accepted that they will retain some of their wildness. The designation
'pet' generally indicates belonging: a placing in the home, either sharing space
with people, adorning the carpet or sleeping on the settee, or confinement within
domestic space as in the case of goldfish, lizards or budgerigars. In the case of
dogs, moving beyond the confines of the home is usually under the control of a
human. However, 'putting the cat out at night' signals incomplete containment in
the home, only a partial domestication, even though cats may be cosmetically
modified to fit conceptions of the homely and domestic. Similarly, the cat-flap is a
breach in the domestic boundary, and cats bringing mice or birds into the home
may still be seen as polluters of domestic space. In this respect, pet cats are transgressive, breaking the boundary between nature and culture.

Feral cats occupy a zone somewhere else on the domestic–wild spectrum. As
members of the same species as pet cats, either ex-pets, dumped cats or born into
a feral colony, they are all potential pets – they could be 'rescued', as the Cat
Action Trust suggests that they should be – but the feral cat also embodies wildness, more so than the domestic version. Some ferals may live in close proximity
to humans, even occasionally occupying domestic space, whereas other feral
groups live entirely apart from people while depending on resources generated
by human settlement such as vermin, scrap food, buildings and waste materials

(the latter two providing shelter). Perceptions of these feral animals then overlap with those of the Wild Cat (*Felinus sylvestris*) from which the domestic cat (*Felinus catus*) is descended. Thus, the pet/feral/wild boundaries are blurred.

'The wild': cats on the margin

In overdeveloped societies, cats as wild animals also have a place on the fringes of civilisation, as a real or imagined and occasional threat to the settled population, and this is probably reflected in attitudes to feral animals in appropriate settings: that is, in marginal spaces characterised by dereliction or uncultivated nature. Berger (1980: 9) has asserted that '[a]nimals have gradually disappeared. Today, we live without them,' meaning that we no longer have that intimate relationship with animals which characterised peasant cultures. Particularly, we could argue either that urban living has resulted in the incorporation of animals into the private sphere (as pets), or that urban culture has removed them to a real or imaginary 'wild' or to some rural past (in Baker's terms, to 'the authentic reality of the good meat-eating peasant': Baker, 1993: 13).

We can see this removal of animals as a loss which might be compensated to some extent by imagining or encountering wild animals on the fringes of cities and in interstitial spaces. There seems to be a need to retain a space for wildness or to recover a time when people had a more intimate relationship with animals. Jackson (1981: 66), in her writing on fantasy and the Gothic, has suggested that people's fears (and desires) are commonly located in unnatural figures, but we could argue that animals which are imagined or whose wildness and danger are exaggerated similarly embody deeply rooted fears and desires. In Jean Genet's portrayal of dusk (quoted by Minh-ha 1996: 101–102), he captures the ambiguity of this transitional time in the expression *entre chien et loup*, between dog and wolf, that is, a time of day when one cannot be distinguished from the other, and he also describes it as 'the hour of metamorphoses when people *half hope, half fear* that a dog will become a wolf' (our emphasis). This quotation exemplifies the ambiguously *unheimlich*, the strangely familiar, which, according to Freud, entailed

> that class of the frightening which 'arouses dread and horror in the old and long familiar' (Freud, 1919 [1990]). [The uncanny] has 'the power to signify the development of meaning in the direction of ambivalence, from that which was familiar and homely [*heimlich*] to that which has become unfamiliar, estranging [*unheimlich*]'.
>
> (Chisholm 1993: 436; see also Wilbert 1998; Wilton, 1998)

Wildness in cats is, on the one hand, 'unsettling'[1] because it contradicts the idea of the cat as a domestic, pet animal close to humans, and, on the other hand, it is a source of desire. In psychoanalytical terms, a fascination with wild cats and with wildness in cats signifies a repressed longing for intimate contact with the natural world.

Representations of the wild cat, often 'sighted at dusk', convey a sense of the *unheimlich*, a repressed relationship with wild nature. There have been hundreds of 'sightings' of large wild cats annually in Britain since the Surrey puma in the 1960s, and some of the evidence – footprints, prey which could only have been killed by a large cat like a puma – is quite convincing. However, many of the sightings are probably imagined or they could be ferals of exaggerated dimensions (Bottriell 1985). Some of these reports convey a strong sense of dread. Thus, in regard to the celebrated Beast of Bodmin:

> Another farmer ... is so convinced that the Beast exists. He told *The Independent* that 'you'll know when the Beast's there. There'll be no rabbits or foxes about and the birds stop singing.' ... And [a local builder] told reporters that he'd had an eerie late night encounter with the Beast near the famous Jamaica Inn.
>
> ('The case ...' 1998)

As sources of fear and excitement, the large feral cat, panther or lynx glimpsed at dusk is akin to the wolf in medieval Europe (Fumagalli 1994), an element of the wild which threatened the precarious hold of settled society over nature (and which has now returned to the fringes of the city in Spain, reforming this boundary between the civil and the wild). The large feral animal, either seen or imagined 'out there', represents both a loss, something now enjoyed only vicariously in TV wildlife programmes, and a threat (to the boundaries of urban civilisation).

Those animals which transgress the boundary between civilisation and nature, or between public and private, which do not stay in their allotted space, are commonly sources of abjection, engendering feelings of discomfort or even nausea which we try to distance from the self, the group and associated spaces (but which we can never banish from the psyche). This is clearly the case with cockroaches and rats which invade public and domestic space, emerging from where they 'belong', out of sight on a stratum below civilised life, and eliding with other cultures in racist discourse to symbolise racialised 'others' (Hoggett 1992; Sibley 1995). But larger animals, like foxes, badgers and wild boar, may also be perceived as transgressive ('out of place') and abject. Thus, while there may be a sense of a loss of contact with animal nature, there is at the same time a sense that the boundary between urban civilisation and animal nature has to be maintained, a fear of the merging of culture and nature. In Britain, this concern is expressed in boundary maintenance and surveillance by environmental health officers, animal welfare organisations and – in the case of sightings of big cats, 'beasts' of Barnet, Brookman's Park, Bodmin and Barnsley, even the Norfolk Gnasher – by the police.

Ideas about which animals belong and which do not belong in urban society result in feral cats, like urban foxes, being regarded as highly transgressive and ambiguous. The feral cat occupies a range of relationships with human populations, from a proximal but non-tactile one to a distant relationship, collapsing in the extreme into these fantastic images of a threatening animal 'other'. They may be perceived as both inside and outside of urban society. This provides one starting-point for understanding feral cat colonies as they vary across the city. The second is a categorisation of urban 'cat spaces'.

Wild things in an ordered urban space

The urban environment is one where nature has been contained and transformed. The city is subject to an ordering process which signals what can be included in urban space and what does not belong, what Foucault referred to as epistemic principles 'defining what objects can be identified, how they can be marked and in what ways they can be ordered' (Bannet 1989: 144). The idea of an ordering system which can be used to characterise a society does, however, need considerable qualification. We have suggested above that people are ambivalent about wild nature, including animals, which can be simultaneously despised and admired. Thus, the place of animals in the city is uncertain and often contested, rather than being determined by epistemic principles. These principles constitute the basis of a power structure or an ordering system, but one which is clearly resisted at the local level.

Some more light is cast on this problem by Peters, who, in a remarkable essay written in 1979, anticipates recent psychoanalytical readings of the city. She uses the metaphors of 'the forest' and 'the clearing' to describe the changing relationship between people and the built environment. The forest, she suggests, 'is still present in our inner lives. It is our pre-conscious stage. Generally, it is so overlaid that we are not aware of it. It surfaces only in dreams and myth' (Peters 1979: 80). The clearing is a place of comfort and security, but also of fear, 'for the security is never secure enough. There is always more work to do, excluding the forest, categorising and creating' (Peters 1979: 80). Peters (1979: 80) then claims that '[we] push back the inner forest by rejecting everything that does not fit into our self-image, pretending these rejected instincts and emotions do not belong to us.' She cites the case of a woman whom she knew who had an intense dislike of dandelions, recognising the dandelions as a scapegoat, like immigrants or dogs:

> [I]t doesn't matter what we fix on. [We] imbue it with all the characteristics which we refuse to recognise in ourselves. Then we spend all our hatred on

it, pushing it outside the city, outside the clearing. We destroy it and, in the process, we destroy ourselves.

(Peters 1979: 84)

This argument anticipates Kristeva's thesis on abjection (Kristeva, 1982), and elsewhere in her book Peters suggests that we can reconnect with nature by purposefully breaking down this urban order, and by accepting 'weeds' and other natural elements that are 'out of place' according to hegemonic ordering principles.

In this vein, we could argue that 'wild spaces' in the city which provide the habitats of feral cats may be particular sites of aversion, signifying 'the forest' intruding into the city, although some people will see the same sites as valuable refuges for those features of wild nature which we have otherwise lost. Thus, such wild spaces may be sites of conflict. Overgrown derelict spaces might fall into this category. Clearly, there will be no necessary tension over the presence of feral colonies in such places if they are avoided or off the map. It is likely that feral cats will be seen as most discrepant where the land is ordered and managed, with pressure from some people to have the cats removed while others enjoy their wildness. Thus, in Palo Alto, California, in 1995, there were complaints about ferals, which suggested that the animals were decidedly 'out of place' in a residential neighbourhood. One of those unhappy about the presence of ferals claimed:

> There are health issues associated with cats going all over in flower pots, sand boxes and yards. ... Granted they were here before us, but how can you have wild animals in a residential neighbourhood?
>
> ('College Terrace ...' 1995)

In support of acceptance of ferals in the wild, Remfry (1996: 520) maintains that feral cats are 'intrinsically interesting and a source of pleasure to many people who feed them and care about them'. In particular, feeders will defend what they see as the interests of cats, perhaps by refusing to give information to either pest control officers or others who may be seen as potentially hostile. A third, probably less common situation is one where feral cats are a familiar part of the urban landscape and are generally accepted as belonging, such as the Fitzroy Square colony in London which was admired by members of the Bloomsbury set, provided the subject for T.S. Eliot's cat poems, was celebrated in music by Andrew Lloyd Webber, and was still in existence in the 1970s (Remfry 1996).

Feral cats in Hull

These arguments about the relationship between people and wild nature, linked to the placing of animals, helped us in thinking through the relationship between people, place and feral cats in the city of Hull. We were able to locate thirty-one feral cat sites in the Hull area. This was done, first, by one of us (IP) contacting environmental health officers in the city who had addresses from which complaints about ferals had been received. As the agency responsible for pest control, the Environmental Health Department of the City Council has an obliga-tion to respond to these complaints. Additionally, information came from a voluntary organisation, Hull Animal Welfare. In some of the locations it was possible to make contact with friends of ferals, including feeders, as well as people antagonistic to the cats, and these contacts provided further information on other sites. Thus, we were able to build up a 'cat map' of the city. Fourteen of the sites were close to houses; five in industrial areas, including derelict industrial sites; and five were in or around institutions, including residential homes for the elderly and a hospital. Three colonies, including one of the largest, were found on allotments, and there were single groups in a cemetery, in the grounds of a church and, allegedly, on grassland next to the River Humber. Residential loca-tions are probably over-represented, clearly because it is in residential areas that the presence of ferals is most likely to seem discrepant, an offence to domestic order, and thus to prompt a complaint to the local Environmental Health Department. Conversely, 'wild' areas, remote from houses, may have feral colonies which are unobserved and unreported.

At the sites we visited, we spent time observing cat behaviour and site charac-teristics, as well as, in a few instances, talking to people with an interest in the cats about the histories of the colonies. It was not possible in every case to contact people who might have had a view on the presence of a feral colony, and in some places identified by the environmental health officers the cats were elusive. However, our observations did allow us to order the sites, this necessarily rough ordering being based on the perceptions of the three of us about what constitutes wildness in urban landscapes. Generally, in an urban context, wildness is suggested by dereliction, a lack of maintenance, with weeds and brambles asserting themselves, or more densely vegetated vacant land. Cats which are perceived as wild may then 'fit' into such patches of urban wilderness, although, as we suggested above, derelict landcapes and feral cats occupying such areas may both be sources of aversion. In other locales, particularly in residential areas, the problem is more complex and more likely to be conflictual than in 'wild' areas. Feral cats may belong according to those who do not discriminate between pets and ferals, but they may be viewed as transgressive and polluting by those whose distinctions between the domestic and the wild consign ferals to the latter cate-gory (as in the Palo Alto, California, case, cited above).

The following notes on a selection of sites give some sense of the range of habitats identified:

1   A completely overgrown garden, next to a Catholic church in a formerly residential street in the inner city, now used primarily for warehousing. This site was the closest approximation to wilderness. The cats stayed hidden in the bushes when people were present (see Figure 3.1). They were fed by the church caretaker, who had once appealed to the bishop to let the cats remain after objections from one hostile priest. The present priest is quite tolerant.

2   An industrial estate where a colony lives on a disused site. The cats are fed by workers from an adjacent factory (see Figures 3.2 and 3.3). Apparently, no-one has objected to the presence of the cats.

3   An allotment, where feeding bowls and shelter are provided for a large feral colony of between twenty and thirty animals. This colony has survived since the 1950s with the support of most of the allotment holders, who are very protective and suspicious of representatives of outside agencies, particularly the City Council. (However, on an allotment elsewhere in the city, feeders were in conflict with other allotment holders who had threatened to kill the cats.)

4   A cemetery, a well-ordered and manicured site where cats stay among the bushes and trees around the edge (see Figure 3.4). A grave digger at the site when we visited seemed unconcerned about the presence of the cats, although a local resident claimed that the cemetery was overrun with about twenty cats. Someone, she said disapprovingly, was obviously feeding them.

5   Outside a shoe repairer's in a small suburban shopping precinct (see Figure 3.5). The shop owner feeds the cats and has provided shelter and bedding. No-one complains, and they have become an attraction for shoppers and children.

Of these sites, the ones where there was a high level of management, and where people felt a sense of attachment and responsibility for the place, were the ones where opposing views on the feral colonies were most clearly articulated. These were the allotments and the cemetery, where the presence of wild nature was seen both as a good thing – the cats were admired and, in the case of the allotments, seen as useful in keeping down vermin – and a bad thing – they were 'out of place' in a cultivated and ordered space (particularly in the cemetery). Thus, in these spaces, the presence of feral colonies served both to fracture and to cement social relationships. Even on the wildest site, next to the church, there was a history of conflict. Only in the left-over space on the industrial estate, colonised by weeds and cats, and in the familiar space of the shopping precinct, where the cats were decorative and pet-like (so long as no-one attempted to touch them), was there an absence of conflict. Interestingly, another (active) industrial space, a British Petroleum (BP) oil refinery, was also in a sense a part of the 'wild', being a habitat for foxes, deer, rabbits and rats as well as feral cats, and the cat colony here was tolerated by the company, which paid for neutering to control the population.

Figure 3.1   A wild space in the city: invisible feral colony with feeding bowl, lower left.
Source: Authors' photograph

Clearly, the local environment is important in relation to the acceptance or disapproval of animals in the city, but attitudes to feral cats are also affected by the ambiguous position of the species in its relationship with people — an object of affection and a wild animal, admired or feared for its wildness. Our project hints at another interesting aspect of human–animal relations: namely, that those people who are close to ferals or who try to engage with ferals, particularly the feeders, may themselves be viewed as discrepant by both council officers and neighbours. See p. 68 for two examples from our field notes.

Figure 3.2   Feral cats on an industrial estate.

Source: Authors' photograph

Figure 3.3   Shelter and feeding bowls provided by factory workers on an industrial estate.

Source: Authors' photograph

Figure 3.4   Feral cats in suburban cemetery.

Source: Authors' photograph

Figure 3.5   Feral colony in small suburban shopping centre: feeding bowls and shelter provided by owner of hardware store.

Source: Authors' photograph

Lady, used to cycle round and pick cats up and take them home in her basket [according to an environmental health officer]. About a dozen cats. Became a member of the CPL [Cats Protection League] but wouldn't let anyone take a cat if they [the CPL] came to re-home them. Extension of house solely for cats, built wire fence for them. 'Roger' [environmental health officer] heard she'd moved house and left the cats, so he referred the case to the RSPCA.

Private house with large walled garden, lady had a cat, adopted a stray, had lots of kittens which started a colony. Neighbours complained … can't go and visit as lady is a bit schizophrenic [sic], although 'Steve' [environmental health officer] will talk to her.

These comments are suggestive, with their intimations of peculiarity, eccentricity and 'madness', even maybe echoing a pre-modern stereotype of 'witch with cats'. While it would be easy to read too much into these observations, and we certainly do not have enough material to draw any firm conclusions about perceptions of cat feeders, it would be interesting to explore the idea that perceptions of feral supporters are gendered and that feeders dangerously transgress domestic norms.

Conclusion

Returning to 'little Darcy', it is evident from our study that feral cats are not necessarily the 'inhabitants of dereliction', as the Cat Action Trust advertisement asserted. As well as retaining their freedom in the wild, feral cats may be well looked after. This view is supported by Remfry (1996: 523), who has argued that 'feral cats can thrive in a free-living state, usually enjoying more interesting lives than those pets confined indoors, so long as certain criteria [relating primarily to shelter, feeding and supervision] are met.' This is also now the position of the Royal Society for the Protection of Cruelty to Animals (RSPCA). However, it is apparent that human responses to feral cats are affected by constructions of the built or cultivated environment, which render cat spaces either discrepant or acceptable features of the urban scene, and by representations of ferals as wild or potentially domestic. The interplay of imagined cats and imagined landscapes contributes to the definition of places where ferals belong or are discrepant. The many associations of animals and place, between cat-friendly Fitzroy Square or the shoe repairer's and the mini-safari park of the Catholic church garden or the oil refinery, are reflected in complex human responses.

The problem could be seen as an instance of the much wider issue of what, in the constitution of space in a modern urban society, is defined as belonging or not belonging. At the level of the home, the neighbourhood and the city, visions of strong and clear boundaries and predictable, recognisable geometries can be seen as a response to heterogeneity, with the consequent uncertainties about social relationships, and the precariousness of urban living: that is, the precariousness of the control of nature within the city. Wild nature might just reassert itself and disturb the urban order. In a reflection on early human efforts to gain security through the medium of architecture, Le Corbusier (1927: 71) suggested that

> in order to construct well and distribute his [*sic*] efforts to gain advantage, to obtain solidity and utility in the work, [the builder] has taken measures, he has adopted a unit of measurement, he has regulated his work, he has brought in order. For, all around him, the forest is in disorder with its creepers, its briars and the tree-trunks which impede him and paralyse his efforts.

The desire to eliminate nature, nature as abject and inimical to domesticity, is still there in the (post)modern city. At the same time, though, wild nature is desired. Its absence in the city figures as a loss, and there is a need to reconnect, if only vicariously or in the imagination. Just as nomads transgress and disturb the urban order but, at the same time, represent a romanticised freedom, so uncontrolled nature is a source of both anxiety and desire. Thus, people 'need' to imagine marginal places, both as abject spaces to which feral cats, Gypsies and other elements of 'the wild' can be consigned and as spaces of desire where wildness can be recovered.

The politics of modernity, however, have been concerned with the *elimination* of both the nomadic (an almost taken-for-granted violation of human rights) and unregulated nature, or their containment and transformation through the imposition of order (zoos, English and Welsh Gypsy sites, nature reserves). But the elimination of the abject, as Kristeva reminds us, is impossible. These removals and purifications can never be more than temporary or partial, and hence the recurrent tension between transgressive nature, resisted and desired, and the ordered city. The presence of wild nature in the city signals a loss of control, just as nomads cross lines which mark secure, predictable and commodified categories of urban space. The realisation of an ordered city, like removing bodily odour or staying young, is an impossible project.

Acknowledgements

We would like to thank the Hull Cat Protection League, officers in the Environmental Health Department of Hull City Council, and Diane Green. We are also very grateful for the help of allotment owners and other cat supporters in the city.

Note

1   Erik Swyngedouw helpfully suggested 'unsettling' as a translation of *unheimlich*, in preference
    to 'uncanny'.

References

Agee, J. and Evans, W. (1965) *Let Us Now Praise Famous Men*, London: Peter Owen.

Baker, S. (1993) *Picturing the Beast: Animals, Identity and Representation*, Manchester: Manchester
    University Press.

Bannet, E. (1989) *Structuralism and the Logic of Dissent*, London: Macmillan.

Berger, J. (1980) 'Why look at animals?', in *About Looking*, London: Writers and Readers.

Bottriell, L (1985) 'Review of Francis's *Cat Country: The Quest for the British Big Cat*', *Cryptozoology* 4:
    80–84.

'The case for the beast of Bodmin Moor' (1998) <http://www.eduweb.co.uk/news/april/beast2.
    html> (accessed April 1998).

Chisholm, D. (1993) 'The uncanny', in E. Wright (ed.) *Feminism and Psychoanalysis: A Critical Dictio-
    nary*, Oxford: Blackwell.

'College Terrace: feral cats evicted from neighborhood' <http://www.fluffynet.com/feral.htm>
    (published 18 October 1995).

Freud, S. (1990) 'The "uncanny" ', in *Penguin Freud Library*, Vol. 14, pp. 335–376.

Fumagalli, V. (1994) *Landscapes of Fear: Perceptions of Nature and the City in the Middle Ages*, Cambridge:
    Polity Press.

Hoggett, P. (1992) 'A place for experience: a psychoanalytic perspective on boundary identity and
    culture', *Environment and Planning D: Society and Space* 10: 345–356.

Jackson, R (1981) *Fantasy: The Literature of Subversion*, London: Methuen.

Kristeva, J. (1982) *Powers of Horror*, New York: Columbia University Press.

Le Corbusier (1927) *Towards a New Architecture*, London: Architectural Press.

Minh-ha, Trinh T. (1996) 'Nature's r: a musical swoon', in G. Robertson, M. Mash, L. Tickner,
    J. Bird, B. Curtis and T. Putnam (eds) *FutureNatural*, London: Routledge.

Perin, C. (1988) *Belonging in America*, Madison: University of Wisconsin Press.

Peters, H. (1979) *Docklandscape*, London: Watkins.

Remfry, J. (1996) 'Feral cats in the UK', *Journal of the American Veterinary Association* 208: 520–523.

Sibley, D. (1995) *Geographies of Exclusion: Society and Difference in the West*, London: Routledge.

Thomas, K. (1996) *Man and the Natural World: Changing Attitudes in England, 1500–1800*, Oxford:
    Oxford University Press.

Tuan, Y.-F. (1984) *Dominance and Affection: The Making of Pets*, New Haven: Yale University Press.

Wilbert, C. (1998) *The Love of Trees: Concepts of Place, Origins and Roots in Economies of Society and Nature
    from Linnaeus to Ecological Restoration*, Unpublished PhD thesis, Chelmsford: Anglia Polytechnic
    University, School of Design and Communication Systems.

Wilton, R.D. (1998) 'The constitution of difference: space and psyche in landscapes of exclusion',
    Geoforum 29: 173–186.

4 Constructing the animal worlds of inner-city Los Angeles

Jennifer Wolch, Alec Brownlow and Unna Lassiter

Introduction

Metropolitan populations of the United States are becoming increasingly diverse. This growing urban demographic diversity is due to high levels of immigration from Latin America, Asia and other parts of the world over the last two decades, as well as to longstanding patterns of rural to urban migration among native-born African-Americans and Hispanic Americans. One seldom-noticed aspect of this rising diversity relates to nature–society relationships. How do urban immigrants think about nature, and how, if at all, do their attitudes change with duration of residence in US cities? Are these attitudes different from those held by native-born urban residents, who themselves are differentiated by race/ethnicity, class and gender? And do variations in nature–society relations play any discernible role in precipitating urban social conflict so prevalent in metropolitan areas today?

Such issues are critical to forging pathways towards human–animal coexistence and strategies for sharing space in an era of rapid urbanisation and related habitat loss/fragmentation, species endangerment and escalating conflicts along the human–animal borderlands of the metropolis.[1] Long-entrenched dualisms between nature and culture, and city and country, are breaking down both on the level of epistemology (especially due to the influence of feminist, post-colonialist and postmodern perspectives), and on the concrete level of everyday urban management of human–animal relations and city space. Traditional, anthropocentric urban theory is inadequate to the task of conceptualising the city as part of nature, or of animals as urban residents (much less providing policy prescriptions). Rather, we need a 'transspecies urban theory' that can make sense not only of cities as spaces of political-economic power and cultural difference, but also as places characterised by particular constellations of animals. For it is within this transspecies urban context that finely differentiated attitudes towards animals are shaped and expressed in the form of human–animal relations played out with particular consequences for both people and animals.

The desire to flesh out such a new, less anthropocentric urban theory provides the motive and context for our current investigations. This research seeks to clarify relationships between cultural background, linked to race/ethnicity or national origin,[2] and urban nature–society relations. In particular, we wish to explore in detail the ways in which attitudes towards animals in the city are formed, and the role of cultural difference in shaping attitude formation. Ultimately, we expect the insights gained from such finely textured analysis to inform understandings of both human interactions in the city (among groups with divergent attitudes towards animals and nature) and the full spectrum of attitudes towards animals, with implications for coexistence.

To this end, we designed and conducted a series of focus groups in Los Angeles, California. Focus groups allow participants to express their ideas in a relatively unconstrained fashion, and to react to one another's statements. Thus this technique is well suited to the task of clarifying issues of culture and race/ethnicity, and the socio-economic contexts of attitude formation. Since prior research indicates significant gender and class differences in attitudes towards animals, we restricted our focus groups to relatively homogeneous participants: within each group, members were low-income inner-city women. Groups differed from each other primarily along lines of race/ethnicity and immigrant status. This enabled us to focus on the nature of attitudes towards animals within relatively homogeneous groups. Comparative analysis of findings from the series of focus groups will allow us in the future to explore race/ethnic and other cultural variations in attitudes towards animals, and the role that cultural background may play in attitude formation and everyday transspecies practice.

In this chapter, we describe and analyse a single focus group involving eleven low-income African-American women living in central Los Angeles. Particular effort was devoted to eliciting detailed accounts from these women, to enable us to put attitudes into cultural and social class context, delineating linkages between attitudes and behavioural interaction patterns. In addition to drawing on prior studies of African-American attitudes towards animals and the environment, our interpretation of the focus group discussion was based on identifying attitudes, the arguments that participants selected to support or denigrate them, and on clarifying how these arguments fit into the elaboration of distinct cultural models of attitudes. Participants also provided us with narratives and anecdotes about the cultural and social meanings of animals and of the human activities that relate to them. Altogether we obtained a rich sampling of attitudes and practices. The main themes to emerge from our discussion with the women in our focus group highlight the roles of generational and class position, urban/rural background, and membership in a historically oppressed and currently marginalised social group. These socio-cultural contexts shape their personal identities, everyday practices and values/attitudes, including their perspectives on animals and appropriate human–animal relations. Anthropocentric and biocentric attitudes quickly emerged in the discussion, along with a number of other, less commonly articulated attitudes.

The remainder of our discussion is organised into four sections. First, we describe our focus group participants and the logistical and analytic procedures used. Next, we characterise participant practices and interactions with animals, and their knowledge and perceptions of animals in order to understand the role of animals in their lives. Third, we explore their values towards nature in general and animals in particular, through questions about how animals should/should not be treated by people. Fourth, we examine the justifications given for attitudes expressed, and their connections to particular socio-cultural contextual variables. On the basis of these findings, we suggest a cultural model of attitude formation that includes the role of racialisation in shaping attitudes towards animals.

The focus group analysis

Most studies of attitudes towards animals have utilised standard survey techniques to determine basic attitudinal dimensions, and to collect information on respondent characteristics, such as gender or income, that past research suggests may be systematically related to attitudes. While informative, such surveys tend to oversimplify attitudes and fail to capture the complex and often contradictory ideas and feelings that people have about animals. Moreover, for respondents whose own thinking on this topic is relatively unexplored, initial responses often reflect stereotypes or the most current news story, and thus can be misleading. Because of these shortcomings of standard survey techniques, and because our goal was to learn more about the ideas of groups seldom if ever studied in the attitudes towards animal literature, we selected focus group methods as an alternative. Focus groups provide an effective means of eliciting fine-grained expressions of attitudes towards animals, and allowing the focus group's inter-group dynamics to help distil participants' positions, stimulate recollections and crystallise alternative perspectives within the group. Moreover, this sort of research approach reveals how participants negotiate, both discursively and in concrete everyday ways, their inter-subjective understandings of animals as friends or food.

All focus group participants were residents in the same affordable housing project,[3] whose manager assisted in our volunteer recruitment efforts.[4] The African-American women who ultimately volunteered spanned a broad age range (from 22 to 75 years old, with a majority in their twenties) (see Appendix A). They also varied widely with respect to educational attainment (three had not completed high school, four were high school graduates and four had either completed some college or had a college degree). Most had lived in Los Angeles for quite some time, with the oldest resident having lived there for thirty of her

75 years. About half had been born either in Los Angeles or another part of California, with the remainder being from other states. One participant, however, had been born in Belize; despite the fact that we had restricted our recruitment to native-born African-Americans, this respondent had a barely detectable accent and we were not aware of her heritage until some time into the focus group. But, although her presence constituted a methodological mistake, she played a very useful role in stimulating discussions about animal rights and environmentalism, topics about which she had forthright views.

Participants were asked a range of questions about their general environmental beliefs, traditional forms of human–animal interaction, attitudes towards animals, and knowledge, perceptions and behavioural interaction patterns. A set of questions specifically regarding one group of animals (marine wildlife) was also posed, as a way to focus on attitudes towards a specific subset of wild animals with whom residents of coastal southern California would likely interact. (In addition, participants completed a brief questionnaire regarding demographic characteristics, their experience with different sorts of animals, and their participation in animal-oriented organisations.) Two warm-up questions focused on the constructions and perceptions of animals in the city, especially 'pets' and 'wildlife',[5] and the animals deemed appropriate for each category. Then, questions were grouped around three general topics. Topic One questions looked at the practices, interactions and perceptions of the focus group members towards, as well as access to, marine wildlife in southern California. Topic Two questions focused on family, social and cultural practices, customs, perceptions and attitudes towards animals. Members were asked to reflect on their own familial and cultural practices and histories as well as those within the many other cultures found throughout the greater Los Angeles area. This series of questions accounted for the bulk of the discussion and what may be considered the most intense period of the focus group. Topic Three questions targeted, respectively, animal welfare values and gendered attitudes towards animals. Finally, the discussion ended with a question posed by one focus group member concerning a very particular type of human–animal interaction.

Focus group members responded in an uneven fashion depending on question and trajectory of discussion. While Carla, Denise and Frankie remained conspicuously absent from most of the conversation, others, such as Georgia, Alice, Vivian, Bernadette and Irene, tended to be well represented throughout.[6] Extended narrative (that lasting longer than a couple of sentences) was regularly provided by Alice, Susan, Georgia and Irene. The group discussion was tape-recorded, transcribed and analysed.[7] Text was 'coded' for a particular conceptual 'node', with nodes collectively constituting an 'index tree' (see Appendix B). In our particular case, nodes tended to fall into one of three 'umbrella' categories: (1) practices, (2) perceptions and knowledge, and (3) values and attitudes.

People and animals: the everyday experiences of focus group participants

Participants had a wide variety of experiences with animals, both direct and indirect. While only five of the women had ever held jobs that involved animals, such as farming, virtually all presently keep, or have in the past kept, pets. From experience, information handed down to them from others, and formal education, the women developed knowledge and perceptions about animals as well as preferences for some sorts of animals over others. Interestingly, only one of the participants mentioned watching nature programmes on TV as a source of either experience with or knowledge about animals, despite being directly asked (although they were familiar with TV animal characters such as Lassie and Benji). The recently opened Long Beach aquarium was the only animal display institution mentioned specifically by members of the group. None belonged to voluntary environmental or animal rights organisations.

Types of interactions with animals

The two extremes with respect to experience with animals were Georgia, who was raised in Belize, where her family owned a wide variety of both farm animals and pets, and Carla, who had never owned or raised an animal. Most participants, however, had experienced animals in either home or work settings. In addition, many had encountered animals in the wild. With one exception, participants had broad experience with pet ownership, either as a result of their own pet-keeping (8) or membership in a pet-keeping family (7). Over one-third had worked or were raised on a farm. Most described having parents who fished and/or hunted. Many recounted vivid experiences with and recollections of animals. One participant's father had worked in a slaughterhouse; another father once killed a neighbour's pesky opossum and served him up for supper. A grandmother regularly wrung the chickens' necks and served the birds up for family dinner. There were tales of summer camp and horseback riding, and memories of reading nature books or having parents who read animal stories aloud. Such recollections were rich and redolent of the human relationships that surrounded these animal practices.

Differences in the nature of relationships between people and pets, on the one hand, and people and farm or service animals (such as guard dogs), on the other, emerged during the discussion. Alice's response to our question about experience with pets, for example, displayed an interesting change in verb: she began 'my parents raised', but, as if catching herself, immediately changed course, to 'my parents *had* two dogs and a cat' (emphasis added). The reason for this shift became clear later when she detailed the deaths of the two dogs, one at the hands of her father, the other at the hands of her family's neighbour. This change in word use, from 'raised' to 'had', appears to constitute a correction to the *meaning* applied to the two dogs by her family (especially her father). To 'raise' implies some form of emotional connection or investment between the dogs and the family, while to 'have' objectifies the 'animal' as just another material commodity, one easily disposed with; in this case, violently ('daddy took a two-by-four and hit … the little dog').

Our questions about participants' experiences with marine wildlife drew many and varied responses, identifying not only animals encountered at the beach but, furthermore, the context within which they were encountered and the perceptions that resulted from these experiences. Animals encountered ranged from 'little itty bitty crabs' (Laura) and jellyfish to more charismatic fauna, including whales, dolphins, seals and swordfish, the latter group (whales, in particular) being described with phrases like 'so pretty' or 'so cool'. Remarkably, except for swordfish, nobody cited 'fish' as being an animal encountered at the beach, despite the fact that four of the participants had fished at some point in their lives. Of special note was a conversation between Georgia, Bernadette and Susan on the question of which fish were best to eat. This dialogue emerged towards the centre of the discussion about fishing, and produced a general consensus that red snapper was 'the tastiest'. This suggests that the primary way in which participants encounter fish from the ocean was as food, likely in 'ready to eat' or prepared (dead) form. This, of course, is true of most Americans today.

Knowledge and perceptions of animals

Rather than asking direct questions about participants' knowledge or perceptions of animals, their responses and spontaneous discussions provided clues about the extent of their knowledge and, frequently, its socio-cultural context. The women often relied on first-hand experience for understanding animals as sentient beings, as well as on second-hand information from parents, family or friends, and, interestingly, even urban folktales. Competing knowledge claims were strategically used to convince others of particular attitudes (such as why people should not eat animals). Knowledge about wild animal habitats also emerged, along with a basic understanding of ecosystem relations. Participants gained some of their knowledge from childhood experiences with animals, such as horseback riding (Bernadette), picnicking in parks (Vivian and Bernadette) and (for Irene and Susan especially) going on field trips or to summer camp. Susan is an example: her mother sent her to summer camp, and 'we learned about all the wild. You know, we used to go hiking and fishing and we learned about the flowers, the birds, all that.'

The sorts of animals typically seen on local beaches were identified, and participants volunteered specific urban sites where animals such as opossums could be seen in numbers, suggesting a familiarity with urban wildlife distributions. In addition, rudimentary knowledge of animal behaviour emerged; for example, in talking about a pet snake that had eaten a neighbour's dog, one respondent argued:

> They're used to liv[ing] in the wild. That's where they was born so it's like very hard for them to be, you know, locked up and as soon as they get out they do what's natural for them … look for prey.

Similarly, discussions about El Niño indicated general ideas about natural system linkages, based on popular conceptions of ecosystems and species diversity. And at least one participant (Alice) was somewhat familiar with recent shifts in predator management (especially wolves in the Rocky Mountain region).

Lively discussions on the topic of animals as food (detailed in the next section) revealed culinary knowledge, but also clarified what one participant, Susan, knew about animal pain and its similarity to human suffering on the basis of her witness of an animal slaughter. She linked the suffering of stray companion animals to human material and emotional needs: 'I've seen stray dogs and I said, "well, oh, that dog needs to be taken out of his misery," but actually the dog just wants some love and attention and a home.' Participants also knew about various foreign practices towards animals, as Irene related: '[T]hey have farms out there in the Philippines where they raise dogs just for the meat … they raise these special kind of dogs and they slaughter [them] … and they sell the meat in the grocery stores out there.'

Knowledge was profoundly uneven, however, and there were many examples of ignorance of basic information. The most spectacular example cropped up at the end of the focus group:

SUSAN: Does anybody have any stories or has anyone ever heard of people having sex with animals? In [a rural town], we used to hear stories about people having sex with sheep.

IRENE: Yeah, you can get a sheep pregnant, but the lamb won't live once it's born.

LAURA: [mumbling] Half-human, half-animal.

Animal practices: socio-cultural context, family histories and individual behaviour

Discussions about animal practices generated the richest commentary, the most intense debate, and the most in-depth historical narratives of the focus group. Interestingly, most of the discussion revolved around animal (meat) consumption, including individual and family practices and perceptions, practices of other cultures and the perceptions of those practices, and histories of animal consumption.

Within the general rubric of animal consumption, the conversation ranged from individual and family consumption practices to those of other cultures, from gendered and cultural knowledge to consideration of cultural and ethnic survival, from the most mundane and innocuous family narratives (for example, raising worms) to historical accounts of the significance of animal consumption to household and cultural persistence. The strongest theme that emerged was the necessity to eat meat – often of animals or animal parts devalued by mainstream white society – in order to survive. The general rationale that meat is necessary to survive is a common psychological mechanism for enabling humans to harm animals while still seeing themselves as compassionate (Plous 1993), but one that carries particular force for African-Americans given their historical and contemporary oppression. Their perspectives from the margin also allowed them to view the animal practices of other cultural groups with understanding rather than condemnation, despite the conflict between such practices and their own forms of animal consumption.

Individual practices: influences of generation and place

The practice of meat consumption by focus group members was framed in the following ways: (1) what was explicitly or implicitly *eaten*; (2) what was explicitly or implicitly *not eaten*; (3) what constitutes an 'appropriate' meat and/or consumable animal in an (African-)American context; and (4) what does not. The social and cultural construction of (food) animals and of the parts therein (such as gizzards, tripe and the like) emerged as a fascinating topic and one that the discussants themselves dealt with thoughtfully and critically. Individual eating practices were often offered in list-like fashion, with discussants presenting experiences they considered most extraordinary, surprising (to the other group members and to themselves) or daring. The following is an example:

BERNADETTE: Frog legs ... I've eaten snake ... shark.
VIVIAN: Shark. I don't care too much for octopus.
IRENE: Shark. Shark is good.
SUSAN: I've had octopus before.

In this exchange, snake, shark, octopus and frog legs appear to be constructed in ways emphasising their qualities as fashionable consumption items similar to swordfish, duckling and, increasingly, deer meat (venison). Perhaps significantly, the four women engaged in this discussion were either Los Angeles natives or had spent the greater part of their adult lives in the Los Angeles area.

Interestingly, this list of exotic animals was immediately preceded by another very different list, older and more 'rural' in character. The women raised in rural settings initiated this discussion. Frankie, who spent most of her life in Alabama, introduced the subject of subsistence hunting, and Alice, raised in rural Texas, joined in a discussion of the hunting, preparation and consumption of jackrabbits, squirrels and opossum. These animals are commonly eaten by many African-American families in the rural South. In a fascinating study of hunting culture in rural North Carolina, Marks (1991) finds that such so-called 'trash' animals have long been a default form of protein for many local African-Americans due to the sequestration of 'legitimate' game animals (for example, deer, partridge or quail) by local, often wealthy, white hunting clubs. Animals like opossum, 'coon' and squirrel are generally considered vermin by the dominant Euro-American ideology, more often recognised as roadkill than as something prepared for dinner.

The younger participants, raised in cities, clearly did not identify with those women whose families hunted and/or consumed such animals. They consistently used distancing mechanisms in their speech (for example, '*they* eat opossum': emphasis added). What appears, then, is a rural–urban split defined on the basis of animal consumption: opossum, 'coon' and squirrel are conceptually linked to an impoverished, rural, African-American community whose dietary needs are supplied by hunting certain game animals and whose access to a normative urban diet is limited. The African-American 'move' to the city and subsequent development of an urban consciousness have broken the rural shackles, however, such that these women are no longer obliged to participate in eating habits that may be considered culturally backward. This emancipates them from the diet of a violent, impoverished and oppressed rural past when 'we' were poor, 'we' were rural, and 'we' were slaves, forced by necessity to eat 'trash' animals.

Socio-cultural practices: legacies of slavery and marginalisation

Just as opossum, 'coon', jackrabbit and squirrel were conceptually linked to a rural, impoverished African-American diet and socio-cultural context, so too were organ meats (tripe, heart, chitlins, liver and others). Alice, especially, emphasised the link between eating animal organ meats and survival. She talked about her father, who worked in a rural California slaughterhouse, and his contribution to family well-being. 'Nearly every day he would bring home a bucket of meat that the slaughterhouse didn't want: chitlins, tripe, the things they charge you a fortune for now. Chitlins, tripe, the stomach, sweetbreads … .' These organs were (and in some cases still are) considered 'trash', vulgar and disgusting by white America, much like the 'trash' animals hunted in the rural South. And yet, organ meat became a dietary staple – indeed a necessity for survival – for black slaves seldom provided with sufficient food.

ALICE: The reasons blacks know how to eat chitlins and all the things that are repulsive in an animal is that we learned it from slavery because we had little to eat ...

VIVIAN: We took whatever was left, and what they [white slaveowners] didn't know how to prepare or they couldn't stomach, that's what they gave the slaves to eat. So we acquired that taste.

Despite its obvious significance to African-American culture and African survival during the period of slavery in this country (as described by Vivian and Alice), the consumption of organ meat, like hunting 'trash' animals, seems to be losing some of its cultural significance in today's largely urban context. Indeed, many of the participants were completely unfamiliar with certain popular organ meats: Bernadette admitted to not knowing what 'sweetbreads' were, while Laura mistakenly described them as pastry ('it's bread. That's all it is'). Nevertheless, organ meat and animals such as opossum, 'coon' and squirrel ('all the things that are repulsive' to white society) appear to maintain an important and critical place in the history of African-American culture, not only as cultural artefact ('we acquired that taste') but as symbols of survival, resistance, perseverance, ingenuity and cultural/ethnic pride. As Alice claimed: 'We [African slaves] were creative, we learned ... and we ate what the white people didn't want.'

Cross-cultural animal practices

The dietary practices of other cultures in the Los Angeles area were debated, critiqued and defended by the focus group members. Focusing almost entirely on dog-eating practised by many Southeast Asian cultures, this topic, interestingly and significantly, emerged within the context of 'trash' animal consumption by blacks in US rural landscapes. Indeed, the topic was initiated by Vivian with the phrase, '[s]ince you mentioned the opossum ...'. Discussion of dog-eating primarily focused on the revulsion of one young woman, Susan, on the one hand, and on an effort to contextualise the practice as culturally legitimate by Alice, Frankie, Vivian and Georgia, on the other:

SUSAN: That's horrible. That's disgusting/

FRANKIE: ... that's the same way we feel about eating, slaughtering goats and cows and chickens. The same way we feel; it's all meat and we gonna eat anything.

SUSAN: That ain't no meat. That's not meat for us to eat/

FRANKIE: That's not what we see it as/

ALICE: That attitude is part of their culture.

SUSAN: Dogs is not meant for us to eat.

GEORGIA: ... they eat dogs because that's a part of their culture, but it's very hard for us to adapt to because we have always considered dogs as pets ...

SUSAN: Dogs *are* pets [emphasis in original]. That's wrong, that's wrong.

GEORGIA: ... if you were brought up eating [dog], it would be just fine ... It's like if you would eat a chicken ...

VIVIAN: But you know ... the difference [between chicken and dog]. I mean, you were born in the United States.

The arguments of Susan, a young Los Angeles native (whose economic means may be greater than that of the other focus group members), centred entirely on the Western construction of 'dog' as pet not intended for human consumption. Juxtaposed to her arguments were those of Alice, Frankie, Georgia and Vivian, who comprehended the Asian construction of 'dog' not as pet but (in certain instances) as food. By providing Susan with examples of how Western society constructs animals as food (chickens, goats, cows), these women shed a more culturally empathetic light upon Asian consumptive practices. Vivian went so far as to detail the consumption of horse meat in parts of the US, consciously using an animal whose popular construction in American culture maintains a privileged position along with that of the dog. Susan herself reinforced this contradiction, using the example of kangaroos ('they had kangaroo meat at Jack-in-the-Box before ... I swear. It was on the news a long time ago'), and Alice mentioned ostrich consumption.

Unsurprisingly, Alice, Frankie and Vivian maintained positions of cultural empathy, understanding and contextuality in the case of Asian dog-eating practices; to do otherwise would have fundamentally undermined their earlier statements concerning the consumption of repulsive 'trash' animals and organ meat within their own culture. Again, this position appears to revolve around the notion of cultural survival, in which there is an historically negotiable 'place' for pets as well as for consumable animals that is culturally, socially, politically, economically and geographically mediated. The cultural placement of animals into one category or another (or, for that matter, into any of several other categories of animal: beasts of burden, wildlife, game, and so forth) is an historical process that, in the African-American context, was powerfully mediated through the processes of oppression, violence and neglect.

Cultural categorisation is, therefore, *a priori* legitimate and demanding of 'our' appreciation, understanding and acceptance lest racism based on consumption become manifest (Wolch and Emel 1995). During a discussion of Mayan practices of eating iguanas, turtles and other animals uncommon in Western diets, Alice went so far as to suggest that cannibalism could be legitimate if cultural or individual survival is at stake. This led Susan to claim that food is entirely relative to culture and situation:

ALICE: I don't want to offend any of my people by reminding them but in Africa they used to [eat] human beings ... and they ate their enemies. When they killed their enemies they ate, they took great pleasure in eating their enemies ...

SUSAN: Anything is edible. Everything, even you all. Everything [is] edible. We're edible, animals are edible, everything is edible, huh? People will eat just any old thing.

Gendered practices

One portion of the focus group discussion, about the existence of an urban poultry market, illuminated the existence of gendered (as well as generational) animal practices among the city's African-American community. Again, it was Vivian who initiated this discussion:

VIVIAN: When I was growing up, we had what they called poultry markets. They still have a few around ... where we used to go ... and buy a whole chicken/

CARLA: Kill it and skin it/

VIVIAN: ... we would take it home and my grandmother would have a big ... tub and dip it in boiling water. First she would wring the neck.

These statements by Vivian (age 48) and Carla (age 50) exemplified the urban persistence of a gendered cultural practice whose roots likely lay in a more rural context, or in a time prior to large-scale urban supermarkets. From purchase to pot, chickens appear to fall into the woman's work space. However, much like the consumption of particular animals and organ meats, the following comments indicate that this gendered/cultural institution may be on the decline among the younger members of the urban African-American community:

SUSAN: [disgusted in response to the process of chicken decapitation] the head gone!? Unh uhhh ... I couldn't do all that.

IRENE: ... well after all that, I've lost my appetite.

SUSAN: ... you all can buy one for me and ... cook it up for me and I'll come get it.

At first glance, it would appear as if the passing of time has been adequate to separate the practice of poultry purchase and butchering from the daily routine of the younger women, so that it is now perceived as an odd, even grotesque, practice. The purchase of pre-butchered, pre-packaged chicken in grocery stores lessens the need to participate in what appears to be an outdated, grisly and messy practice. Space, too, may play a role here, especially the location of the remaining poultry markets (for example, in Chinatown). However, as two of the women indicated, poultry markets *do*, in fact, still exist in downtown Los Angeles and remain within access to the city's black communities. Therefore, two questions arise. First, why is this cultural practice being lost on younger women? Second, why do the older women continue to participate in the seemingly undesirable chore of buying and butchering live chickens when, indeed, 'ready to eat', pre-cleaned, pre-butchered chicken is easily available at most, if not all, super-markets?

To the latter question, clues may be found in the responses of Vivian and Bernadette, each patrons of a local poultry market in the south central district of Los Angeles:

VIVIAN: I went up there Sunday before last and I think I bought about four chickens/

BERNADETTE: The live chicken is so good.

VIVIAN: Isn't it good?! It's fresh/

BERNADETTE: It is so good! And tender ... I had never bought one from [a particular poultry market] and I went down there Easter and got some.

Two clues may be found here. The first is simply one of taste: the chicken is 'fresh', it is good and perhaps of better quality than that bought in a supermarket. The second clue refers to days of occidental religious (Judeo-Christian) observation and ceremony, the Sabbath and Easter. Religious sacrifice still occurs within ethnic communities throughout urban America (for instance, the slaughter of Easter lambs by Greeks in urban New England; personal observation A.B.). Although not explicitly examined here, the potential religio-cultural role(s) of poultry markets (or any market selling 'live' animals for consumption) in the observation of the Sabbath or other holy days is intriguing.

Reasons for the decline of poultry market use among younger African-American women in the city remain obscure and in need of further investigation. However, the convenience of supermarket shopping, alongside the perception of this grisly and messy practice that creates (now) unnecessary conflict between the desires to be compassionate towards animals yet still eat them, are likely candidates. Animal slaughter no longer appears a welcome event in urban American culture, a fact that is likely true across racial, ethnic and gendered boundaries (Philo 1995).

Attitudes towards animals: consumption and competition, coexistence and care

Attitudes towards animals were expressed throughout discussions of many topics, most linked to normative questions of how humans should or should not use

animals, but also in descriptions of animal practices, noted above, that were justified on the basis of particular normative values and attitudes. Two basic attitudinal
(or cultural) models emerged from the discussion (Kempton *et al.* 1995). One
was an *anthropocentric* model that emphasised utilitarianism ('humans must use
animals to survive'), which has been revealed in many attitudinal studies. This
model was linked not only to individual and household survival by certain of our
participants, but also to the cultural survival of African-Americans as an historically oppressed minority. Other anthropocentric views also emerged, in
particular negativism ('some animals are pests'), the spiritual value of animals to
humans ('animals have supernatural powers to help or harm people'), and animal
welfare, recognising the contribution of animals to human emotional well-being
('animal companions make people happy'). The second model was a *biocentric* one
that had several variations. These variants included naturalistic perspectives that
stressed animals as part of nature ('animals should not be harmed for their natural
behaviour'), and also one suggesting an animal rights standpoint ('animals have a
right to existence'). A fourth and related biocentric view was that people should
coexist with animals ('the world is big enough for everybody to share').

Because attitudes towards animals have frequently been found to vary sharply
by gender, we asked focus group participants to ponder the issue of whether men
and women had different attitudes towards animals. Responses fell back upon
essentialised cultural models of gender differences ('women are more compassionate than men'), but exceptions revealed the porosity of such distinctions.

Anthropocentrism

The small number of prior studies of attitudes towards animals among African-
Americans, and research on African-American attitudes towards the environment
more generally, suggest that their views are slightly more anthropocentric and
especially utilitarian than those of whites (Kellert and Berry 1980). However,
there is substantial evidence that these attitudes reflect socio-economic and
cultural factors (Caron-Sheppard 1995; Dolin 1988; Kellert 1984), or even the
bias of survey questions (Dolin 1988), rather than a lack of concern for nature or
animals (Caron 1989). Moreover, there is also evidence of change over time
(Caron-Sheppard 1995: 25).

Perhaps the strongest advocate of an anthropocentric, utilitarian perspective
was Alice, who justified her attitudes in part on the need to use animals in the
competitive struggle for survival. Recalling the deaths of her two childhood dogs,
Benjamin and Bob, at the hands of, respectively, her father and her neighbour's
father, Alice offered a complex glimpse of her beliefs on both the inferiority of
animals and their expendability during times of need:

When we were living in [a rural town] my father found this beautiful black and white dog, Benjamin. And we had chickens, and a chicken coop, a chicken house, and Benjamin would suck eggs. And about the fourth or fifth time daddy caught him sucking eggs, daddy took a two-by-four and hit, hit the little dog with it. ... And the second dog we had, Bob, was a big German Shepherd; had been a police dog. ... One day, the neighbours were playing with us and Bob took that, took the little four- or five-year-old boy and mauled his head. They were Latinos, the little boy, and the parents were good friends of my parents. But about two or three days later, the dog came up missing and we didn't understand it at the time but, of course, that father killed that dog. ... So that's, I think that's why I don't, I'm very sure about animals being inferior to humans and that humans come first because animals have always been our servants. ... I don't remember even ever being concerned with animals.

The story is illuminating on several counts. The perception by Alice of the inferiority of animals, one that emerges in many of her comments, may have come directly from her father. Combined with the earlier discussion of her father's work in the slaughterhouse, a picture emerges of a childhood in which the father's work, both occupationally and at home, and his approach towards animals within a necessarily utilitarian value system resulted in the hierarchical, survivalist approach towards animals that Alice embraces. In Alice's narratives, household survival was of immediate consequence and consumable animals and their products (in this instance, eggs) played a critical role to the family's persistence and well-being. Benjamin, by 'sucking eggs', was perceived as a direct competitor for family food and, therefore, was likely seen as undermining the household's health and welfare. In times of such resource constraints, cultural constructions (in this case, the idea of pet who is not to be eaten) become negotiable and, in certain instances, will likely dissolve altogether. Benjamin was no longer pet but 'pest', and was treated accordingly.

Alice also argued that animals are inherently inferior to humans, and justified her attitudes on the basis of religious beliefs rooted in deeply entrenched notions of a 'chain of being' extending from God, to 'man', and finally down to animals. With this rationale, which implies that human needs come before those of animals, she had no trouble articulating a clear remedy for perceived problems such as predators venturing into the suburbs. Her suggestion was to 'have open season' on the animals in question, so that there would be 'so many men with guns they'd be up there and they'd be rid of 'em in less than a year'. In an interesting contradiction, however, Alice was the most educated of the group, and had lived in at least four distinct regions of the country, two of which are urbanised (New York City and Los Angeles). This warns us that attitudes cannot be 'read off' from a small number of individual characteristics, but instead must always be understood as evolving within a more complex socio-cultural and individual biographical context.

Biocentrism

Biocentric attitudes were not necessarily predicated on the basis of the past research cited above. Such attitudes were found to be uncommon among African-Americans (Kellert and Berry 1980), as were concerns for wildlife or natural habitat (Dolin 1988). But they emerged swiftly in our focus group discussion. For example, Alice was promptly challenged in her resolutely anthropocentric, utilitarian stand by several other participants who articulated biocentric attitudes. These attitudes emphasised the need for human–animal coexistence on the basis of fairness to animals and their prior claims to territory. For example, in discussing what to do about predators in the suburbs, several participants argued that 'people are going more and more up into the mountains and they're invading these animals' territory … they've been there for all these years', and the animals are simply 'invading us back … we should stop building. Start sharing!' In response to Alice's suggested hunt and assertion that humans come first, Vivian, Georgia and Irene responded with calls for fairness and coexistence:

VIVIAN: I don't think we can have that kind of control.

GEORGIA: I don't think we should. That's not even fair.

IRENE: Wasn't they here before us? The animals? … So, you know, actually they were here first. God made them first, all right.

ALICE: No, man is just a little lower than God. Man is just a little lower than the angels!

IRENE: We don't have, God put them animals there. We don't have no right just to take 'em out, because/

ALICE: Man is the ruler of all things/

IRENE: Well, this world is big enough for everybody to share/

VIVIAN: If they would only start sharing.

IRENE: We don't have any right to take out these animals because they were here before we were. Period. They have their life too.

Even though Irene privileged the lives of her own children over those of animals, if it came to a choice, she nevertheless insisted that the animals were only doing what they needed to do to survive – echoing themes of survival in a harsh environment but also situating animals as sentient beings with natural behaviours and needs:

I'm not gonna lie. Some, one of the animals attack my children I'm gonna try my best to kill that animal, you know? … that's because he's harming mine … but I'm not just gonna go looking [to kill the animal] … . He came down there for a reason. He was harmless, probably, and like I said, he was probably, they was in his territory, you know, where he probably hunt for food …

Several women, across all age groups, also articulated biocentric animal rights attitudes that promoted human responsibility to help wild animals in distress, threatened by starvation or other harm, even when this resulted from a natural phenomenon such as El Niño (as opposed to human-induced dangers). The absence of comment here from Susan was surprising, since her earlier comments about the slaughter of animals and the consumption of dogs would seem to place her directly in the animal rights field. Less surprising perhaps was the absence of Alice, whose utilitarian approach to animals was detailed above. Only Vivian voiced a fatalistic position, on religious grounds: 'Well, that's nature. That's an act of God. Man can't do anything about [it]' – and even she considered it appropriate for humans to help relieve animal distress in some fashion. Others, such as Georgia, advocated assistance, not because the animals were important for human use, but because they were other creatures who shared the planet with us and deserved consideration just like other humans:

We can't do nothing about temperature change but we can help them out. It's like if we, you're hiking somewhere and you get stuck in the snow, that's nature but then we run to save you so why can't we run to save the animals?

The six women (Georgia, Vivian, Laura, Bernadette, Norma, Carla) advancing the pro-animal rights commentary overwhelmingly agreed that it did not matter what kind of animal was in distress; that life in distress warrants human intervention. They were willing to dissolve inter-specific differences that might make it more appropriate to help one animal versus another. This perspective appeared to be a projection of anti-racist attitudes into a normative guideline for human–animal relations.

NORMA: An animal is an animal and if they need our help, well, you know, we should help them.
GEORGIA: Right. Just like humans, they're all different, we're all different in some ways/
NORMA: In some ways, but, you know, we stay people/
GEORGIA: In some ways, but I would rush to help you. I would, I mean, you know/
NORMA: Black, white, purple/

GEORGIA: Yeah. So why wouldn't I rush to help like a geese, a lion/
CARLA: Animals are just like humans.

Some participants, such as Susan (who, as described above, had reacted so nega-
tively to dog-eating), went further in their biocentric attitudes, venturing into
animal rights territory. Susan, in her early twenties, had a more affluent childhood during which she regularly watched nature programmes and was taken into
the mountains by her mother to hike and to learn about wildlife and plants.
Looking somewhat embarrassed and nervously laughing, she described the impact
of witnessing an animal being slaughtered on the formation of her attitudes and
her resolve to become vegetarian:

> I don't think that we should even eat animals, no kind of animals. Because
> animals cry, when they slaughter, they cry just like humans. They call out just
> like humans, I heard it with my own ears and it made me look totally
> different on a lot of different things.

Later, she added that her son was staying with his grandmother at her ranch, and
had recently been traumatised by the sight of neighbours slaughtering a goat, and
maintained that:

> If you seen them kill an animal and take the meat out and skin it and all that, I
> swear to you, you would never want to eat another hamburger, you would
> never want to eat another steak. It sound good, it smell good, but you would
> never want to eat another one because that is so awful.

The neighbours who killed the goat happened to be Mexican. In a fascinating
discursive move, Susan used a specific series of strategies to avoid racialisation of
this Mexican family: '[S]ome Mexicans that live next door, *not being funny or
nothing*, but they had *sacrificed* a goat ...' (emphasis added). By saying that she is
'not being funny or nothing', she may be warning listeners against making racist
assumptions about these neighbours (or at least not in front of a group of white
liberal academics!), and she softened the act of slaughter by using the term 'sacrifice'. These strategies may have been used to keep listeners focused on how the
goat's death created a negative emotional impact on her son, and thus on her
rationale for vegetarianism.

Other anthropocentric attitudes: negativism and spiritualism

Less often, two other sorts of anthropocentric attitudes were expressed: negativism,
or the fear, dread, contempt and animosity towards animals; and spirituality, in

which animals are viewed as agents or possessors of supernatural powers, to be respected, avoided or placated. Norma, for example, had some starkly negative sentiments, despite her assertion that animals were just like people and deserved assistance if they were in some sort of trouble (see above). In response to questions about who might be responsible if someone's boa constrictor escaped and ate a neighbour's pet dog, her expression turned mischievous, and she repeatedly proclaimed, *sotto voce*, 'death penalty!' – presumably implying that the snake should be killed. And, in discussing how to handle wildlife in the suburbs, her position was 'some animals deserve to die, honey.'

Similarly, negative attitudes were expressed towards the beach and associated sea life. Three participants (Bernadette, Vivian, Laura) revealed an aversion to visiting the beach, two of them describing the beach as 'nasty'. Frankie and Irene each hinted at a sense of fear towards sea life that is unknown or out of sight (under water), with Irene claiming that 'I can't stand them seaweeds, honey! My Mom used to scare me with them things. ... When I'd see 'em I screamed.' Similarly, Frankie provided a narrative about the drowning of a young boy due to entanglement with seaweed, while Irene revealed anxiety over her children 'playing with' sand crabs. Seagulls, however, were the primary target of negative perceptions and feelings among these women, demonstrated by the following discussion:

LAURA: I don't like … seagulls.
ALICE: Well, they make a mess around here and we used to feed them/
IRENE: You can kill them, too/
ALICE: And they hate for us to feed them because they really mess up this building and wherever else they are.
LAURA: I don't like … seagulls.
IRENE: Seagulls are bold birds! …
LAURA: I don't like … seagulls.
IRENE: [seagulls are] bold …
SUSAN: Yeah.
LAURA: I don't like them things … I wish I had a gun to shoot them suckers!

Seagulls emerged from this bit of discussion as both materially destructive (a pest, vermin), as described by Alice, and labelled (in anthropomorphic fashion) with an unattractive characteristic ('boldness') that signified an immediate, almost confrontational, aspect of the animal. Laura, whose contempt for seagulls appeared most intense (indeed, to the point of violence), unfortunately gave no qualification for this feeling.

The normative attitudes concerning human–animal relations, as linked to the spiritual value of animals, arose during talk about childhood memories of animal stories, and religious ideas about animals. These attitudes were primarily expressed by our Belizean participant Georgia, who raised the issue of animals as symbols of bad luck that therefore needed to be removed from the home environs:

GEORGIA: …if for some reason a black cat would be born/
LAURA: It would be bad luck/
GEORGIA: It would be put … out of the yard. You would have to take it like maybe a mile or two miles and just let it go in the woods. … Whatever would happen to it we don't know but we would just get rid of it.

Other animals were never to be harmed, however, due to their role in human salvation. Again, Georgia raised this point, which may have been related to her syncretic religious training in Belize that appeared to blend Catholicism and indigenous animist beliefs:

A dove in Belize … that's the only bird that you would … not [be] allowed to play with, [or throw] stones, you know, because it was a dove that led us back to life. So … they would have eggs, and … I'm tempt[ed] to want to go take the eggs, but I mean, you would never do it cause it's like you were told ever since you were little that, 'never touch the dove, never stone the dove'.

Both the notions that a black cat brings bad luck and doves are somehow a symbol of life and peace are common in Western cosmology. Indeed, cats – especially black ones – have a long history of connections with black magic, witchery and the devil, a history that legitimised torture and death for untold unfortunate felines. Similarly, doves are a potent biblical symbol, but this rarely affords them much protection except, it would appear, in those places such as Belize where older traditions of animism may persist beneath an overlay of Christianity and continue to influence everyday attitudes and practices.

Gendered attitudes

Responses to 'Do men and women vary with respect to their attitudes towards animals?' included perceptions of gendered preferences and attitudes towards animals:

ALICE: Men like horses and dogs … some women like dogs and cats … very few men like cats.
LAURA: Men like snakes.
VIVIAN: The idea that they [men] are better than animals … that they try to have … control.

SUSAN: Men are harsh ... men are harder on animals than women. ... Men are real rough with animals and women are more soft.

Interacting here are the social and gendered constructions of both animals and gender itself. Accordingly, gender seems to structure attitudes towards, and values of, animals, practices involving animals, and animal-specific preferences.

The notion that men are 'better', 'harder', 'harsh' and 'real rough' with animals is consistent with post-colonial and feminist literature on the violent and oppressive history of, generally, white patriarchal societies and value systems towards women, (other) peoples and the environment. The idea of male control, as posed by Vivian, was reflected in Alice's comment on gendered preference, noting that dogs and horses tend to be preferred by men over, say, cats (generally considered an animal with more of a 'feminine' character). That dogs and horses may be considered the two most successful cases of domestication (control) in the course of human history is significant. Although posed separately and with different intent, each of these two comments reinforces the construction of 'male' as dominant, oppressive, controlling.

The idea was found in earlier areas of discussion as well. In various ways, male family members, especially fathers, were the decision-makers on the fate of animals. They were the hunters (Bernadette, Frankie), the dispatchers of family pets (as in the case of Alice's dog, Benjamin), the 'professionals'; in effect, the 'keepers' of animals with all of the power and control that position entails. Consider the following statements:

BERNADETTE: They [father and friends] weren't 'fishermen', they were 'professionals'.

IRENE: [A male] friend came from the islands and [brought] real small monkeys that fit in our pocket ... but they ... recently died. He had a big iguana; jumped off his balcony ... his house was like a zoo.

ALICE: [On wolf eradication] you got so many men with guns they'd be up there and they'd be rid of 'em in less than a year.

FRANKIE: [My father] hunted food and trapped it.

Nevertheless, a gentler side of 'male' attitudes towards animals also appeared, as Susan indicated when recounting her son's distress at witnessing the slaughter of a goat. Even Alice identified a chink in her father's hard emotional armour; discussing her father's experience at the slaughterhouse, she said: 'Daddy couldn't kill the sheep because of ... the sound they would make when you kill 'em. The sheep broke his heart.' But despite these stories of empathy, the gendered construction by women of 'male' and of 'animal' appears to fix in place a certain relationship between the two.

Relative to men, little was explicitly said about women's relationships with animals. Irene noted her mother's sensitivity to animal needs and happiness, contrasting it to the anthropocentric attitude of her father and siblings, as she explained her mother's decision not to grant the children their wish for a dog:

> I couldn't understand why my mom just didn't want to let the dog live back there and let him in the house sometimes. And she used to tell me, 'We're not going to be able to keep the dog because this is mean to the dog. We're gonna give him away to somebody with a big farm so he'll run around.' ... I wanted my dog in the backyard. ... You know, my mom was more sensitive to the animal's feeling ... she did do the right thing, cause ... we lived in a townhouse but ... the backyard wasn't that big. ... I understand now but then I couldn't stand my mother.

But there were plenty of counter-examples. Carla's abrupt and somewhat brutal description of the slaughter of chickens ('kill it and skin it') and Alice's admission of never being 'concerned with animals' both seem to contradict Susan's belief that 'women are more soft' on animals. The abstraction of gender to a totalising 'male' or 'female' was deeply ingrained in participants' minds, despite the fact that their own discussion revealed such essentialised views to be equally as erroneous as the abstraction of animals to a single category. Instead, the discussion of concrete events revealed that human–animal relationships occur within personal/local histories that make use of and illustrate the variety of relationships that both men and women maintain with animals in their various and flexible categories on a day-to-day basis. 'Male'/pet relations differ significantly across both time and space, as much if not more so than 'male'/pet versus 'female'/pet relations. Alice's dog, Benjamin, as we have seen, constituted pet in one set of circumstances yet quickly emerged as 'pest' within a different context, resulting in his death at the hands of her father, who, incidentally, had to change jobs at the slaughterhouse because the slaughtering of sheep 'broke his heart'. In short, despite participants' stereotypical responses when asked about gender differences, their situated stories showed how animal-human relations vary across space and time, across gender, and within context.

Discussion: attitudinal conflict and cultural models of people and animals

How can we make sense of the fundamental conflict between the dominant attitudes articulated by the focus group participants – anthropocentric utilitarianism and biocentric views? For our group, resolution was achieved by segmenting the animal world into three categories: food, pet and wildlife. 'Food' animals were simply necessary for survival; people had to distance themselves from their unfortunate fate. Indeed, animals consumed for food seem to have become the ultimate urban 'other', a group whose plight and suffering could not be recognised without grave risk to longstanding everyday dining practices and patterns. For Susan, such recognition had deeply disturbed her, and led her towards a vegetarian diet, but most of the others seemed resolute in their desire to put animals used for food consumption into a separate category. In the case of pets and wild animals, however, the human–animal divide became permeable, and similarities between humans and animals demanded care and compassion and legitimised animal rights.

These socially constructed categories were not rigidly defined or mutually exclusive, but rather depended upon time, place and situation. Indeed, as the discussion of cross-cultural practices above indicates, these categories were readily discerned as relative and culture-bound. But in general, the suffering of pets or wildlife appears fundamentally different from killing for consumptive purposes, engaging emotions, sympathies and values that become manifest in human–animal analogies. Indeed, the following comment by Bernadette rang of an almost 'maternal' quality: 'Help the animal. … Take care of them. You know, nurse [it] back to health and let [it] go.' The idea of suffering, in general, is a disturbing one, regardless of the causes. But animal suffering, in particular, is likely a rare experience in an urban setting, outside of the suffering of pets. In contrast, the suffering from slaughter, as detailed by Susan, is something few in an urban setting will ever experience. Hence a place-based *distancing* operates to shield people from animal suffering during slaughter in the urban context, working along with packaging and advertising to mute the psychological dissonance involved in killing for food. Although poultry markets continue to exist in Los Angeles, and the variety of live animals for butchering and consumption can be impressive in certain ethnic enclaves, the cultural construction of animals either as food, pet or wildlife serves to shield people from the emotional impact of slaughter-related suffering. This renders livestock animals such as chickens, pigs, sheep and cattle the most 'outsider' of outsider animal groups.

Particularly for urban residents distanced from subsistence hunting, pets and wild animals can more readily be seen as standing on the boundaries of humanity. For example, animals identified in the question about whether humans should intervene to help animals in distress due to natural causes were (in order) whales, seals, birds and fish, each framed explicitly and implicitly as wildlife, a category which carries with it strong socio-cultural, even political, images and understandings. Suffering wildlife may be interpreted or imagined in such contexts as a suffering earth, a suffering innocence, a suffering outsider group (such as African-Americans); its rescue as a chance at redemption.

Such ideas about animals as food, pets or wildlife together reveal the subtle ways in which participants identify with other outsider groups, both human and non-human. Their willingness to tolerate dog-eating among Southeast Asian immigrants, for example, reflects their own experience as an oppressed, marginalised group in American society, and their sensitivity to racialisation based on colour and culture. Similarly, their sympathy for suffering wildlife and their claim that people should help wildlife in distress, just as people should help each other regardless of colour, hints at their solidarity with animals as brethren due to their outsider status. But the history and experience of oppression among African-Americans also creates imperatives to survive in the face of adversity. Thus, when animals are critical to survival, they will be hunted, trapped, slaughtered and served up in ways that may become potent cultural symbols of strength and perseverance. And, when they compete with humans and thus deprive them of needed resources, animals can expect harsh treatment if not extermination.

Thus certain practices are appropriate for specific social, cultural and, significantly, *ideological* constructions, and are mediated by the values and attitudes inherent within such constructions. Suffering wildlife or pets legitimise sympathy and rescue, while the suffering of 'food' animals is considered an unfortunate but necessary externality of insufficient consequence to change behaviour – although over time such cultural constructions can break down or dissolve as circumstances change. In any one time or place, however, our analysis reveals that people are bound – by shared affinity, need or disgust – to particular sorts of animals understood on the basis of historically rooted social constructs. Such human–animal connections will vary in both subtle and startling ways across the diverse cultural groups living side by side in the contemporary metropolis.

Appendices

Appendix A   African-American focus group participant profiles

Participant	Age	LA residency [a]	Education	Animal experience (type)
Bernadette	52	37 yrs (Richmond, CA)	High school	No
Georgia	22	3 yrs (Belize)	High school	Yes (farm, stable, pets)
Vivian	48	Native	Some college	Yes (pet)
Carla	50	Native	High school	No
Susan	25	Native	College degree	Yes (farm)
Denise	75	30 yrs (New York)	High school	Yes (pets)
Irene	28	13 yrs (Colorado)	<High school	Yes (pets)
Frankie	22	9 yrs (Alabama)	Some college	Yes (farm, stable, pets)
Alice	64	33 yrs (Texas)	College	Yes (farm, pets)
Laura	26	Native [b]	<High school	Yes (pets)
Norma	25	Native	<High school	Yes (pets)

Notes:
[a] years of residency in LA followed by (state or country of birth)
[b] moved to LA before her first birthday and therefore considered a 'native'

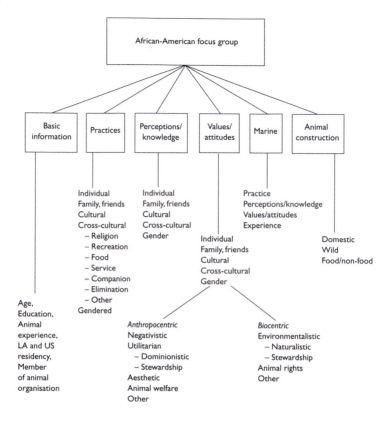

Appendix B   Coding tree.

Acknowledgements

This research was supported by the US National Science Foundation Grant #9605043, Programme in Geography and Regional Science. We are grateful to Sister Diane Donoghue and Wallace Moore for helping us to organise the focus group, and to Robert Wilton for his assistance with qualitative analysis software. We also wish to thank our editors, Chris Philo and Chris Wilbert, for their constructive and thoughtful comments on an earlier version of this chapter.

Notes

1   For further elaboration, see Wolch (1996).
2   In this chapter and in our research more generally, we consider 'cultural background' to reflect the complex of socially constructed individual and group identities linked to colour, nativity and non-US heritage that, along with class and gender and other markers of difference, are used to categorise, legitimise and marginalise people in US society. Such identities can also form the basis, over time, for some degree of shared perspective or standpoint, as well as common place-specific historical experience and material circumstances that we expect to shape attitudes towards animals. For an elaboration, see Elder *et al.* (1998).

3 No non-humans participated. Whereas we considered the inclusion of a friendly canine as an 'ice-breaker', we ultimately decided that such a visible presence would inhibit human respondents from talking about controversial practices – such as dog-eating!

4 Although these participants might have volunteered due to their interest in or affection for animals, creating a self-selection bias, the service provider who assisted in recruitment indicated that our \$25 honorarium would be the primary incentive to join the focus group.

5 The terms 'pet' and 'wildlife' should really be placed in 'scare quotes' throughout the chapter, given that much of the discussion in the focus group swirled around the question of what exactly is meant by pets and wildlife, and about how humans should treat them. In what follows, though, we will avoid quote marks for these terms, which we use repeatedly, to avoid over-complicating the appearance of the text on the page.

6 Participants' names have been changed in order to protect their anonymity.

7 The text was analysed using QSR NUD*IST, a qualitative, non-numerative research programme designed for textual and narrative analysis.

References

Caron, J.A. (1989) 'Environmental perspectives of blacks: acceptance of the "new environmental paradigm"', *Journal of Environmental Education* 20(3): 21–26.

Caron-Sheppard, J.A. (1995) 'The black–white environmental concern gap: an examination of environmental paradigms', *Journal of Environmental Education* 26(2): 24–35.

Dolin, E.J. (1988) 'Black Americans' attitudes toward wildlife', *Journal of Environmental Education* 20(1): 17–21.

Elder, G., Wolch, J. and Emel, E. (1998) '*Le pratique sauvage*: race, place and the human–animal divide', in J. Wolch and J. Emel (eds) *Animal Geographies: Place, Politics and Identity in the Nature–Culture Borderlands*, London: Verso.

Kellert, S.R. (1984) 'Urban American perceptions of animals and the natural environment', *Urban Ecology* 8: 209–228.

Kellert, S.R. and Berry, J.K. (1980) *Knowledge, Affection and Basic Attitudes Toward Animals in American Society, Phase III*, Washington, DC: US Department of the Interior, Fish and Wildlife Service.

Kempton, W., Boster, J.S. and Hartley, J.A. (1997) *Environmental Values in American Culture*, Cambridge MA: MIT Press.

Marks, S.A. (1991) *Southern Hunting in Black and White: Nature History and Ritual in a Carolina Community*, Princeton: Princeton University Press.

Philo, C. (1995) 'Animals, geography and the city: notes on inclusions and exclusions', *Environment and Planning D: Society and Space* 13: 655–681.

Plous, S. (1993) 'Psychological mechanisms in the human use of animals', *Journal of Social Issues* 49(1): 11–52.

Wolch, J. (1996) 'Zoöpolis', *Capitalism Nature Socialism* 7: 1–48.

Wolch, J. and Emel, J. (1995) 'Bringing the animals back in', *Environment and Planning D: Society and Space* 13: 632–636.

5 Taking stock of farm animals and rurality

Richard Yarwood and Nick Evans

Introduction

The study of livestock has long been a part of rural and agricultural geography. However, farm animals have usually been studied in an economistic manner and have been regarded simply as 'units of production' within agricultural systems. As Philo (1995: 657) has stated, animals have only appeared 'in the background of studies in rural geography' and there have been few attempts to view animals as 'animals' in their own right or, at the very least, as animals which have cultural as well as productive value to people in the countryside. Some early effort was expended on mapping and explaining livestock distribution using broad (productive) categories such as 'dairy' cattle, 'beef' cattle, sheep, pigs, 'table' foul or 'laying' foul, as exemplified by Coppock (1964) in his agricultural atlas. As agricultural geography has become more process-orientated, so too have analyses of livestock. For example, Bowler (1975) considered specialisation in livestock enterprises in Montgomeryshire, North Wales, through an examination of 'behavioural influences'. Symes and Marsden (1985) examined pig production in Humberside as part of a trend towards the increasing industrialisation of agricultural systems. Similarly, Halliday (1988) investigated the state imposition of quotas on dairy farmers in Devon, reflecting an increase in interest among researchers on the effects of government intervention in agriculture.

These changes generally reflect paradigm shifts in agricultural geography, except that livestock are always incidental to some other goal, irrespective of theoretical position. Even then, Cloke (1989) has suggested that rural geographers have had an apparent reluctance to engage in new theoretical approaches favoured in other fields of geography. This has certainly been true of the 'cultural turn' which influenced the direction of so much geographical research in the 1990s. Although Cloke (1997) has demonstrated that the application of cultural theory is now leading to new 'excitements' in rural studies, these have yet to impact fully on agricultural geography. Here, studies have remained so firmly underpinned by political economy perspectives that, despite attempts to accommodate greater diversity in analysis through modification (see Marsden 1998; Whatmore *et al.* 1996), agricultural change has barely been mentioned in recent reviews of rural geography (Cloke 1997; Phillips 1998).

Less anthropocentric approaches to geography, as advocated by a growing number of commentators (Anderson 1997; Murdoch 1997; Philo and Wolch 1998; Tuan 1984), present a whole host of possibilities for agricultural geographers and offer one way in which the study of agrarian change can be linked more closely with recent theoretical developments in rural geography. At the very least, there is a need to recognise that farm animals have, quite literally, been constructed by people to fit into particular rural spaces. This emphasis would also advance a 'new' animal geography as, somewhat surprisingly, many studies in this vein have focused on the place of animals in the home or urban environment (Anderson 1995; Philo 1995; Wolch *et al.* 1995) and there is still much work to be done on the place of animals in the countryside.

One way in which this can be achieved is to pay greater attention to different breeds, as well as species, of farm animals. This is because people have deliberately bred animals to meet different environmental, economic and cultural demands over time and space. Livestock breeds are therefore socially constructed, in both physical and cultural terms (see Anderson 1997). As a consequence, there is an astonishing diversity of livestock breeds across Britain specifically and the world generally which can reveal much about society and people's relationship with animals (Yarwood and Evans 1998). Yet, despite such significance, the geographies of livestock breeds have been largely ignored in contemporary study and there have been few attempts to examine the place of farm animals within the geographical imagination.

This chapter demonstrates how a sharper focus on livestock can contribute to current thinking in rural studies and to the development of new ideas in agricultural geography. It begins by examining how rare breeds of livestock have been valued by farmers and how the changing rural economy has affected this relationship. Emphasis is given to the new roles that livestock are finding in a post-productive countryside, to their contribution to landscape in Britain, and to ways in which they have been re-imagined for the purposes of heritage and place marketing. This leads to consideration of the ways in which livestock are culturally constructed to fit with certain discourses of rurality and how these can be expressed through conservation groups. In doing so, the chapter moves from a perspective underpinned by political economy ideas to a more culturally informed standpoint, demonstrating how a greater sensitivity to cultural issues can enrich studies of agricultural change. Given the extensive nature of this review, the reader's attention is also drawn to discussions on these themes which have been published in more depth elsewhere (Evans and Yarwood 1995, 1998, 1999; Yarwood and Evans 1998, 1999; Yarwood *et al.* 1997). This chapter uses a range of empirical evidence, including information from 1,834 interviews with members of the Rare Breed Survival Trust (RBST) conducted as part of a research project[1] by the authors on the 'empirical' and 'imagined' geographies of livestock in Britain and Ireland. The text which follows in italics is taken directly from these interviews.

Agricultural change and livestock

Many studies of livestock within agricultural geography have treated animals as 'units of production'. This is a logical way to start examining livestock as they have long been exploited primarily by people for meat, wool or dairy produce. Over time, centres of domestication emerged where local breeds were developed to match productive need and climatic conditions. Breeds were diffused from these centres by human migration and, as a consequence, the historical geographies of animal breeds reflect the changing geographies of dominant societies (Yarwood and Evans 1998). In the British Isles, for example, a succession of invasions led to the introduction of many new breeds of livestock animals to different parts of the country (Henson 1982; Wallis 1986).

The respective locations of Kerry, West Highland and Welsh Black Cattle in the West of Ireland, Scotland and Wales can be explained by past cultures. These breeds are closely related and have a common ancestor which was kept by some Celtic peoples. After successive invasions, these people were forced to retreat from the lowlands to the periphery, taking their livestock with them (Alderson 1976; Friend and Bishop 1978). The present-day survival of these livestock continues to reflect historic differences and to impact on the landscapes of the British Isles to the same extent as do other features such as field boundaries, settlement morphology and defences (Evans and Yarwood 1995). Clearly, strong associations can exist between breeds, place and culture.

Significant changes in the geographies of livestock occurred during the agricultural 'revolutions' of the eighteenth and nineteenth centuries as a result of 'selective' breeding. This technique used scientific principles to improve the genetic characteristics of livestock, such as their weight, milk yield, leanness and speed of maturity, in order to increase profitability. This did much both to create and to destroy associations between particular places and breeds. On the one hand, livestock improvement led to the establishment of many new breeds which were closely linked with place: '[F]or many breeds their vernacular names ... precisely describe their place of origin' (Clutton-Brock 1981: 30). On the other hand, selective breeding led to the diffusion of livestock breeds across wider areas and diluted association with particular places (Walton 1984, 1987).

In the twentieth century, capitalist farming was characterised by greater intensification, concentration and specialisation, especially in the post-war 'productivist' era (Bowler 1985). As the efficiency and profitability of livestock were given greater emphasis, many new breeds of livestock were introduced to Britain which were suited to capital-intensive farming systems because they could produce more food at lower costs (Evans and Yarwood 1995). Examples include Friesian Cattle, Poland China Pigs and Texel Sheep. This meant that many British breeds were viewed as unprofitable or inefficient and were no longer kept by farmers. Between 1900 and 1973, over twenty-six breeds of farm animals became extinct in Britain alone, including Glamorgan Cattle, St Rona's Hill Sheep and Lincolnshire Curly Coat Pigs (Alderson 1990). Others declined drastically in numbers as they were not suited to technologically dependent conditions of production. Currently, over sixty breeds of farm animals in Britain are classed as having 'rare' or 'minority' status by the RBST on the basis of their pedigree numbers and genetic status. This applies to 25 per cent of all British sheep breeds, 40 per cent of cattle breeds and 70 per cent of pigs. Since 1973, however, no further breeds have become extinct and many have actually increased in number (Yarwood and Evans 1998). This can be largely explained by the significant changes which have occurred in agricultural modes of production.

By the mid-1980s, persistent encouragement of productivist farming methods had created a political 'farm crisis' caused by overproduction of foodstuffs, the immense budgetary costs of agrarian support and environmental damage. These pressures propelled farmers into a new set of circumstances, known generally but uncritically as 'post-productivism'. Three central characteristics of this shift in agrarian priorities away from food production have been evident as 'pluriactivity', a desire for more environmentally friendly farming and consumer concern over food quality. Although many rare breeds had no (profitable) purpose within the productivist era, they have been re-valued and can play important roles within post-productivism.

First, some farm families become pluriactive through an on-farm economic diversification into activities such as farm shops or farm-based tourism (Evans and Ilbery 1993). The keeping of rare breeds has become a major tourist attraction (see Figure 5.1). Many rare breeds, by virtue of their present-day scarcity, have 'unusual' characteristics, such as large horns, capable of attracting paying customers onto the farm. In other cases, pluriactivity is evident where farm households have become 'part-time', one or more members spending a proportion of their time engaged in income-earning activities off the farm (Evans and Ilbery 1993). Part-time farm households value traditional breeds, such as Large Black Pigs, because these animals can be reared outdoors with minimum supervision in contrast to modern breeds which require capital- and labour-intensive support systems. Second, rare breeds can contribute to the wishes of society to create a more sustainable agriculture through agri-environmental policy (Evans and Morris 1997). They are currently used in the conservation of certain endangered habitats because they graze differently to modern animals (Small 1995).

Rare breeds also contain unique genetic material which, if lost, is lost for ever. Their conservation therefore contributes to national and global biodiversity. Third, although many rare breeds cannot produce the same *quantity* of food as their modern counterparts, many farmers believe that their produce is superior. Consequently, some farmers have engaged in the niche marketing of rare breed food and are keeping rare breeds to supply a growing demand for *quality* food caused by recent food scares.

As these three post-productive priorities become entrenched, rare breeds are being kept by more and more farmers. The survival and status of livestock breeds relies largely, therefore, on how farmers perceive their usefulness as items of productivity or profitability to the wider farming economy. Thus, economistic approaches offer a useful introduction to the study of farm livestock and go some way towards explaining temporal changes in livestock numbers.

However, many rare breeds are only found in specific localities. The RBST membership survey noted a strong degree of local loyalty to particular livestock. In Scotland, for example, there were higher than average rates of ownership of Shetland Cattle (kept by 27 per cent of Scottish RBST owners), Belted Galloway Cattle (kept by 24 per cent of Scottish RBST owners) and Clydesdale Horses (kept by 43 per cent

Figure 5.1   Different breeds of livestock can be a central attraction in many farm tourism enterprises. Here the 'Live Sheep Show', part of the Lakeland Sheep and Wool Centre in Cumbria, introduces visitors to the diversity of British sheep breeds in an entertaining manner.

Source: Authors' photograph

of Scottish RBST owners). In Wales, higher than average rates of ownership were recorded for Hill Radnor, Llanwenog and South Wales Mountain Sheep, all Welsh breeds. Furthermore, 60 per cent of Dorset Down Sheep, 57 per cent of Cotswold Sheep (see Figure 5.2), 56 per cent of Greyfaced Dartmoor Sheep and 48 per cent of Wiltshire Horn Sheep were found in the south of England, clustering most strongly in those places that the breed names suggest.

In many of these cases, economic sense would suggest that other breeds could be used in a much more profitable manner. Economistic approaches cannot, therefore, fully explain the *geographies* of farm animals, and more culturally sensitive approaches are needed to understand links between livestock and place. The following sections offer ways of doing this and consider in more detail how livestock have been re-imagined to fit within new modes of accumulation in rural places. As well as allowing some consideration of structural change in the countryside, this discussion starts to consider how value and identity are placed on particular breeds of livestock in different localities. By adopting this approach, it is possible to see how different livestock and their roles are culturally constructed to fit into different human spaces. This fundamental goal of animal geography has not previously been attempted in the rural field.

Market(ing) places

Some commentators have suggested that the countryside is changing from being a place of production to a place of consumption and, as part of this process, many areas have been commodified to attract new forms of capital (Cloke and Goodwin 1992). Central to this process has been the use of rural heritage to sell places and products associated with an idealised rural lifestyle (Bunce 1994; Thrift 1989). Just as features of the industrial past have been commodified in order to re-image post-industrial cities (Hall 1995), so too has there been a re-imagination of livestock in the post-productive countryside.

An obvious example of this is in the growth of farm parks which seek to create hyper-real versions of agricultural heritage to attract the paying public. Of those farmers who had on-farm tourist activities in the survey, 94 per cent kept rare breeds as part of their venture because they were '*local breeds*', '*part of the region's agricultural history*', '*relevant to the area*' and '*for history and heritage*'. By way of example, the survey identified a part-time farm manager in Buckinghamshire who ran an 18 hectare open-air museum and who kept Berkshire Pigs, Shorthorn Cattle and two local breeds of poultry because of their '*local, historical relevance*'. Another full-time farmer in Oxfordshire kept Oxford Down Sheep, Cotswold Sheep, Berkshire Pigs and Gloucester Old Spot Pigs because they allowed him to re-create '*a local Victorian farm*'. Such breeds remain popular in their places of origin as they can be used to construct versions of local history for wider consumption.

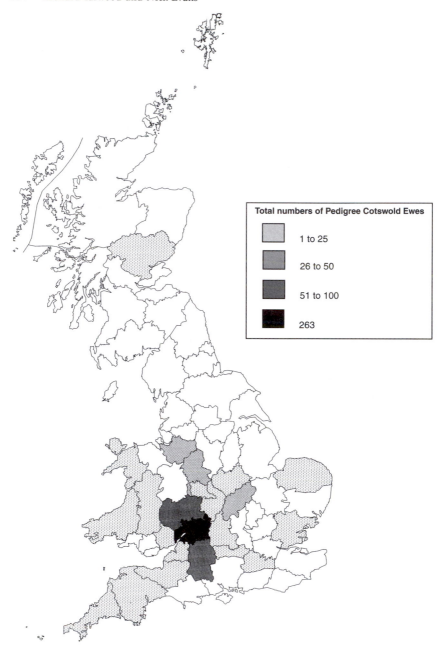

Figure 5.2   The geographical distribution of Cotswold Sheep in Britain.

Source: Authors' research, derived from information in Rare Breeds Survival Trust (RBST) (1994) *Combined Flock Book, 1994*, National Agricultural Centre, Stoneleigh Park, Warwicks: RBST

The use of local livestock in this manner extends beyond farming enterprises to the marketing of place and its local products (Yarwood and Evans 1998). Local historic livestock animals are also used in the iconography of particular places (Higgins 1993). The Cotswold Sheep is used in the emblem of the Cotswolds Area of Outstanding Natural Beauty (AONB), as is a leaping Dartmoor Pony in the Dartmoor National Park and the Swaledale Sheep in the Yorkshire Dales National Park. The unique and wild appearance of Highland Cattle are used in tourist adverts to market similar qualities in the Scottish Highland landscape. In Gloucestershire, the Uley brewery produces a beer called 'Old Spot' which is served in many local hostelries, including the 'Old Spot' public house in Dursley, Gloucestershire. Affection for the Gloucester Old Spot Pig is used to sell local products and the pub itself uses images of the breed in much of its internal decor. When particular breeds of livestock are used to market particular places and experiences in these ways, they both reinforce and exploit existing local loyalties to a breed. Such marketing can be beneficial to a breed as it sustains local interest in it and further encourages people to keep it for its 'heritage' value. This helps ensure the survival and continuation of many breeds in their places of origin, and farm parks are valued for their contribution to national breeding and conservation programmes. However, constructions of animals in this way can be problematic, for two main reasons.

First, many rare breed centres keep animals which appeal to the public more through their long horns, shaggy coats, unusual size or curious markings than their historic connection with locality. Many farm parks keep animals such as White Park Cattle, Leicester Longwool Sheep, Shire Horses or Gloucester Old Spot Pigs for these respective reasons. Although 'unusual' animals have a novelty value that appeals to visitors, they can shatter local landscape coherences if they are kept in places which have no history of farming with these breeds (Evans and Yarwood 1995). Furthermore, this focus on 'unusual' animals has meant that plainer livestock have been lost, for, as Alderson (1990: 54) laments: '[M]aybe if the Irish Dun and Suffolk Dun [cattle] had been a more spectacular colour they might have escaped extinction.'

Second, rare breed centres are aimed at a lay public who have little experience of farm animals. Consequently, the animals are clean, healthy, docile and even have 'pet' names. This has contributed to a sanitisation of livestock, even in RBST-recognised farm-parks. This is furthered by the use of livestock caricatures to promote a range of products, including toys (Houlton and Short 1995), holidays, television programmes and food (see Evans and Yarwood 1995). It is necessary to consider how these anthropogenic constructions affect the ways in which animals are perceived by different groups of people and how these are expressed in various discourses of rurality, especially given recent reports of conflict between farmers and non-farmers over the place and role of livestock in the countryside

(Cloke 1993; Yarwood and Evans 1998). The following section begins to address these issues by considering the ways in which farm animals are culturally constructed to fit into particular rural spaces.

Livestock, landscape and rurality

The impact of modern farming methods on elements of the British landscape such as field trees, drainage features, topography and buildings has been well documented (see Rackham 1987; Westmacott and Worthington 1984). However, it has rarely been questioned whether livestock can contribute towards localised landscape distinctiveness (Evans and Yarwood 1995). This is a serious omission as livestock are very common in many rural landscapes and bear strong testimony to historic and current cultures. White Park (see Figure 5.3) and Chillingham Cattle provide a good illustration of this. There is some controversy over the exact origins of these animals, but they are believed to be Britain's oldest breed of cattle and it would appear that they are descended from a breed of white cattle which had been roaming freely in Scotland and northern England since the fifth century. Some of these animals were enclosed in parks in the thirteenth century for hunting purposes. These were allowed to live in a relatively feral state, although they may have been subject to some forms of management, including artificial

Figure 5.3   White Park Cattle.
Source: Authors' photograph

selection through culling. One feral herd survives today at Chillingham Park, Northumbria (Latham 1998). These Chillinghams have been in-bred to such an extent that many regard them as a separate breed from the White Parks. The animals have been the subject of many myths (including claims that they are direct descendants of Auroches [wild oxen], or that they were used by Romans and Druids in religious ceremonies), yet their distinctive presence in the landscape reflects much about people's historic links with livestock and landscape (see Alderson 1998 for a full account of this breed).

Today, livestock changes have become more rapid and widespread based on technological advances. These account for the dominance of Friesians, which, with their exceptional milk yield, supply 90 per cent of milk in the United Kingdom. The impact of such distinctive black and white cattle on the landscape has been similarly influential, and there has been concern that the widespread proliferation of this breed has eroded local landscape distinctiveness (Evans and Yarwood 1995).

At a time when there is political willingness to conserve the appearance of local landscapes (Evans and Morris 1997), livestock remain surprisingly neglected in study and policy. Greater attention to them would be beneficial because, as Small (1995) notes, some breeds can contribute actively towards preserving sensitive landscapes because they graze differently to modern animals. For example, Hebridean Sheep exhibit a marked preference for grazing tough species of upland grass, encouraging the regeneration of heather. Longhorn Cattle have been employed on pastures in the Derbyshire Dales overrun by tough plant species which are inedible to softer-mouthed modern cattle. Small (1995) advocates that grazing with local breeds is the best way of preserving local habitats and points to the successful use in the Cotswolds of Gloucester Cattle on Minchinhampton Common and Cotswold Sheep on Cranham Common.

To date, this use of livestock in creating or diluting landscapes has only been pursued by agents such as the National Trust or members of the RBST on an *ad hoc* basis. National conservation schemes, such as Countryside Stewardship, do not, as yet, offer incentives to use local rare breeds to preserve landscape distinctiveness (Evans and Yarwood 1995). Such policies would not only aid local ecology and the preservation of breeds, but would also help in the (re-)creation of local identity. Ironically, there is provision in the European Union for member-states to encourage the conservation of rare breeds under the agri-environmental 'accompanying measures' Regulation 2078/92. However, the UK (along with Denmark) has decided not to fund rare breeds through this mechanism. The 'official' reason is that no one area in the UK is entirely dependent on a specific rare breed for the maintenance of its landscape or habitat. Given the lack of attention that has been paid to such relationships, a more convincing explanation is that the Ministry of Agriculture, Fisheries and Food (MAFF) does not wish to fund a process of rare breed conservation which is already occurring through the charitable actions of the RBST. Further investigations of these assertions are necessary.

By contrast, lay agencies and people have recognised the importance of livestock in the creation of place identity and local culture, as Clifford and King (1995: 14) recognise in the 'Common Ground' campaign:

> [T]he cows of this place, eating the grass of this valley, with the expertise built up over generations ... breed culture and identity which has meaning for the people who live and work here and for those who chance upon it or make it their destination.

This assertion was supported by the RBST survey. When asked to comment on the general nature of rural change, as many as 18 per cent of respondents referred to changes in livestock. Farm animals, it was felt, added '*to the general rural scene and are important for urban dwellers and us countryfolk*', and it was important to '*preserve breeds for the enhancement of rural life*'. However, many respondents bemoaned losses in breed diversity and the replacement of 'British' cattle with continental breeds. In these cases, respondents noted that '*the introduction of foreign breeds was leading to the destruction of landscape*' and that there had been a '*continual loss of traditional breed types in favour of black and white cattle and white pigs, hybrid poultry and the introduction of more continental beef stock*'. These comments reveal that farm livestock are important within the geographical imagination of the countryside, and that interest in them extends well beyond their productive value to include their contribution to landscape identity and biodiversity. Indeed, only 11 per cent of respondents feel that rare livestock are important economically, and, of those who keep rare breeds, less than a third said that they do so for their produce. Instead, three-quarters (responses were not mutually exclusive) said that they are supporting rare breeds to maintain '*unique sources of genes*' and two-thirds to preserve '*part of British heritage*'. This contrasts starkly with suggestions that farmers are often reluctant to engage with the ideas and practice of conservation (Carr and Tait 1991; Morris and Potter 1995). Thus, the membership and aims of the RBST raise interesting questions because it is a conservation organisation which draws from both the farming and non-farming community (Evans and Yarwood 1999).

The RBST was formed in 1973 and, like many conservation charities (Thrift 1989), has experienced a surge of interest over the past twenty-five years. Charities have been criticised because all too often an overt desire to conserve or to preserve the countryside masks a desire to protect property prices, cultural capital and elitist visions of the countryside (Murdoch and Marsden 1994). It is assumed that farmers, acting as entrepreneurs, either oppose or adopt the ideas of the conservation lobby depending on economic circumstances or class interest.

Thus, they will oppose conservationists if production is threatened, but will adopt agri-environmental schemes if suitable compensatory payments are offered by the state or Europe (Wilson 1996). Despite calls for greater sensitivity to 'intra-class' division (Cloke and Thrift 1987) and a well-established behavioural tradition in agricultural geography (Gasson 1973; Ilbery 1978, 1979), scant coverage has been given to the role of farmers in contributing to conservation debates compared with their well-reported propensity to engage with conserving actions on the ground (Potter 1986). Yet, as the results of the survey show, many farmers appear to keep rare livestock for the sake of conservation alone, even though they have no direct monetary incentive to keep rare farm livestock or to join the RBST (although its services are of some commercial value). Further investigation is necessary into the reasons why this is the case and why farmers have responded so well to this aspect of conservation. On one level, it might be suggested that farmers are keeping rare breeds in the hope that they may possess economic advantages in future productive systems and that conservation is a fortuitous by-product. But, on another level, it might be suggested that this reflects a genuine concern about farm animals which transcends their productive value. Specific questions also arise about farmers' deeper motives for belonging to the Trust, whether the RBST supports class interest and the influence it has had, or could have, on policy-making. The preceding discussion obviously reflects the view-points of RBST members, who have an interest in farm livestock, and so apparent interest in livestock may be restricted to a minority of people – although minorities are important, as Philo (1992) points out. Research which addresses these issues and considers the place of animals in farmers' constructions of rurality would clearly be worthwhile.

Attention also needs to be focused on how these constructions mesh with those held by other social groups. Certainly, the constructions of livestock discussed above are not unproblematic. As Agyeman and Spooner (1997) indicate, discourses which suggest that flora and fauna are 'native' or 'foreign' to Britain, and 'do' or 'do not' belong in rural places, can contribute towards feelings of exclusion felt by some people of colour in the countryside. Caution and sensitivity must therefore also be exercised when referring to the 'origins' of livestock and to whether or not they 'belong' in Britain.

Other commentators have reported conflict when the place of livestock in the countryside has been contested by different groups. These conflicts have been fuelled by the anthropocentric commodification of livestock to sell products and places. This has led to a significant disparity between the sanitised expectation of livestock and the ways in which they are used within farming practice (Yarwood and Evans 1998). Cloke *et al.* (1997: 218) have noted that some migrants to rural areas were surprised to find 'manure smells, cows mooing'. Such attitudes have led to disputes over the right of cocks to crow (Cloke 1993) and cattle to roam

(Yarwood and Evans 1998). Central to these arguments, Cloke (1993: 120) suggests, is cultural conflict over the place of the 'sights, smells and sounds of agriculture' in the countryside, not to mention the right of animals to behave in a natural way. At another extreme, the idea that livestock are simply 'units of production' can lead to the mistreatment of some farm animals through intensive breeding and feeding programmes (Yarwood and Evans 1998). As recognition of animal rights has grown, some animals have benefited from improved living conditions through more extensive, organic methods of farming.

These examples represent quite exceptional ways in which animals are constructed, contested and subsequently placed within the countryside. It would be simplistic to think that all discourses about livestock centre on debates over their sanitised expectations or their productive value. For most people, livestock are simply 'there' in the countryside and make up part of the rural scene. Indeed, Halfacree (1995) has noted that animals can be important within individual constructions of rurality. It is unlikely, though, that breed type matters for most people, even though the existence of animals might as an entity. Nevertheless, it has also been shown that, for these people, individual breeds can play a significant role in the creation and dilution of place identity, and are hence an important facet of rurality. It is nonetheless clear that farm livestock are more than the simple 'units of production' so commonly portrayed in geographical literature.

Conclusions

This chapter has sought to move analysis in agricultural geography away from political economy-based approaches and towards greater cultural sensitivity. Using the example of livestock, it has demonstrated that by taking a more cultur-ally sensitive approach to agricultural geography, new directions for research become apparent. It should be stressed, however, that abandonment of political economy approaches is not being advocated. They are helpful to explain why certain breeds survive or how a breed is commercially useful (even if it is not necessarily for agricultural output), but, as the survey of RBST membership demonstrated, political economy approaches provide only a partial understanding of the geography of livestock. Even in the modified format frequently employed in agricultural geography, they are unable to cope with the complex spatial rela-tionships that have developed between animals and people or groups of people. It cannot be denied that economic factors have influenced abundancy and scarcity of breeds in the countryside, but it is too deterministic to rely upon these explana-tions. It is here that culturally informed perspectives become relevant to help interpret and reinterpret facets of agricultural change (Morris and Evans 1999). As this analysis of livestock breeds has demonstrated, cultural influences can help to provide a better understanding of spatial and temporal variations in animal distributions.

These considerations (re-)engage agricultural geography with some of the debates affecting (post-)rural geography. On the one hand, it is possible to treat cultural approaches as 'stirring some additional ingredients into the mix of rural studies' (Philo 1992: 193). Certainly, the main purpose of this chapter has been to outline a few, but by no means all, of the ways in which agricultural geography can be enriched by such cultural 'flavourings'. However, others (Murdoch and Pratt 1993) have argued for a paradigm shift which takes as its starting-point an understanding that 'the rural' and everything within it is socially constructed. There have also been suggestions that to ignore political or economic circumstances would lead to studies which lacked critical, explanatory power. Thus, Murdoch and Pratt (1997) have advocated a more flexible approach to rural study which recognises the role of local 'actor networks' within fluidly constructed rural spaces. Such 'a focus upon network construction and mediation thus begins to provide a way of breaking down the inevitable rigidities in conceiving rural space from its physical composition or its strictly internal or external definition alone' (Marsden 1998: 25). Taken to its logical conclusion, these networks should include not only people, but animals and flora too (Murdoch 1997). This will lead to a more symmetrical, and therefore better, understanding of rural change. Yet, whether viewed as a paradigm shift or as just a new ingredient, the study of livestock beyond productivism can contribute much to improving the understanding of rural change, the development of agricultural geography and reducing the anthropocentric human–animal dualism in geography.

Acknowledgements

This work was funded by a grant from the Nuffield foundation. We would also like to thank Dr Julie Higginbottom for her work on this project.

Note

1 The project had two main aims. The first was to examine the relationship between particular breeds and places. Using data from herd books, the current distributions of individual breeds were mapped across the British Isles. These maps reveal that, in many cases, a breed has a strong relationship with a particular place and is often clustered in a particular region (see Figure 5.2, for example). Frequently, it is the rarest breeds which have the strongest association with place, and, by contrast, the more commercial is a breed, the more uniform it is in its distribution (Yarwood and Evans 1999). The second stage of the project examined the reasons for these distributions and, in particular, why rare breeds remained important in particular rural places. This was done through a survey which targeted members of the Rare Breeds Survival Trust (RBST), a charity dedicated to the preservation of rare livestock. From the 1,834 successful returns (a 23 per cent response rate which included farmers and non-farmers), it is possible to begin exploring how livestock are placed within discourses of the contemporary British countryside.

References

Agyeman, J. and Spooner, R. (1997) 'Ethnicity and the rural environment', in P. Cloke and J. Little (eds) *Contested Countryside Cultures: Otherness, Marginalisation and Rurality*, London: Routledge.

Alderson, L. (1976) *Farm Animals*, London: Frederick Warne.

Alderson, L. (1990) *A Chance to Survive*, London: Helm.

Alderson, L. (1998) *A Breed of Distinction*, Kenilworth: Countrywide Livestock Ltd.

Anderson, K. (1995) 'Culture and nature at the Adelaide Zoo: at the frontiers of "human" geography', *Transactions of the Institute of British Geographers*, 20: 275–294.

Anderson, K. (1997) 'A walk on the wild side: a critical geography of domestication', *Progress in Human Geography* 21: 463–485.

Bowler, I. (1975) 'Factors affecting the trend to enterprise specialisation in agriculture: a case study in Wales', *Cambria* 2: 100–111.

Bowler, I. (1985) 'Some consequences of the industrialization of agriculture in the European Community', in M. Healey and B. Ilbery (eds) *The Industrialization of the Countryside*, Norwich: GeoBooks.

Bunce, M. (1994) *The Countryside Ideal: Anglo-American Images of Landscape* , London: Routledge.

Carr, S. and Tait, J. (1991) 'Farmers' attitudes to conservation', *Built Environment* 16: 218–231.

Clifford, S. and King, A. (1995) 'Losing your place', in S. Clifford and A. King (eds) *Local Distinctiveness: Place, Particularity and Identity*, London: Common Ground.

Cloke, P. (1989) 'Rural geography and political economy', in R. Peet and N. Thrift (eds) *New Models in Geography*, London: Unwin Hyman.

Cloke, P. (1993) 'On "problems and solutions": the reproduction of problems for rural communities in Britain during the 1980s', *Journal of Rural Studies* 9: 113–122.

Cloke, P. (1997) 'Country backwater to virtual village? Rural studies and the cultural turn', *Journal of Rural Studies* 13: 367–375.

Cloke, P. and Goodwin, M. (1992) 'Conceptualising countryside change: from Post-Fordism to rural structured coherence', *Transactions of the Institute of British Geographers* 17: 321–336.

Cloke, P. and Thrift, N. (1987) 'Intra-class conflict in rural areas', in T. Marsden, P. Lowe and S. Whatmore (eds) *Rural Restructuring: Global Processes and their Responses*, London: David Fulton Publishers.

Cloke, P., Milbourne, P. and Thomas, C. (1997) 'Living lives in different ways? Deprivation, marginalisation and changing lifestyles in rural England', *Transactions of the Institute of British Geographers* 22: 210–231.

Clutton-Brock, J. (1981) *Domesticated Animals from Early Times*, London: British Museum and Heinemann.

Coppock, J. (1964) *An Agricultural Atlas of England and Wales*, London: Faber.

Evans, N. and Ilbery, B. (1993) 'Pluriactivity, part-time farming and the farm diversification debate', *Environment and Planning A* 25: 945–959.

Evans, N. and Morris, C. (1997) 'Towards a geography of agri-environmental policies in England and Wales', *Geoforum* 28: 189–204.

Evans, N. and Yarwood, R. (1995) 'Livestock and landscape', *Landscape Research* 20: 141–146.

Evans, N. and Yarwood, R. (1998) 'Rare breeds, livestock and the post-productive countryside', in A. Ontiveros and F. Molinero (eds) *From Traditional Countryside to Post-productivism: Recent Trends in Rural Geography Research in Britain and Spain*, Murcia: Associacion de Geografos Españoles Grupo de Trabajo: Geografia Rural.

Evans, N. and Yarwood, R. (1999) 'Charity begins at Linga Holm: the Rare Breeds Survival Trust', in A. Ferjoux (ed.) *Environment et nature dans le campagnes: agriculture de qualité et nouvelles fonctions*, Nantes: Actes du colloque Franco-Britannique de geographie rural tenu à Nantes.

Friend, J. and Bishop, D. (1978) *Cattle of the World*, Poole: Blandford Press.

Gasson, R. (1973) 'Goals and values of farmers', *Journal of Agricultural Economics* 24: 521–542.

Halfacree, K. (1995) 'Talking about rurality: social representations of the rural as expressed by residents of six English parishes', *Journal of Rural Studies* 11: 1–20.

Hall, T. (1995) '"The second industrial revolution": cultural reconstructions of industrial regions', *Landscape Research* 20: 112–123.

Halliday, J. (1988) 'Dairy farmers take stock: a study of milk producers' reactions to quota in Devon', *Journal of Rural Studies* 4: 193-202.

Henson, E. (1982) *Rare Breeds in History*, Cheltenham: Olivant and Son Ltd.

Higgins, D. (1993) *The Lake District: The Contemporary Representation of a Valued Place as a National Institution*, Unpublished PhD thesis, Exeter: University of Exeter.

Houlton, D. and Short, J. (1995) 'Sylvanian families: the production and consumption of a rural community', *Journal of Rural Studies* 11: 367–387.

Ilbery, B. (1978) 'Agricultural decision-making: a behavioural perspective', *Progress in Human Geography* 2: 448–466.

Ilbery, B. (1979) 'Decision-making in agriculture: a case study of north-east Oxfordshire', *Regional Studies* 13: 199–210.

Latham, H. (1998) 'Chillingham cattle – extracts from historical records', *The Ark* 25: 141–143.

Mardsen, T. (1998) 'Economic perspectives', in B. Ilbery (ed.) *The Geography of Rural Change*, Harlow: Longman.

Marsden, T., Murdoch, J., Lowe, P., Munton, R. and Flynn, R. (1993) *Constructing the Countryside*, London: UCL Press.

Morris, C. and Evans, N. (1999) 'Research on the geography of agricultural change: redundant or revitalised?', *Area* 31: 349–358.

Morris, C. and Potter, C. (1995) 'Recruiting the new conservationists: adoption of agri-environmental schemes in the UK', *Journal of Rural Studies* 11: 51–63.

Murdoch, J. (1997) 'Inhuman/nonhuman/human: actor-network theory and the prospects for a nondualistic and symmetrical perspective on nature and society', *Environment and Planning D: Society and Space* 15: 731–756.

Murdoch, J. and Marsden, T. (1994) *Reconstituting Rurality*, London: UCL Press.

Murdoch, J. and Pratt, A. (1993) 'Rural Studies: modernism, post-modernism and post-rural', *Journal of Rural Studies* 9: 411–428.

Murdoch, J. and Pratt, A. (1997) 'From the power of topography to the topography of power: a discourse on strange ruralities', in P. Cloke and J. Little (eds) *Contested Countryside Cultures: Otherness, Marginality and Rurality*, London: Routledge.

Phillips, M. (1998) 'The restructuring of social imaginations in rural geography', *Journal of Rural Studies* 14: 121–153.

Philo, C. (1992) 'Neglected rural geographies: a review', *Journal of Rural Studies* 8: 193–209.

Philo, C. (1995) 'Animals, geography and the city: notes on inclusions and exclusions', *Environment and Planning D: Society and Space* 13: 655–681.

Philo, C. and Wolch, J (1998) 'Through the geographical looking glass: space, place and human–animal relations', *Society and Animals* 6: 103–118.

Potter, C. (1986) 'Processes of countryside change in Lowland England', *Journal of Rural Studies* 2: 187–195.

Rackham, O. (1987) *The History of the Countryside*, London: Dent and Sons.

Small, R. (1995) 'Rare breeds in habitat conservation', *The Ark* 22: 246–250.

Symes, D. and Marsden, T. (1985) 'Industrialisation of agriculture: intensive livestock farming in Humberside', in M. Healey and B. Ilbery (eds) *The Industrialization of the Countryside*, Norwich: GeoBooks.

Thrift, N. (1989) 'Images of social change', in C. Hamnett, L. McDowell and P. Sarre (eds) *The Changing Social Structure*, London: Sage.

Tuan Y.-F. (1984) *Dominance and Affection: The Making of Pets*, New Haven: Yale University Press.

Wallis, D. (1986) *The Rare Breeds Handbook*, Poole: Blandford Press.

Walton, J. (1984) 'The diffusion of the improved Shorthorn breed of cattle in Britain during the eighteenth and nineteenth centuries', *Transactions of the Institute of British Geographers* 9: 22–36.

Walton, J. (1987) 'Agriculture 1730–1900', in R. Dodgshon and R. Butlin (eds) *An Historical Geography of England and Wales* (First Edition), London: Academic Press.

Westmacott, R and Worthington, T. (1984) *Agricultural Landscapes: A Second Look*, Cheltenham: Countryside Commission.

Whatmore, S., Munton, R., Marsden, T. and Little, J. (1996) 'The trouble with subsumption and other rural tales: a response to critics', *Scottish Geographical Magazine* 112: 54–56.

Wilson, G. (1996) 'Farmer environmental attitudes and ESA participation', *Geoforum* 27: 115–131.

Wolch, J., West, K. and Gaines, T. (1995) 'Transspecies urban theory', *Environment and Planning D: Society and Space* 13: 735–760.

Yarwood, R. and Evans, N. (1998) 'New places for "Old Spots": the changing geographies of domestic livestock animals', *Society and Animals* 6: 137–166.

Yarwood, R. and Evans, N. (1999) 'The changing geography of rare livestock breeds in Britain', *Geography* 84: 80–87.

Yarwood, R., Evans, N. and Higginbottom, J. (1997) 'The contemporary geography of indigenous Irish livestock', *Irish Geography* 30: 17–30.

6 Versions of animal–human

Broadland, *c.* 1945–1970

David Matless

Introduction

> [Y]ou need one thing above all in order to practise the requisite *art* of reading, a thing which today people have been so good at forgetting – and so it will be some time before my writings are 'readable' –, you almost need to be a cow for this one thing and certainly *not* a 'modern man': it is *rumination* ...
>
> (Nietzsche 1994: 10)

This essay considers material from the Norfolk Broads in the decades after the Second World War. The Broads, a region of shallow lakes and interconnecting rivers in eastern England, have been a key site for the development of conservation practice, and the disputed nature of conservation, and the human–animal relations that it presupposes, runs through the paper. The animal here marks out different cultures of nature in the region, each of which entails different claims to speak for the place and rests variously on practices of watching, listening, recording and killing. My concern is for how animal and human became enfolded as subjects and objects; while the final part of the essay focuses on specific species, the starting-point is not a concern for animals *per se*. A relational sense of the subject–object, with human and non-human as objects and subjects in the story, generates various senses of 'animal geography'. First, the assumption is that for all of these subjects–objects life entails a becoming-through-environment (on debates concerning the nature of animality and humanity, see Ingold 1988). Second, we find moral geographies of human and animal conduct in terms of appropriate and inappropriate ways of being in the region. Third, we encounter animal–human geographies in terms of a spatial zoning reflecting and reinforcing differences in human activity, habitat, property relations and regimes of management.

The first part of the essay outlines two ways in which animals act to define a relational human: first as visceral observers, second as reserved watchers and listeners. Each version of animal–human produces and is produced through a different mode of human authority, each involves a negotiation of different scales of being – local, regional, national – and each works through different technologies of nature: the gun, the camera, the recording instrument, the radio. The second part takes two creatures, the bittern and the coypu, to show how particular animals act as refractors of cultural–ecological value. If the first part of the essay concerns the definition of the human through the animal, here the pattern is reversed. The focus is on how humans act to define a relational animal, the presence of the bittern and coypu in the region being figured through particular facets of their being in a world subject to human valuation and intervention. The two parts of the essay take different routes into the same regional animal–human material.

Two versions of animal–human

Visceral eyes

> A peasant becomes fond of his pig and is glad to salt away its pork. What is significant, and is so difficult for the urban stranger to understand, is that the two statements in that sentence are connected by an *and* and not by a *but*.
>
> (Berger 1980: 5)

Berger's analysis of peasant love and death in 'Why look at animals?' holds for the first version of animal–human considered here. While our human subjects are not peasants, their pursuit of animals for pleasure, profit or food entails both a regard for the animal and a desire to see it dead. Animal flesh and blood are foregrounded, whether to be admired and preserved, or admired and shot, and such acts of love and death are held to embody a local, rural outlook distinct from the urban. The same ideological trajectory has been followed in recent years by groups such as the Countryside Alliance in their defence of blood sports.

I use the term 'visceral eyes' here to highlight a particular conjunction of body and sight in this culture of nature, which differs markedly from the second version of animal–human considered below. If all those considered in this chapter use their eyes to guide their relationship to landscape and nature, none profess a visual detachment; rather all of these ways of seeing are self-consciously embodied and embedded. Different cultures of nature are here distinguished less by the use of different senses, or by varying degrees of attachment to place, than by different metaphoric connections of the eye to the rest of the body. If some, as we shall find below, seek a balance between alert, independent senses in a body moving through landscape, this first version of animal–human rests on a metaphoric connection of eye and guts, in three senses: that one sees in order to kill in order to eat; that the deadly dead-eye of a good shot brings the visceral human closer to the visceral animal through the act of killing; that guts signify a courageous and hardy masculinity in the field.

In Broadland this specific nature–culture was articulated by author, journalist and broadcaster James Wentworth Day, author of around fifty books on shooting, royalty, speed, nature, ghosts, gambling and more. Day, who died in 1982, pronounced on and shot over the region from the 1930s, notably in *Marshland Adventure* (1950), *Broadland Adventure* (1951), *Norwich and the Broads* (1953) and *Portrait of the Broads* (1967), all produced by educated-mass-market publishers such as Batsford and Country Life. Day also broadcast on the area as 'The Man from the Country' (1946), and through his work becomes the voice of one version of the region. A deliberately maverick figure, who acted as political assistant to the equally maverick Lady Houston in the 1930s (Day 1958), Day worked his hard Right political views into mainstream topographic commentary, claiming Broadland, Norfolk and East Anglia as a whole (Day 1949a, 1954) as places for which such views were appropriate. Commenting on Norfolk's topographic variety in *Norwich and the Broads,* Day (1953: 73) suggested that 'Norfolk has almost everything except mountains which produce rain and coal which breeds factories, Socialism and dirt.' Day determined through environment his political claim to the regional ground: Norfolk is the kind of place over which his kind of politics should have authority. In similar vein Day (1953: 7) opposed post-war proposals for a Broadland national park:

> Whether the threatened creation of a National Park will enhance or improve the scene is, to my mind, doubtful. I do not believe in the dead hand of remote bureaucratic control. Those who dwell in and by the English country-side are the best guardians thereof.

Day's Broadland, one of dwelling 'in and by' the region, starred local humans, local animals and Day himself. The dustjacket of *Broadland Adventure* (see Figure 6.1) showed a drawing of Day, face lowered, gun over his shoulder. The frontispiece photograph showed 'The author in a flight pit on the Waxham marshes', Day in trilby, jacket and tie, gun ready, dog ready, and with a serving man holding already dead ducks by his side. Elsewhere we see Day and colleagues 'In the reeds on Martham Broad', head and shoulders showing, guns pointing up in a visual extension of the reed bed. Day was well known as a sporting writer, having edited *The Field* and *Illustrated Sporting and Dramatic News*; his shooting books included *Sporting Adventure* (1937), *Sport in Egypt* (1938) and *The Modern Fowler* (1949b). As a royal biographer of the most sycophantic order, Day produced *King George V as a Sportsman: An Informal Study of the First Country Gentleman in Europe*, presenting the King as 'The Squire of Sandringham' knowing 'the woodcraft of the sport', and 'no lover of big bags': 'He never takes an easy shot and never risks a long one'

(Day 1935b: 5, 23). Gunning is here a sign of both masculine action and ecological connection, of conduct becoming the locality.

The stress on action is common in this period on the political Right, and it is in this vein that in the 1930s Day could write not only eulogies of sport but eulogies of speed, producing biographies of the racing drivers and speed-record breakers

Figure 6.1   Broadland Adventure: an image of J. Wentworth Day.

Source: Day, J.W. (1951) *Broadland Adventure*, London: *Country Life*, dustjacket illustration

Kaye Don, Henry Segrave and Malcolm Campbell (Day 1931). Campbell became motoring correspondent of *The Field* while Day was editor, and echoed Day's hard Right political philosophy, in his case by supporting Sir Oswald Mosley. Active masculinity, whether on the track or the fen, whether on the sands in a speed car or on the sands with a gun, bound together such Right engagements with technology and nature. Day trumpeted the maleness of gunning, ending his 1948 *Wild Wings* with some 'Gunroom Philosophy':

> [A]n immortal temple of memories, a snug house-corner forever masculine, a room whose four walls are an enclave of enchanted memories of marsh and mountain, river and tidal mudflat ... nothing keeps the domestic female more effectively at bay than a half dozen stuffed birds.
>
> (Day 1948: 211)

Day's Broadland commentary policed the region for non- or anti-masculine action, behavioural or architectural traces of decadence, femininity or effeminacy: '[T]he banjo boys of the Broadland summer season' who would hardly be up to holding a gun (Day 1951: 57); the 'little standardised boxes of mock Tudoresque which are as much out of place in the countryside as a woman with fox furs and high red heels in a muddy lane' (Day 1967: 191); the 'scantily clad summer beauties from Leeds and Wigan' who frivolously disregard local nature (Day 1953: 87). Day salivated (with anger) at such scenes.

Day set various forms of gunning against the scanty banjos. Wildfowling was upheld as a free, authentic and embedded local practice:

> I may be prejudiced, but somehow I think that wildfowling is one of the last forms of strenuous relaxation left to the man whose heart and soul are so essentially masculine that he must by necessity escape from the shams, conventions and orthodoxies of modern life.
>
> (Day 1935: 25)

Day was hardly against arranged shoots, though, especially when these were deemed part of local tradition. He celebrated the annual or biannual Hickling coot shoot: thousands of birds driven to one end of the broad, to panic and fly back over the guns: 'Then the fun begins. The punts rock, the gunners load and fire, the coots fall and fly. ... The average bag is about 500, but it varies enormously' (Day 1949b: 236–7; 1950: 46–8). In 1927 twenty-one guns killed 1,175 birds. If coots are so common as to be fit for carnage, rare birds are to be watched, preserved and hopefully not shot, although Day could not always resist the latter: '[T]he land-rail or corncrake is, alas, nowadays a comparative rarity. I shot one on a fen adjoining Calthorpe Broad in the autumn of 1945, whilst walking up snipe' (Day 1953: 147; 1967: 168). For Day, such birds could survive because of the care given to wildlife by sporting estates. Shooting and naturalism were linked through this socio-economic institution, the decline of which, he argued, had allowed destructive birds such as crows and little owls to flourish (Day 1951: 47–8). Care of the Broads by those dwelling in and by them implies a bolstering of the estate. In *Norwich and the Broads,* Day praised the former owner of Catfield Hall, Lord William Percy:

> [T]hat good naturalist and wonderful shot ... Lord William is not one of those eager 'bird-watchers' who dash into print, burst upon the air and smirk at us from the television screen on every possible occasion in order to impress upon the world that they, and they alone, are experts.
>
> (Day 1953: 126–127)

Never averse to recycling his own words, Day (1967: 100) repeated this passage fourteen years later in *Portrait of the Broads*, replacing 'bird-watchers' with 'birdy-boys'. Day's dismissal of 'expert' bird-watchers was echoed in his *Poison on the Land*, an early critique of the use of pesticides in agriculture as a misguided application of expert knowledge. Day (1957) reflected on '[t]he war on wild life, and some remedies', highlighting the impact on game birds and placing the 'revolution in farming' alongside myxamotosis in its adverse effects. Day critiqued scientific authority to uphold landed authority, declaring war on the war on wildlife in order to maintain shooting.

Day enrolled other Broadland writers and artists into his particular nature–culture. Several of his books were illustrated by bird artist Roland Green (e.g. Day 1948, 1960), whom he first met after he shot a swan which came to ground on the roof of Green's hut. In *Marshland Adventure,* Day visited Green at Hickling, finding a 'hermit-artist of the Broads' with the same relation to land and the outside world as Lord Percy: 'Roland Green is no drawing-room publicity-hunter. ... He is neither photographed nor paragraphed.' Day then devoted several paragraphs and a picture to him (Day 1950: 48; 1967: 159–161). Day also made alliance with Alan Savory, author of *Norfolk Fowler* (1953) and *Lazy Rivers* (1956); three chapters of *Broadland Adventure* describe a cruise down the Yare with Savory from his home at Brundall (Day 1951: 42–73). Savory shared Day's agenda against 'bird protectionists and anti-sports cranks' (Savory 1953: vii), presenting wildfowling as a pursuit working in the grain of local ecology, only succeeding if the sole disturbance to the animal was its death. Savory's fowling was functional: 'I do not personally shoot birds that one cannot eat, other than vermin, for they cannot be put back like fish' (Savory 1953: 48). 'Vermin' for Savory also included animals such as the otter, on whom he declared 'war' after the death of some ornamental duck. Savory captured an otter: 'He was a beautiful animal, and he stared me out as I shot a .22 into his brain. I felt like a murderer until I found my last red-crested pochard duck snatched off her nest and chewed to pieces' (Savory 1953: 118–119). Savory shared Day's enthusiasm for fishing as a field sport, *Lazy Rivers* giving anecdote and technical advice, and showing '[t]he author with a 20-pounder from the Broads.' Day presented Broadland as an 'anglers' paradise' (Day 1967: 140–154), with the angler, like the fowler, not detached from but a 'part of the landscape': 'Therefore he sees a lot that is denied to the chattering hiker, the petrol-propelled motor-boat fiend, the moronic Beatle-disciple with his long-player and longhair. These people make a noise. Therefore they repel nature' (Day 1967: 150).

For all his local rhetoric, Day was neither a native nor a resident of Norfolk. Day did not 'dwell in and by' the region, having his permanent home in Essex, and closeness to Green and Savory helps head off any sense that *his* might be the intrusive voice of an outsider. Day also aligned himself with local fowlers and keepers, who might not write their own story but could be represented through his pen. The cover of *Marshland Adventure* shows a Breydon fowler at night, bending to pick up a downed bird, a 'Moonlight trophy on the mud'. Day's eulogy of free-shooting over marshes, mudflats and tidal shore is also a eulogy of local humans, independent and resourceful, out with nature in all weathers. The tidal flats of Breydon, unlicensed and beyond human ownership, with matching human and animal freedom, sit alongside the shooting estate as signifying different yet, for Day, complementary ways of being in Broadland. Day also gave local wild-fowlers genetic distinction as part of the local fauna: a photograph of Stanley Nudd of Hickling is captioned: 'One of a Famous Family of Broadsmen, said to be of Norse Descent' (Day 1950: 15). He found another variant of human fauna on the Horsey estate, captioning a photograph in *Marshland Adventure*: 'Half Bird, Half Man! Crees, the Horsey Hall Gamekeeper, an Authority on Harriers, Bitterns and Other Rarities' (Day 1950: 64). Day was having a serious joke here, tapping a consistent enfolding of local human and animal fauna, the authority of one on the life of the other deriving from a long and close sharing of space.

This version of human-as-visceral-animal-observer rests then on an alliance of elite humans, local serving humans, and animals, united in experience of land and water, and united through the shooting/hunting and eating of animal flesh. And set against this, but always in relation to it, are ways of not-being in Broadland, personified by the figure of the tripper (Matless 1994). In Day's work the tripper emerges as a kind of polluting pond skater, slipping across the surface of local ecology and having no meaningful connection to it, but managing to destroy it in the process. The image of trippers as literal scum is stated most clearly in *Portrait of the Broads* in 1967, where Day remembers a Broadland not 'exploited, capi-talised, vulgarised, transistorised': 'A pleasanter picture I suggest than that of a boatyard yacht-basin littered with ice-cream cartons, candy floss sticks, empty potato crisp bags, cigarette ends and the thin grey scum of contraceptives on the water' (Day 1967: 36). These people only consume, can never produce. We now turn to a second version of animal–human which, if also set against the tripper, offered a more benign moral geography of Broadland.

Weather eyes and ears

Alongside this visceral sense of the animal–human we find a reserved mode of watching and listening, committed to conservation and education; the term 'reserved' refers here to both the practice of establishing nature reserves and the reserved engagement with nature which they are held to demand, and to cultivate. This sense of the animal–human forms part of a 'new naturalism' that emerged before, during and after the Second World War (Marren 1995; Matless 1998), embodied at a national scale in the New Naturalist book series produced by Collins, the establishment of the Nature Conservancy and National Parks, and the general promotion of citizenship through popular natural science (Matless 1996). The key Broadland organisation here is the Norfolk Naturalists Trust (NNT), founded in 1926, and the most prominent written product is the New Naturalist volume on *The Broads*, published in 1965 under the editorship of Ted Ellis (1965).

This culture of nature presumes a different authority to that of Day and company. As R.H. Mottram (1952: 235) expressed it in his *The Broads*, from this perspective the three alternatives for 'the future of the Broads' were 'National Park, Private Reserve or Commercial Exploitation', and the former was preferred. This is a long way from Day's sense of the national park as remote bureaucracy, and while shooting is not necessarily opposed here, and the national park is indeed put forward in part as a public adaptation of the estate ethos, there are key departures from Day's version of animal–human. The sense of visceral connection through killing is not evident, and new naturalist cultural authority rests on a very different form of expertise to that of the modern shooter. Day typically had no doubt about his distinction from this other version of animal–human. In *The Modern Fowler*, discussing Scolt Head on the north Norfolk coast, a key site for new naturalist study through the work of, among others, geographer J.A. Steers, he considered bird-watching:

> Today the Head is the annual summer rendezvous of a great many earnest people in spectacles who appear to derive infinite joy from watching its shel-duck, terns and ringed plover at their connubial business. In winter it is more interesting. … Then, snow and gales and bitter cold conspire to keep the fine-weather naturalists safely tucked away among their text-books.
>
> (Day 1949b: 87)

Day (1951: 7) contrasted the outlook of the sporting naturalist to the 'vulgar attentions' given by bird-watchers; the latter become nature-trippers in their lack of deep connection to animals, place or weather.

Day typically overdid it. Setting up a choice of either visceral animal–human connection or dry, bookish detachment served to isolate him at a time when a new naturalism fitting into neither category was gaining momentum through a different kind of bodily engagement with the region. A visceral sense of connection to the blood and flesh of the animal may have been absent from this new naturalism, but this new expertise was less dry and detached than damp, gathered not just from the lab and library but through keeping expert eco-canny weather eyes and ears to the field, as well as talking with the same locals highlighted by Day for their half-human qualities. As in Day's work, there arises a clear sense of how not to conduct oneself in nature, but from this perspective the errant tripper is not to be expelled from the place as alien scum but re-embedded into local ecology through education. Naturalist and audience here gain a different relationship, linked to a sense of the expert not as remote, impersonal and apart from the local scene but grounded, individual and of necessity a touch eccentric. The new naturalist explores nature by bordering on the eccentric, far from the dry authority of the laboratory, damp in his or her chosen field. The naturalist's delicate balance of attachment and detachment regarding nature is matched in his or her relationship to society.

It is again appropriate to approach this version of animal–human through key individuals who became known for a particular voicing of the region. The key local figure in this case was Ted Ellis, natural history keeper at Norwich Castle Museum from 1928 till 1956 and resident at Wheatfen on the River Yare from 1946 until his death in 1986; the surrounding marshes are now in the care of the Ted Ellis Trust. Ellis was prominent in the Norfolk and Norwich Naturalists Society and Norfolk Naturalists Trust, contributed nature notes to the *Eastern Daily Press* and the *Guardian*, and edited the New Naturalist volume on *The Broads* (Ellis 1965, 1998; Stone 1988).[1] He was also something of a radio and television personality, making regular appearances on BBC TV's regional *Look East* programme.[2] Ellis began a 1952 essay on 'Wild Life in Broadland' by addressing different possible readers:

> You may be a summer voyager upon the waterways of this North Sea lowland, ... You may be a week-end angler and dreamer, ... Perhaps you journey in search of the fantastic bittern, to catch a glimpse of the swallow-tail butterfly in all its glory, ... Belonging, you may keep cattle on our vast green levels or be concerned with the mowing of ronds and the slubbing of dykes in season. Whether you find yourself here for a day or a lifetime, mysteries will crowd upon you and the peace that is felt in solitude will be

yours in the midst of this land of bright waters, jungle wildernesses and twinkling facets of life on every side.

(Ellis 1952: 185)

Here, as elsewhere in his work, Ellis addressed different modes of being in Broadland as equal forms of access: cruising, fishing, bird- and butterfly-watching, farming. Shooting is a notable absence, and if farming denotes 'belonging', even that does not give it cultural priority. It is on 'our', not the farmer's, green levels that cattle are kept; Broadland is a place of common cultural ownership, and ideally of common access, and if some do conduct themselves inappropriately, then they might be educated out of it to take their place in the landscape.

Ellis's public role was furthered by appearances on radio; by 1957 he was experienced enough to do a series of live early evening talks on the Midland Home Service, *Through East Anglian Eyes*, telling of the Broads by night, of harvest time and beach life, and of the undergrowth. Ellis, 'probing the undergrowth', encouraged his listeners on an intrepid and educational journey of reserved wonder into an 'Aladdin's cave', a 'jungle' of marsh, bog and carr: 'I don't suppose you've even thought of looking closely into the depths of a reedswamp, with your nose only a few inches off the mud. It isn't likely to appeal to many people, I suppose' (Ellis 1957b). Ellis's Broads appear a strangely animate realm, where even the plants sing:

A warm summer night has a special music. The thing to listen for is the singing water-weed. You almost have to hold your breath to hear it. ... It's a queer experience; a sort of fairy music rises from the water. If your ear is not attuned, you might compare it with the 'muzz' in sound broadcasting before VHF came along; but it's really rather like a faint jifflin of cymbals, or the hiss of rain on water. Only one sort of weed performs in this way; the hornwort.

(Ellis 1957a)

Ellis, making analogy with a then popular musical movement, suggested that as the leaves scratch together they 'skiffle. ... Very much the same sounds are made by musical water-bugs in our marsh ditches' (Ellis 1957a; see also Ellis 1952: 197, 206, on 'Reed-mace music' and 'When waterweeds sing'). The message is that to connect with the animal and the vegetable most effectively you need to get off the beaten tripper track, into Ellis's Wheatfen or the NNT reserves, or into the dark of a Broads night, where you can exercise this way of being human, with eyes and ears open, with neither disturbance nor ridicule from those who might find nosing in a swamp bizarre.

Ellis, with his specialist knowledge of insects and fungi as well as mammals and birds, invited the listener into a total ecology. Birds remained the most prominent wildlife in new naturalist public discourse, however, helped along by the increasing use of radio to bring distant nature into the home. Radio acts as an introductory space for the possible naturalist, nurturing weather ears in the living room, and the Broads became part of educational radio's common ground, the subject of travelogues, schools programmes, talks, adventure stories and outside broadcasts, such as that from Hickling Broad in May 1947. Central to this radio nature was the transmission of a convincing aural ecology, produced through mimicry and recording; such fields have been ignored as yet in discussion of sonic geographies. If John Betjeman in his 1937 poem 'Slough' had mocked the inhabitants of a new commercial England who 'do not know / The bird song from the radio' (Betjeman 1975: 22–23), new naturalists inverted Betjeman's caustic commentary by getting people to *listen to* bird song from the radio, transmitting calls into the home.

Imitation was the province of Percy Edwards, who worked for the BBC on nature and country programmes from soon after the war, and remained a celebrity until his death in 1996. In 1948 Edwards could impersonate 153 birds. Edwards's public persona shuttled between that of expert and variety turn, but he and others took care to present his bird calls as the work of a field naturalist. Brian Vesey-Fitzgerald prefaced Edwards's 1948 *Call Me At Dawn* by stressing that this was more than just entertainment:

> [T]hough there are thousands of people who regard Percy Edwards, quite rightly, as a 'star,' there may be many who do not regard him as a naturalist. Let me say straight away that Percy Edwards is a naturalist – and a very good one.
>
> (Edwards 1948: 9)

Edwards's language was close to that of Ellis in calling the listener or reader to adopt a childlike wonder in the face of nature, to see and hear again with supposedly unprejudiced childlike senses. This is a wonderland of vivid accuracy and immediacy rather than fantastic dreams, and radio is its appropriate medium. Edwards's light voice, with a hint of a Suffolk accent, and his courteous and genial persona, offered a naturalism far removed from Day's gunning masculinity. In November 1959, on an edition of *Woman's Hour* from Norwich, Edwards took listeners into 'A Fairyland', describing and mimicking the scene at Ranworth, going through each bird and doing their calls:

> Those rocket-winged little ducks, the teal, they flash past, but no protests, just dainty call notes. [Example]./ Then the inevitable mallard, always

breaking out of cover, and always sounding vulgar. How else can be describe [*sic*] such a voice? [Example]

(*Woman's Hour* 1959)[3]

Alongside such imitation came recorded animal sound, pioneered by Ludwig Koch, a refugee from Nazi Germany. Koch's 1954 *Memoirs of a Birdman* recalled making the first ever bird song recording aged 8 in 1906 in Germany, producing 'sound-books' of animals, birds and cities, subsequently destroyed by the Nazis, and planning a Sound Institute to explore the rendering of life on disc. Moving to Britain in 1936 Koch produced more bird sound-books, sponsored by Parlophone, combining gramophone discs with text by leading new naturalists. Max Nicholson provided words for *Songs of Wild Birds* (Nicholson and Koch 1936) and *More Songs of Wild Birds* (Nicholson and Koch 1937); Julian Huxley wrote the text for *Animal Language* (Huxley and Koch 1938). Koch formed the aural wing of new naturalism; images in his *Memoirs* show him setting microphones up trees to record birds, and 'interviewing a young grey seal on the beach at Skomer' (Koch 1954: 96). A sea-lion in Regent's Park Zoo pops its head up by a gramophone on the edge of the pool to hear a recording of its own voice. There is evident humour here, a humour derived from a sense that catching and replaying sounds in this way is a strange disruption of the boundaries of nature and culture, with a part of the animal becoming part of technology. The animal may not understand the process, but it knows something is going on, which, as it involves neither death nor food, remains a puzzle. Broadland's birds similarly encountered men with microphones. Another *Memoirs* image shows Koch in a punt, guided along a dyke by Crees, carrying his equipment into the marshes: 'In search of the rare bittern on Horsey Mere, Norfolk. Its booming was successfully recorded in this floating "studio"' (Koch 1954) (see Figure 6.2). We return to the sound of the bittern below.

We have, then, identified two contrasting modes of being human in relation to the animal, one working through visceral eyes and embracing death, the other through a reserved watching and listening with weather eyes and ears. These versions of animal–human mark out and shape two nature–cultures contesting the definition of the region in the post-war period, and both work in relation to another nature–culture, that of the tripper, presented as merely skimming the surface of the place. The latter is of course as much a culture of nature as new naturalism or gunning masculinity; surface skimming is as much an engagement with material, a presence in ecology, as deep-rooted belonging.

Two more versions of animal–human

For the remainder of this essay we address versions of the animal–human through the place of two animal species in Broadland's cultural ecology. Both show the

Figure 6.2 'In search of the rare bittern on Horsey Mere, Norfolk': Koch is punted through the
reeds with his floating 'studio'.

Source: Koch, L. (1954) *Memoirs of a Birdman*, London: Scientific Book Club/Phoenix House, between p. 64
and p. 65

animal as a refractor of cultural value, and both make sense in terms of the
versions of animal–human already considered. One is a bird, the other a rodent.
One is deemed native, the other alien. The native bird is unreservedly valued, but
we also find some welcome for the alien rodent. Categories of native and alien do
not play here in the way in which we might expect from recent conservationist
arguments against invasive plant and animal species.

Bittern: Broadland bird

The bittern, a bird which nests in reedy places, has long been presented as a symbol of Broadland; a rarity, a talisman, a curiosity, an indicator of ecological health.

We begin by returning to Day. In 1935 Day turned his hand to urban ecology. His *A Falcon on St Paul's: Being a Book about the Birds, Beasts, Sports and Games of London* received an excruciating introduction from Viscount Castlerosse:

> My friend Mr James Wentworth Day has, after much close study, discovered the fact that since time immemorial there are other sports in London besides the hunting of rich men by poor women. Jimmy Day has turned day into night. Read his book and you will be bitten.
>
> Bitterns boom. I have been bitten by Jimmy Day, that is why I am booming his book.
>
> Jimmy Day knows his subject, which is rare amongst authors.
>
> Castlerosse, Brook Street, W.1
> (Day 1935a: Introduction, no pagination)

In Day's writing, as in Ellis's and many other commentators on Broadland, the bittern is a symbolic bird of the region. In 1954 there were estimated to be sixty 'booming males' in Norfolk, concentrated around Barton and Hickling Broads and Horsey Mere (Ellis 1965: 202). The bittern's chosen habitat matches those zones upheld by conservationists for ecological and cultural value, areas spatially close to but culturally far from the tourist centres of the upper Bure and lower Thurne; the heart of what today's Broads Authority term the 'Last Enchanted Land' (Matless 1994). The bittern's distribution helps it stand for one culture of nature against another, and its habits and demeanour also lend symbolism. The bittern is presented as a curious, vulnerable bird at one with an environment itself secretive, enchanting, enigmatic and fragile. Blending in with its reed bed habitat to the point of invisibility – 'A bittern when it assumes the freezing attitude is extraordinarily invisible in reed stems' (Buxton 1946: 155) – the bird is ungainly on dry land but nimble in the marsh. And it is valued as much for its sound as its look, a weird booming call deemed strangely to fit a place of mists, mystery, enchantment:

> Whoever has heard the occult, resonant gong of this baroque fowl beaten from the reeds on a still May evening, as I have heard it, will realize that a new mastery of interpretation has been given to the watery landscape of Broadland.
>
> (Massingham 1924: 116)

Since the early twentieth century efforts have been made to restore or to maintain the bittern as a breeding rather than simply visiting bird, the assumption being that it belongs, and that the place lacks something without it. Jim Vincent, keeper at Hickling, successfully reinstated the bird after it chose to nest in 1910 after decades absent from the region (Matless 1994; Vincent 1926). Anthony Buxton, owner of the Horsey estate adjacent to Hickling, recorded the bittern's habits in prose and photography, his 1946 *Fisherman Naturalist* describing the bird's movements and noises, and showing the bittern feeding its young, emerging from reeds, and going through 'various stages of powder puffing' (Buxton 1946: 150–155; 1948). This is not, however, a straightforward story of a bird happy to be in its place. The unpredictable bittern is liable to fly off elsewhere, as birds do, leaving the region without its elected symbol for a year or more, and the creature is hard for even the most dedicated conservationist to cater for; as Buxton (1950: 109) put it: 'There is no fathoming a bittern's brain.' The bittern is symbolically eccentric, like some of its local caring humans.

If there is unanimity concerning the bittern's regional value, there are telling differences between figures such as Day and Ellis in their presentation of the bird. Day, for whom the bittern was 'the bird-spirit of these haunting solitudes' (Day 1953: 150; 1946: 164–7), hailed its restoration as a vindication of the sporting estate. The threat to the bird comes in part from misguided gunners – Day told of sportsmen mistaking it for a goose and taking aim (Day 1953: 150) – but more from bird-watchers. Day (1967: 186–187) offered a warning story of a bittern in Holland having its nest trampled by bird-watchers whose 'sheep-like mentality' drew them to gawp at the creature; the bird chooses to go elsewhere. The relationship of bird to holiday-maker in Day's work is more complex: '[T]wo summers ago, I saw a bittern rise from a reed-bed opposite ... and fly with the utmost unconcern, over a whole flotilla of yachts and motor-boats, which were taking part in an extremely noisy regatta' (Day 1953: 97). While this anecdote could be taken to imply a happy coexistence of bird and boaters, in Day's terms the interpretation is more likely one of a resilient animal, part of a deep local ecology, enduring despite the messy and passing surface ecology of cruising. The bittern becomes a figure of ecological stoicism, akin to the cow noted by Day in *Portrait of the Broads*: 'I have known a cow swallow a plastic mackintosh and pass it entire out of her system' (Day 1967: 208).

If for Day the bittern was threatened by an ignorant audience, for Ellis and others the task was to take the bittern, and especially its symbolic sound, to new audiences, in part via radio. The bittern was one of Percy Edwards's 153 talents, and on *Woman's Hour* he mimicked the bird: '[P]erhaps the strangest of all bird voices [Example]' (*Woman's Hour* 1959). The distance in radio here reinforces the sense that this is a creature not to be disturbed, but given space in which to resound. If there is little hope of seeing a bittern, here is a chance to hear the bird *in situ*; the bird in its reedy nest, the human in its domestic armchair. Koch's *Memoirs* describe the first recording of the bird. Invited by Anthony Buxton, Koch brought heavy gear up from London by train: 'It would of course be a great achievement to record the boom of the bittern for posterity and to give millions of people a chance of hearing a sound which they have never heard before' (Koch 1954: 74). An account of the expedition was published in *The Times*. Koch and Crees punted around Horsey Mere for several days and evenings with a floating studio, sometimes hearing the bird's call and not being near enough to record it, sometimes disturbed by aircraft drowning out the boom. Koch worked out that booming was preceded by four very low clicks, 'a sort of prelude', followed by an intake of air and three to six booms. At last he got his noise: 'To get ten to fifteen seconds of booming we had had to be on the alert for about a hundred hours and have discs running for about 130 minutes altogether' (Koch 1954: 77–78). The strangeness of the sound, 'very uncanny when one is alone in the dark', and the effort of getting it, gave it 'a place of honour in my collection of nature sounds' (Koch 1954: 78).

Such recording efforts helped to cement the bittern as a regional symbol in an age of radio nature. On a Home Service programme in April 1959 on '*The Naturalist' in East Anglia*, chaired by Ellis, naturalist R.P. Bagnall-Oakeley introduced the bird to the listeners: '[T]he obvious one to begin with is the bittern – the one which the locals call the "bog-bumper" – and the one which has had this delightful North American name introduced – the thunder pumper'; a twelve-second recording followed. Just as the bird flies in and out of the region, so nomenclature migrates: 'Well, there it was "bog-bumping" or "thunder-pumping", have it how you like' ('*The Naturalist' in East Anglia* 1959). Broadcasting from Hickling on the Midland Home Service in 1947, Buxton described the sound:

> [T]hree noises, clicks, … intake of air, a sort of gasp, and expulsion of air which makes the boom. The boom can be heard on a still day inside my house with all the windows closed, and the nearest booming place is 700 yards away. And yet it is not a very loud sound even at 10 yards.
>
> (*Hickling Broad* 1947; also Buxton 1946: 151)

After six seconds of bittern noise on a *Saturday Review* programme of 29 November 1952, presenter Doreen Pownall interjected: 'Don't be alarmed – it isn't your radio set that's gone wrong – that's the booming of a bittern.'

Coypu: unexpected guest

> Visitors to the Broads are sometimes amazed to see enormous rat-like creatures with reddish-brown fur and orange-coloured front teeth. They look rather like giant hamsters and are called coypus.
>
> (Althea 1975: 15)

The coypu has a rather different presence in the landscape. A native of South America, this 'Great South American Marsh Rat', as Day (1967: 181) termed it, was brought to the UK from Argentina in 1929 for fur farming. Several farms were established in the Yare and Wensum valleys, one by Alan Savory at Brundall. Most closed by 1939, but by this time some coypu had escaped, most notably in 1937 from a farm at East Carleton during a heavy rainstorm, from where the animals moved down a tributary stream into the Yare. The coypu thrived out of captivity, and established itself throughout the Broads by 1948. Some trapping was carried out by the War Agricultural Committee, and the population was further checked by the harsh winter of 1946/7 and local trapping for skins, but numbers continued to increase. The market for pelts collapsed in 1958 (Ellis 1960; Norris 1963), and the late 1950s saw a dramatic increase in the population to a maximum of around 200,000 (Fitter 1959: 103–104; George 1992: 231–234; Gosling and Baker 1989; for a related study, see Sheail 1988). A lack of adequate food supply led animals to forage more widely in farmers' fields, and in 1960 the government authorised rabbit-clearance societies to kill coypus, with 50 per cent of the costs covered by the Ministry of Agriculture (Ellis 1960). The coypu was added to the list of animals proscribed under the 1932 Destructive Imported Animals Act in 1962, and at the same time the Ministry established the Coypu Research Laboratory (CRL) in Norwich. A combination of severe winter weather in 1962–3, when 90 per cent of the population perished, and a centrally organised cull from 1962 using cage trapping and shooting reduced numbers dramatically. Cage trapping would sometimes catch, although not harm, other species: hares, otters, pheasants, lambs, even bitterns (Norris 1967). The cull was scaled down in 1965, but the animal survived to increase again with a run of mild winters in the early 1970s. A further eradication campaign in 1981, guided by CRL biologists, was deemed to have been successful by January 1989, and terminated by MAFF (Gosling and Baker 1989).

Like the bittern, then, the coypu has persisted in unpredictably variable numbers over decades, although preyed upon rather than preserved by humans. In considering why the coypu has been killed in various ways at various times, it is important here not to assume a story of alien victimisation. As we shall see below, many welcomed the coypu, but even those engaged in its trapping or shooting have not tended to present this as the pursuit of an animal because it is out of place. There are a range of concerns regarding the damage done by coypus: burrowing in banks which might lead to flooding of adjacent land; destruction of reed beds which might threaten the local thatching industry; erosion of bird habitat; destruction of plant species; transmission of disease to grazing animals; destruction of crops (Davis 1963; Laurie 1946). Animal biologists and Ministry of Agriculture pest control officers produce scientific studies of the coypu population, detailing the possible methods of trapping and control, but even where the latter is outlined in the manner of a military campaign, with details of the 'Plan of Campaign' and 'Operations' (Norris 1967), the argument has seldom been made that the coypu should be destroyed because it is alien. In a *New Scientist* article by Regional Pests Officer J.D. Norris, a photograph is captioned 'an alien rodent from South America', but Norris does not develop the theme in his text (Norris 1963). Alienness is recognised in the coypu, but it is not held against the animal as an inherently negative characteristic.

In an essay on the coypu in the journal *Agriculture* in 1956, R.A. Davis of the Ministry of Agriculture wrote of the rabbit, grey squirrel, mongoose and muskrat that '[t]he introduction of any new species into an environment not its own may be dangerous', but he concluded that the coypu 'forms an interesting addition to our fauna' and 'need not cause "alarm and despondency"'. Davis concluded here that the coypu should not be 'proscribed as a pest' (Davis 1956: 127, 129, 127), but later would change his view: 'By 1957 it had become apparent that the feral coypus were not an unmixed blessing' (Davis 1963: 346). It is the classification of the coypu as a crop-damaging pest, rather than its status as an alien, that is the key issue in the instigation of a systematic anti-coypu campaign. Norris recorded that Ministry posters appeared in police stations and post offices across East Anglia asking people to report sightings, and presented the animal as a quiet, even sly creature demanding vigilance: 'Coypus are largely creatures of the twilight, and, moving about singly or in small numbers, leave few signs behind. It therefore requires a considerable alertness to spot the first colonists' (Norris 1963: 626). The coypu may be being othered here as a crafty twilight creature which might creep up to colonise a new environment, but the striking thing about such discourse is how its South American origins are not given prominence. It would be easy for such writing to play on fears of the foreign, particularly in a period when that argument was being made in human cultural politics, but the native/alien polarity which has come to the fore in more recent ecological conservationist discussion (Agyeman and Spooner 1997) did not drive this campaign. It might also be argued that earlier local trapping of coypus for fur was not in itself an anti-coypu action in terms of antipathy to the species as a whole. It would seem, paradoxically, that it was the collapse of the fur market, reducing the incentive for trapping and thereby allowing numbers to increase, which generated the conditions for a state-sponsored eradication campaign. Commentators such as Davis noted a very different coypu story in countries such as the USSR, where coypus were deliberately released into the wild and regularly culled as a source of fur through live trapping (Davis 1963: 347).

We can further explore the cultural ecology of the coypu by considering those who, in various ways, positively welcomed the animal. In their review of coypu policy, Gosling and Baker (1989) suggest that the coypu escaped systematic culling for twenty-five years due to a lack of knowledge of its destructive potential, and a belief that numbers could be easily kept down. One should also though attend to a complex cultural–ecological story of welcome for rather than resistance to the coypu, which we can trace in the work of Day, to whom we return below, and in that of Ellis and his new naturalist colleagues. The new naturalist movement indeed displayed a general openness to alien species as an inevitable element in an evolving ecology. Richard Fitter's urban ecology, *London's Natural History*, offered one example of such an outlook (Fitter 1945; Matless 1998); likewise James Fisher's Presidential Address to the Norfolk and Norwich Naturalists Society in 1950 emphasised the need to welcome any future new bird 'invaders' (Fisher 1950). In his review of *Britain's Nature Reserves*, Nature Conservancy director Max Nicholson (1957: 72) reflected on the coypu:

> While opinions are divided, there seems so far to be no evidence that the Coypu is anything but beneficial, and while there has been considerable discussion about increasing the area of open water for navigation in the Broads, the Coypu can claim to have been so far the only creature to do anything effective about it.

The extent of open water was a key economic and ecological issue: the Nature Conservancy's 1965 *Report on Broadland* recorded increases in open water of between 14 per cent and 40 per cent in broads on the upper Bure between 1946 and 1952, crediting this to the coypu (Nature Conservancy 1965: 20). Many feared the ultimate disappearance of the Broads through ecological succession, a fear perhaps accentuated by the discovery in the 1950s that the Broads were the product of medieval peat-digging (Lambert *et al.* 1960; Williamson 1997). If the Broads could be made so swiftly by human action, and were not natural elements in the landscape, then might they quickly disappear in a natural loss of unique ecology? And if this was an artificial landscape, then was its maintenance by an imported animal against the grain of the place?

Writing on 'The Coypu Threat in East Anglia' for *Country Life* in December 1960, Ellis (1960: 1591) noted that 'for a long time the conclusions reached were that coypus in moderate numbers had a beneficial effect in keeping local waterways clear'. Ellis weighed up pro- and anti-coypu arguments here, on radio, and in the *Broads* New Naturalist volume, welcoming their clearing of vegetation but wary of erosive burrowing in great numbers, the threat to 'characteristic plants of the broads' (though those resistant to coypu were thriving), and the consumption of crops (Ellis 1965: 225–228). Ellis (1960: 1591) considered that control was needed, but that eradication would be impossible 'without annihilating almost everything else in the area'. The alienness of the coypu is not in itself a problem here; indeed, in Ellis's writing this alienness is recognised, pondered and enjoyed. The existence of 'moderate numbers' of coypu would be no problem; they could be ecologically beneficial, provide a source of local income through trapping, and could even be enjoyed for food. Ellis, who ate the animal regularly and wore a hat of coypu fur, secretly served the beast to a British Association for the Advancement of Science dinner in Norwich in 1961 (Stone 1988: 102); the Association had made excursions to Hickling, Horsey and Wheatfen (British Association for the Advancement of Science 1961). Ellis's *Country Life* article concluded:

> The meat of coypu has long been appreciated abroad and, incidentally, consumed under various pseudonyms in this country. A few Norfolk people are beginning to overcome their scruples and are eating casseroled coypu regularly now, but there is no organised demand for the local product so far.
>
> (Ellis 1960: 1591)

In Ellis's work the coypu comes to signify a happy foreignness in the landscape:

> A warm summer day will sometimes bring them forth to frolic in quiet waters. On such occasions they may be seen floating with their tails stiffly erect, or wallowing for pure pleasure, like hippopotamuses in an East African river. They tend to remain mute during the day, even when taking part in a communal water frolic.
>
> (Ellis 1965: 226)

As with the bittern, the sound of the coypu plays a key role, although this is a rather differently meaningful sonic ecology. Ellis's July 1957 radio talk on *Night Excursions and Alarms on the Broads* used sound to negotiate the coypu's alienness and its appeal. Ellis had been comparing the vegetation of the Yare valley to jungle:

> We have South American music on the Broads now: genuine jungle stuff – not the sort they imported from Spain. It echoes across the valley when the moon is full. If you've ever heard a cow in trouble at any time, imagine listening to one lamenting the loss of a calf in the middle of every reed-swamp. That's the nearest description I can give of a coypu's love-call. … I wouldn't say they make the night hideous with their cries, but they're very mournful under the moon's influence. [Give the coypu's cry here – MNAAHW – MNAAHW –]
>
> (Ellis 1957a; also 1965: 226)

Sound suggests the coypu as authentically lost, its authentic 'jungle' cries provoking sympathy. If the Yare valley vegetation is like a jungle, although not a jungle, then the coypu's jungle cries almost belong, but not quite. The animal seems melancholy, and its sound provokes a melancholic and sympathetic curiousity in the local human listener.

If the coypu was a matter of shifting opinion from the 1930s to the 1970s, one strand of local thinking consistently favoured the animal, admiring and valuing its resilience, its fur, its meat. Alan Savory ran a coypu farm at Brundall from 1933 until he joined the RAF Air–Sea Rescue in 1941, showing his 'nutria' fur at a Crystal Palace fur show in November 1934. Savory, who was keen to stress that it was not his farm from which coypus escaped (Savory 1956: 46, 83), presented the coypu as a gentle beast, wary even of the ducks which got into its farm pens (Savory 1953: 90–93). He held the sporting rights on a Brundall marsh, and on returning from the war found that the invading coypu had done a good job. What was once a boggy thicket, difficult to shoot over, was now 'Coypu Marsh', with only reeds and yellow iris thriving around open water:

> The coypu had ... turned the marsh into the most delightful little duck swamp with little pools leading into big pools. ... It was the perfect teal shoot. ... [T]he coypu had done a good job, and they kept on doing it and keeping the open water from growing up.
>
> (Savory 1956: 83–84)

Savory was writing here in the mid-1950s, and his positive view of the coypu was consistently echoed by Day, who supported this 'Great South American Marsh Rat' as a good ecological influence and an ally against officialdom. The coypu's vegetation clearing was to be welcomed – 'It is a thousand shames that whenever one appears some lout with a gun immediately shoots it' – and Day (1953: 154) hoped that it would become 'a permanent and truly wild Broadland animal'. Coypus deserved 'every protection. They can certainly become a charming, useful and attractive addition to our wild life' (Day 1951: 56). The creature added 'a primeval echo of an older and larger, but vanished, fauna' (Day 1951: 181), restoring rather than invading the spirit of the region. When the official cull began, Day took the side of the coypu as both a victim of officialdom and a thorn in its side. The coypu, and Day, become part of a roguish ecology resistant to the bureaucrat:

> The coypu became not only a menace to the farmer, but a little godsend to the chair-bound 'rural experts' of Whitehall, who never by any chance get mud on their boots. The coypu was a wonderful excuse for orders, instructions and memos in quadruplicate.

Day (1967: 182) gleefully noted that 'the coypu beat them every time. He bred, increased and travelled!' The coypu knew Broadland better than the officials, could move ahead of their men, and if the 1962 campaign killed many, Day reported in *Portrait of the Broads* that he had seen them breeding again in 1965–6. For Day the cull was less about environmental management than an excuse for officialdom and bureaucracy to interfere in local ecology, and he applauded the coypu's stubborn survival, allying himself and the animal in a maverick ecology, refusing to be properly subject to others:

> He is probably here for good. He is harmless to man and will only attack you if you attack him. Incidentally he is good to eat and, when roasted, very much like roast pork. If you attempt to skin one, cut the skin down the back-bone, not on the belly. Thus you will obtain the best fur.
>
> (Day 1967: 183)

Concluding remarks

In front of me, pinned to an office shelf, are seagulls, two adults and three young, stuffed and photographed against a rocky background to make a postcard: 'No. 233. Great Black-Backed Gull and young' (see Figure 6.3). The postcard is a curiosity in itself, and a memory of a curious place, the Booth Museum of Natural History, Dyke Road, Brighton, built by Edward Booth in his garden in 1874 to house a collection of stuffed birds, mounted in over 300 display cases showing 'natural' settings. Cases of birds go up to the ceiling, many shot on Breydon Water before making a last posthumous migration to the south coast for stuffing. Day, not surprisingly, liked this place, although he missed the name of the founder:

> Probably the best collection of Breydon birds can be seen in the Dyke Road Museum at Brighton. They were collected by the late R. Fielding Harmer, a noted 'gentleman gunner', who regularly punted on Breydon some sixty years or more ago. The collection is quite outstanding.
>
> (Day 1953: 102)

Forty years further on, visiting the museum, I too rather liked the place, although I do not much like Day. Perhaps these stuffed seagulls have migrated in meaning from being simple trophies of gunning men. Taxidermy a hundred years on, still in its original case and multiplied 300 times up to the ceiling, carries a pathos in relation to both the doomed-to-be-ever-static animals and their killers. The museum was no longer, for this visitor at any rate, a celebratory exhibition of naturalist shooting. Rather the birds' mute presence triggered some reflection on the different reasons for which they and other birds like them might have died. If these birds ended up as south coast trophies, others not here were shot for the

Figure 6.3   A postcard of stuffed gulls from the Booth Museum, Dyke Road, Brighton.

Source: Postcard in author's possession

pot, others were sold on for a living, and others died just for the fun of their killers, landing dead on the marsh and staying there. And still others ended up in the Castle Museum, Norwich, to be kept by keeper Ted Ellis and installed in diaromas of the Norfolk landscape created between 1931 and 1936 under his direction (Stone 1988: 52). I remember being taken as a child to this 'Norfolk Room', with its scenes of coast and marsh populated by local birds I had neither seen nor looked for; Ellis had set up the displays as a site for the education of young and old local citizens about their local nature.

These differential fates of dead birds, migrating even after death through space and time, reflect and reproduce different human cultures, different versions of the animal–human. If one does not fundamentally object to the killing of animals, then such reflections might form part of an animal geography which does not necessarily begin with matters of rights or welfare. The latter are important driving forces for much work in this field, but they do not exhaust the motivation or ambition of such inquiry. This chapter has taken one regional culture of the animal, where regional nature is formed in relation to other scales of being and authority; different forms of and claims to national and local expertise. Different versions of the animal–human reflect and shape contesting cultures of nature and region, with specific animals taking particular symbolic roles, their nativeness or alienness refracting different normative senses of cultural ecology. Subject and object are enfolded here in ways which make it impossible to figure ways of being human outside of relations to the non-human environment. This may be an obvious point to make from the Broadland material in this chapter, where human conduct is a matter of self-conscious self-formation through relations with environment, but one could extend this general argument to anyone and anywhere, including situations where human self-formation claims a subjectivity formed outside of or despite environment. Such claims remain a matter of relational geosubjectivity, even if the relation is that of rejection or retreat. In Broadland, though, we have found figures such as Day and Ellis consciously and enthusiastically making themselves, and the nature–culture of the region, through their relation to the non-human, especially the non-human animal but also the vegetable and mineral; looking, touching, testing, killing, stuffing, listening, tasting, via various technologies and techniques, all operating through and producing claims to authority within and over the region. If the Norfolk Broads are a site where such matters have been especially acute given the tight interweaving of local ecology and economy, parallel stories might be found in any region, with different animals and different humans in conjunction, making for different versions of animal–human.

Acknowledgements

Earlier versions of this essay were presented to the annual Institute of British Geographers conference at Exeter in January 1997, and to a workshop in the School of Geography at the University of Nottingham. Thanks to all those who provided comments on each occasion. I would also like to thank the extremely helpful staff of the BBC Written Archives Centre, Caversham Park, Reading, for providing access to transcripts of radio programmes. Much of the research for this essay was carried out with the help of a University of Nottingham research grant (94NLRG454).

Notes

1   Day typically sought to claim that Ellis, like every other local figure, was his friend, following an initial encounter at Wheatfen where he had to convince Ellis he was not doing anything wrong there (Day 1951: 56–57; 1953: 82; 1967: 60).
2   I recall sitting with friends outside Norwich Library *c*.1982 and seeing Ellis walk up to a tree and contemplate it in great detail; we thought he was a very curious man.
3   The programme also featured an early performance by Allan Smethurst, later (in)famous as 'The Singing Postman', singing 'Mind Your Head', detailing the risks of decapitation when travelling by train.

References

Agyeman, J. and Spooner, R. (1997) 'Ethnicity and the rural environment', in P. Cloke and J. Little (eds) *Contested Countryside Cultures: Otherness, Marginalisation and Rurality*, London: Routledge.
Althea (1975) *Exploring the Broads*, Ipswich: East Anglia Tourist Board.

Berger, J. (1980) 'Why look at animals?', in *About Looking*, London: Writers and Readers.

Betjeman, J. (1975) *Collected Poems*, London: John Murray.

British Association for the Advancement of Science (1961) *Excursions Guide: Norwich*, Norwich: British Association for the Advancement of Science.

Buxton, A. (1946) *Fisherman Naturalist*, London: Collins.

Buxton, A. (1948) *Travelling Naturalist*, London: Collins.

Buxton, A. (1950) 'Wild bird protection in Norfolk', *Transactions of the Norfolk and Norwich Naturalists Society* 17: 109.

Davis, R.A. (1956) 'The coypu', *Agriculture* 63: 127–129.

Davis, R.A. (1963) 'Feral coypus in Britain', *Annals of Applied Biology* 51: 345–348.

Day, J.W. (1931) *Speed: The Authentic Life of Sir Malcolm Campbell*, London: Hutchinson.

Day, J.W. (1935a) *A Falcon on St Paul's: Being a Book about the Birds, Beasts, Sports and Games of London*, London: Hutchinson.

Day, J.W. (1935b) *King George V as a Sportsman: An Informal Study of the First Country Gentleman in Europe*, London: Cassell.

Day, J.W. (1937) *Sporting Adventure*, London: Harrap.

Day, J.W. (1938) *Sport in Egypt*, London: Country Life.

Day, J.W. (1946) *Harvest Adventure*, London: Harrap.

Day, J.W. (1948) *Wild Wings*, London: Blandford Press.

Day, J.W. (1949a) *Coastal Adventure*, London: Harrap.

Day, J.W. (1949b) *The Modern Fowler*, London: Batchworth Press (First Edition 1934).

Day, J.W. (1950) *Marshland Adventure*, London: Harrap.

Day, J.W. (1951) *Broadland Adventure*, London: Country Life.

Day, J.W. (1953) *Norwich and the Broads*, London: Batsford.

Day, J.W. (1954) *A History of the Fens*, London: Harrap.

Day, J.W. (1957) *Poison on the Land*, London: Eyre and Spottiswoode.

Day, J.W. (1958) *Lady Houston OBE: The Woman Who Won The War*, London: Allan Wingate.

Day, J.W. (1960) *British Animals of the Wild Places*, London: Blandford Press.

Day, J.W. (1967) *Portrait of the Broads*, London: Robert Hale.

Edwards, P. (1948) *Call Me At Dawn*, Ipswich: East Anglian Daily Times.

Ellis, E.A. (1952) 'Wild life in Broadland', in R.H. Mottram (ed.) *The Broads*, London: Robert Hale.

Ellis, E.A. (1960) 'The coypu threat in East Anglia', *Country Life*, 29 December: 1590–1591.

Ellis, E.A. (ed.) (1965) *The Broads*, London: Collins.

Ellis, E.A. (1998) *A Tapestry of Nature*, Norwich: Eastern Daily Press.

Fisher, J. (1950) 'Bird preservation', *Transactions of the Norfolk and Norwich Naturalists Society* 17: 71–83.

Fitter, R.S.R. (1945) *London's Natural History*, London: Collins.

Fitter, R.S.R. (1959) *The Ark in Our Midst: The Story of the Introduced Animals of Britain*, London: Collins.

George, M. (1992) *The Land Use, Ecology and Conservation of Broadland*, Chichester: Packard.

Gosling, L.M. and Baker, S.J. (1989) 'The eradication of muskrats and coypus from Britain', *Biological Journal of the Linnean Society* 38: 39–51.

Huxley, J. and Koch, L. (1938) *Animal Language*, London: Country Life.

Ingold, T. (ed.) (1988) *What is an Animal?*, London: Unwin Hyman.

Koch, L. (1954) *Memoirs of a Birdman*, London: Scientific Book Club.

Lambert, J.M., Jennings, J.N., Smith C.T., Green C. and Hutchinson J.N. (1960) *The Making of the Broads*, London: Royal Geographical Society Research Series No. 3.

Laurie, E.M.O. (1946) 'The coypu (*Myocastor coypus*) in Great Britain', *Journal of Animal Ecology* 15: 22–34.

Marren, P. (1995) *The New Naturalists*, London: Collins.

Massingham, H.J. (1924) *Sanctuaries for Birds and How to Make Them*, London: Bell.

Matless, D. (1994) 'Moral Geography in Broadland', *Ecumene* 1: 127–156.

Matless, D. (1996) 'Visual culture and geographical citizenship: England in the 1940s', *Journal of Historical Geography* 22: 424–439.

Matless, D. (1998) *Landscape and Englishness*, London: Reaktion.

Mottram, R.H. (ed.) (1952) *The Broads*, London: Robert Hale.

Nature Conservancy (1965) *Report on Broadland*, London: Nature Conservancy.

Nicholson, E.M. (1957) *Britain's Nature Reserves*, London: Country Life.

Nicholson, E.M. and Koch, L. (1936) *Songs of Wild Birds*, London: Witherby.

Nicholson, E.M. and Koch, L. (1937) *More Songs of Wild Birds*, London: Witherby.

Nietzsche, F. (1994) *On the Genealogy of Morality*, Cambridge: Cambridge University Press.

Norris, J.D. (1963) 'A campaign against the coypus in East Anglia', *New Scientist* 17: 625–626

Norris, J.D. (1967) 'A campaign against feral coypus (*Myocastor coypus Molina*) in Great Britain', *Journal of Applied Ecology* 4: 191–199.

Savory, A. (1953) *Norfolk Fowler*, London: Geoffrey Bles.

Savory, A. (1956) *Lazy Rivers*, London: Geoffrey Bles.

Sheail, J. (1988) 'The extermination of the muskrat (*Ondatra zibethicus*) in inter-war Britain', *Archives of Natural History* 15: 155–170.

Stone, E. (1988) *Ted Ellis: The People's Naturalist*, Norwich: Jarrold.

Vincent, J. (1926) 'The romance of the Bittern', *Country Life*, 28 August: 303–307.

Williamson, T. (1997) *The Norfolk Broads: A Landscape History*, Manchester: Manchester University Press.

Radio programmes

Day, J.W. (1946) *The Man From The Country: Call of the Broads*, Light Programme, 23 February, 20.10–20.15.

Ellis, E.A. (1957a) *Through East Anglian Eyes: Night Excursions and Alarms on the Broads*, Midland Home Service, 23 July, 18.50–19.00.

Ellis, E.A. (1957b) *Through East Anglian Eyes: Probing the Undergrowth*, Midland Home Service, 6 August, 18.50–19.00.

Hickling Broad (1947) Midland Home Service, 9 May, 21.30–22.00.

'*The Naturalist' in East Anglia* (1959) Home Service, 24 May, 13.10–13.40.

Saturday Review (1952) London Calling Area, 29 November, 13.25–13.45.

Woman's Hour (East Anglian Edition) (1959) Light Programme, 6 November, 14.00–15.00.

7 A wolf in the garden

Ideology and change in the Adirondack landscape

Alec Brownlow

Introduction

In late March of 1995 a group of imported Canadian gray wolves (*Canis lupus*) exited their makeshift pen and dashed hurriedly into the Lamar Valley of Yellowstone National Park. The first wolf to trip the signal wire indicating to Park Service wildlife officials that the wolves were 'free' was the capstone moment following years of planning, lawsuits, death threats and general hostilities between 'pro' and 'anti' wolf groups around the region and nationally. It was an historic and symbolic event, one that captivated the attention of the international public and media. Experimental reintroductions[1] have now been repeated elsewhere in the US (red wolves [*Canis rufus*] in North Carolina; Mexican gray wolves in Arizona).[2] Indeed, at the moment, stretches of 'wilderness' across the nation are being eyed by wolf advocates as potential restoration sites.

The 'experiment' of wolf restoration into Yellowstone, however, has broader social implications and exposes a set of complexities unidentified and unarticulated in the general understanding of the term as it is defined legally. For this restoration fundamentally (re-)created a space to be shared and occupied by an animal long absent in the national landscape. Indeed, wildlife restoration, in general, constitutes the material 'on the ground' manifestation of Wolch and Emel's (1995) recent call to 'bring the animals back in': that is, (re-)creating a place for non-human 'others' in the social realms of theory and space. As I show below, the restorationist's project of 'bringing the animals back in' presupposes *necessarily* an *appropriate* (ecological, social, political) place for animals to be brought back *into*. This assumption maintains the potential to be quite problematic when considered out of its historical context and/or when considering the restoration of an animal as culturally and symbolically embellished as the wolf is in North America.

In this chapter, I examine the recent designation of the Adirondack Mountains in upstate New York as a potential release site for the gray wolf. This designation marks a clear departure from past site selection processes, for the region is not maintained as a national park, wilderness or other area commonly understood as 'protected' (void of human settlement and most human activity), but is, rather, a region heavily settled with human communities maintaining intimate ties to the local landscape and its history. From this, the fundamental question guiding what follows becomes: is the contemporary Adirondack landscape an appropriate 'place' for wolves to be brought back into? In order to address this question contextually, creatively and thoughtfully, it is necessary to consider how and when these animals 'lost' this (their) place to begin with. Further questions emerge from this. For example, how, when and why did the wolf become an animal 'out of place' (Cresswell 1996) in the Adirondacks?[3] Were wolves *re*-placed in the socio-physical landscape, and if so by what? How did these factors interact with and/or dictate the wolf's extermination from and its potential restoration to the Adirondack landscape? Questions of historical and geographical context appear fundamental to any consideration of wildlife restoration today. Any attempt to address them requires the location and consideration of those past and present social, economic and cultural processes which are fundamental to the construction of current landscape meaning and perception. I therefore leave the exposition of the ecological appropriateness of the Adirondack landscape to wolf restoration to others (indeed, ecological appropriateness is too often all that is considered), and will instead focus upon the landscape's social and cultural appropriateness.

This chapter traces the evolution of landscape meaning in the Adirondacks as it has been represented since its settlement by Euro-Americans in the early 1800s. In particular, I focus upon that fifty-year period between the mid-nineteenth century until the turn of the twentieth century when the Adirondacks and the surrounding areas experienced two very distinct 'waves' of white settlement and exploitation, each resulting in a very different use of and, subsequently, application of meaning to the local landscape. This period witnessed the parallel processes of wolf extirpation (local extinction) and the marginalisation of the region's early settlers, each resulting from an urban-based change in landscape meaning in which both wolves and settlers found themselves effectively 'out of place'. These historical and concurrent processes, I argue, are imperative to any discussion concerning wolf restoration to the Adirondacks today. They constitute the historical context, the socio-cultural foundation upon which this discussion must be based. Furthermore, they remain at the fore in terms of current landscape meaning and use.

Examination of the historical changes in landscape meaning and the place of the wolf in the Adirondacks is accomplished through the analysis of the myriad narratives of the region as they appear in documents, accounts and artwork. It is within and through these narratives that the *imaginative repertoires*[4] of wolf, people and place are constructed, promoted and contested, and within which landscape meaning emerges.[5] It is within these narratives that the evolving ideologies applied to the Adirondack landscape are identified and their inclusive and exclusive capacities exposed (Willems-Braun 1997). Finally, it is within these narratives that we are able to track the wolf's loss of place, as well as the place of the region's early settlers and their subsequent re-placement in the Adirondack landscape. They provide invaluable insight into and clarity regarding the context within which wolf restoration into upstate New York is currently being debated.

Making 'place' for animals

In order to identify better the past and any future place of the wolf in the Adirondacks, it is first necessary to examine the place of the wolf as it has existed in the American imagination. Only recently have the social sciences begun seriously to consider animals as significant and symbolic agents within the socio-cultural milieu. An introduction to the novel 'place' that geographical and social theory are currently making for non-humans is, therefore, an appropriate entry-point into the following discussion.

Until very recently, there was no 'place' for animals in social theory, discourse or analysis. Their contribution to landscape meaning and definition, and to our understanding of 'nature' in general, tended to be overlooked or dismissed entirely. When mentioned at all, animals were often framed as 'conscienceless' actors in the capitalist chain of production (e.g. Smith 1984), or, more recently, lumped together within the ambiguous and increasingly contentious term 'biodiversity' (e.g. Katz 1998). Reflecting on this phenomenon, Shepard (1996: 9) posits that the current absence of (wild) animals from social science discourse, in general, and from American consciousness, specifically, is attributable to their 'physical absence (in the landscape) and our shifted attention (to more pressing problems), as though we had lost both the opportunity and the ability to see them'. In their absence, we have been forced to draw upon representations of (wild) animals as they are depicted by and through various media (e.g. zoos, books, television, film, etc.) for our construction of and application of meaning to these non-human 'others' (Anderson 1998; Wilson 1992).

Recently, however, the inability or unwillingness on the part of the social sciences to acknowledge or to consider animals as legitimate components and constitutive agents of the socio-physical landscape has been problematised and addressed by human, especially feminist, geographers and theorists who find this oversight indicative of deeper, more insidious biases within the social sciences generally (e.g. Anderson 1995; Whatmore and Thorne 1998; Wolch and Emel 1998; see also Shepard 1996). These writers expose animals as fundamental to our socio-cultural constructions of self and 'other' (Anderson 1995, 1998; Elder *et al.* 1998; Emel 1995; Haraway 1989; Ohnuki-Tierney 1991; Shepard 1996), and identify them as fellow landscape participants that are central to our perception of and application of meaning to place and space (Gullo *et al.* 1998; Philo 1995; Whatmore and Thorne 1998; Wolch 1996).

Paralleling the emergence of animals within social discourse has been a fundamental reshaping and reconsideration of the physical and imagined landscape by a conservation ideology currently attracted by the idea and the potential of wildlife 'restoration'. This emerging paradigm (see Baldwin *et al.* 1994) is premised upon the ecologico-political process of 'bringing the animals back in' (cf. Wolch and Emel 1995), both materially (physically) and figuratively (symbolically). However, as stated previously, this 'bringing back in' or 'reintroduction' of animals into the socio-physical landscape presupposes the existence of a 'place' where both the physical (embodied) and cultural (symbolic) animal may be welcome, and where its survival may be realised.

The place of the wolf in America

Despite its disappearance from the physical space of the US,[6] the wolf continues to occupy a contested and emotionally laden symbolic space within the American imagination, prompting one author recently to ask, 'what do wolves mean?' (Scarce 1998). Shepard (1996: 291) contends, 'every culture has its ultimate animal ... [one that] implies a unity – the most powerful effect of the animal on the mind.' In that collective culture of 'American', surely there exists no more likely candidate for this post than the wolf. Few other animals have been the focus of such an extensive campaign of cultural representation (in art, science, film, audio, photography) or have been so intensively pursued (by bounty hunters, photographers, film-makers, biologists) as has the wolf in the American context.

Over the past several decades, the wolf has experienced a remarkable transformation in the American imagination. It has emerged from an absolute fear and loathing that resulted in its violent extermination from the greater part of North America to occupy a position of admiration and fetish-like devotion among scores of enthusiasts that has resulted in its recent return to the physical landscape. Aided in no small part by the writings of Aldo Leopold (1949) and Barry Lopez (1978), the photography of Jim Brandenburg and the research of biologist David Mech, the wolf has, to a large extent, shed the loathsome and evil character once associated with it and today embodies metaphorically concepts such as 'wilderness' and 'wildness' (Brick and Cawley 1996: 1; Mech 1995), 'paradise lost' (Brick and Cawley 1996: 2) and the 'wild, the free, the uncommodifiable' (Emel 1995: 709).

For its advocates, this radical change in the symbolic space of the wolf in America signals an opportune and unparalleled moment in American history to restore this animal to the physical space from which it was removed so long ago.

Yet, despite its successful and relatively rapid restoration within the American imaginary, the wolf's restoration to the physical landscape will not prove to be as simple. The embodied, sentient wolf remains an emotional and political lightning rod. Among its enthusiasts, wolf restoration and protection programmes have acquired an intensity and a symbolism that appear more apologetic than scientific, more emotional than ecological, often appearing to parallel an emergent shame over 'historical' brutality towards Native American populations (see, e.g., Wishart 1994).[7] Restoration opponents, on the other hand, have shed the wolf of its 'old world' demonic qualities (see Lopez 1978), replacing them with a more insidious and politically powerful set of metaphors. To its opponents, the wolf is the current embodiment of a conservation discourse perceived as unyielding and relentless, an agenda promoted by an urban-based environmental agenda considered elitist, illegitimate, hegemonic and exclusionary (Arnold 1996; Brick 1995; Heiman 1988: 210–216; Nixon 1992; Snow 1996).

In the Adirondacks, local changes to the meaning of the wolf emerged from changes in landscape description and meaning, resulting in its repositioning as 'in' or, more commonly, 'out of place'. Changes in the descriptions and meanings of place in the Adirondacks precipitated the wolf's extermination from the region and its subsequent re-placement by more socio-economically and culturally 'legitimate' animal types. Remarkably similar changes in the meaning of place may also precipitate its return.

Of wolves and humans in New York

Livestock arrived along with the early settlers to the New Netherlands territory, the region that eventually would encompass the state of New York, and wolf depredations were an immediate concern among the new residents. James Sperry, an early chronicler of the region, notes, 'among the trials of the first settlers, there were none more irritating than the destruction of sheep and swine by ... wolves. ... [O]ften whole flocks of sheep would be slaughtered in [a] night' (in Mau 1944: 136). In his discussion of early Adirondack settlement, Byron-Curtiss (1976: 10) tells a similar story:

> [O]ften after a settler had, by several seasons' patient breeding, obtained quite a flock of sheep, or a number of cows ... his plans and calculations were upset by some nocturnal visitor from the woods.[8]

Wolf depredations on livestock began taking an economic and symbolic toll early on, and anti-predator policies and campaigns were quickly initiated. Bounties and community 'wolf drives' were initiated with a fervent militancy and fanaticism once considered capable only in the American west.[9] Combined, these anti-predator campaigns accomplished the rapid extermination of the wolf from greater New York. By 1800 it had disappeared from the area around Lake Champlain and by 1830 from the western part of the state, a region described as 'much infested' with wolves only thirty years earlier (John Maude, English traveller, in Mau 1944: 112). One New York historian, commenting on a particular wolf bounty of the early nineteenth century, remarked that 'nothing in zoological history ever wiped out wolves as fast as that law' (White 1967: 84). This sentiment would, of course, be repeated time and again across the continent.

As the wolf population dwindled throughout the rest of the state, the remote and mountainous Adirondack interior,[10] with a relatively small Euro-American population at the time, became a haven of sorts for wolves and other species suffering the expansion of white settlement and exploitation elsewhere[11] (Terrie 1992). A visiting New York City journalist, Joel Headley, writing in the mid-nineteenth century, noted that wolves 'swarm[ed] the [interior] forest' (Headley 1849: 85; also Foster in Byron-Curtiss 1976: 180). However, not forty years later, mention of the wolf was noticeably absent from the notes of George Washington Sears, a field columnist for the New York City-based outdoor journal, *Forest and Stream*, who spent months at a time travelling the Adirondack interior (Brenan 1962).

As the above accounts reflect, at the same time that the Adirondacks were becoming a safe haven for animals retreating under heavy pressure from human settlement outside of the region, the area was being heavily promoted by New York City-based journalists, artists and enthusiasts. Writing for an urban audience eager to escape the rapidly expanding pressures of late nineteenth-century city life, a desire facilitated by rapidly increasing industrial wealth and mobility,[12] these boosters lured wealthy urbanites by the thousands to the 'wilderness next door'. The region was triumphed in the journalism of Sears, Headley and Sylvester (1877), the paintings of Winslow Homer, Asher Durand and Homer Martin, and the prose of Emerson, Longfellow and James Fenimore Cooper. Soon, travel guides began to appear *en masse*, expediting the area's emergence as 'one of the most thronged pleasure routes on the continent' (Williams 1871: 81). An 1864 *New York Times* editorial proclaimed the Adirondack region 'A Central Park for the World'. By the turn of the twentieth century, the Adirondacks were popularly considered a 'playground' for New York City's vacationing elite, one

> large enough to afford all the teeming population of the east an opportunity to find place within it for the restoration of health, the pursuit of game in season, and all the delights that a summer paradise and an autumn hunting ground can offer.
>
> (Kitchen 1920: 111; Miller 1917: 17)

Within a period of roughly two decades the place had fundamentally changed, both physically and imaginatively. The Adirondacks were no longer the remote and untamed wilderness that instilled its earliest explorers and chroniclers with wonderment and fear (see below), but a place tamed of its 'wild' and 'dangerous' past, a landscape of leisure and the pursuit of sport, health and recreation. This change in the meaning of place was accompanied, necessarily, by a commensurate change in the place of wolves, assuring their ultimate disappearance from the region. In New York, therefore, the wolf was the victim of two back-to-back changes, pastoral followed by playground, in landscape meaning. Through these cultural changes, the wolf not only lost its place in the landscape but, indeed, found itself *re-*placed by animals more culturally acceptable, more economically viable.

Re-placing the wolf in New York

With the settling of New York, wolves and other 'loathsome animals'[13] were quickly re-placed in the socio-physical landscape by animals more culturally suitable, more economically valuable. The introduction of domestic livestock was among the first and the most significant steps taken by early settlers to 'tame' the New York landscape. Not only was ridding the region of wolves demanded, but, perhaps more importantly, the very presence of livestock hastened and facilitated a domesticated and pastoral characterisation of the region very different from the 'wild' and 'savage' landscape that predated their arrival. Combined, the physical presence of livestock and the tame, pastoral landscape that these animals presupposed necessarily precluded the place of those animals deemed 'threatening' to each (cf. Crosby 1986; Gullo *et al.* 1998). Wolves, cougars and other 'vermin' were constructed as fundamentally 'out of place' in this 'new' landscape, physically and symbolically *dis*placed then *re-*placed by a regionally novel group of domesticated animals. Demands were placed not only upon the defence of livestock from depredation by wolves, but, significantly and perhaps of greater consequence, upon the defence of this new meaning of place from any 'out of place' heresy (Cresswell 1996: 9). Livestock symbolised and signified a new ideological landscape within which wolves had no place. As such, they had to go. Subsequently, bounties and wolf drives became the standard policies and practices for addressing these heretical 'visitors from the woods' (Byron-Curtiss 1976: 10). Combined, they nearly succeeded in the complete extermination of the wolf throughout greater New York state.

In the Adirondacks, wolves were under pressures similar to those described above, especially along the region's outer, less mountainous and more heavily settled periphery. However, the ruggedness of the mountainous terrain of the Adirondack interior, the High Peaks region, precluded any significant presence of livestock and heavy settlement and, for a period, wolves remained relatively abundant there. It was only a matter of time, however, before the Adirondack wolves became victims to a process remarkably similar to that described above yet displaced via a thoroughly different set of conditions, re-placed by an entirely different kind of animal.

By the mid- to late nineteenth century, local economies based upon small-scale timber and mineral extraction were on the decline. They were quickly being replaced by an emerging low wage, service-based economy that catered to the ever-increasing numbers of vacationers and recreationalists from New York City. Out of this resource-based past, the Adirondacks emerged into what one author described as a 'sportsman's [*sic*] paradise' (Murray 1869: 19). Requisite to this transformation of place was the continued physical presence of a new kind of animal resource, a particular 'class' of animal whose presence in the landscape was demanded by the thousands of urban 'sportsmen' that flocked to the region. The pursuit of 'game', above all other Adirondack leisure opportunities (health, painting, hiking), emerged as *the* lure that attracted the (male) urban adventurer to the region. Hunting and fishing rapidly became the area's premier attraction, promoted enthusiastically by travel brochures, guides and catalogues targeting this class of urban recreationalist. Deer, in particular, became the target of choice for these 'gentleman' hunters whose passion for 'the kill' was often ruthless and fanatical.[14] The following journal entry by a pair of vacationing New York City hunters on the pursuit and killing of a deer is typical:

> We rowed up to her and when within a rod or two, I fired and wounded her. … At length I entreated Tim to knock her in the head. … He forthwith took her by the tail and I placing the muzzle of my gun under her ear and firing she finally bade adieu to the upper world sank for a moment in the water and then rose *annihilated*.
>
> (Colles and Johnston [1851] in Adirondack
> Museum 1961: 17, emphasis in original)

While fish and 'game' (deer) had been a regular source of sustenance to local Adirondack families and communities, the arrival *en masse* of the urban sportsman provided these animals with an entirely different set of meanings, meanings that would herald the end for the state's remnant wolf population.

In order to ensure a steady flow of urban-based revenue into the Adirondack region, the maintenance of a healthy deer population was absolutely necessary. Regional economic growth depended, literally, upon the living bodies of the region's 'game' animals. Deer, thus, became *the* critical component of the promotional landscape. The image of the deer appeared regularly in the paintings, poetry and prose of the region; it was *the* selling-point to the wealthy New York City sportsman, attracting the cash flow to the region necessary for continued economic expansion and development. For each of these reasons, the continued physical presence of deer in the region (i.e. their 'management') was fundamental. Any 'threat' to deer implied much more: a threat to the 'new' landscape of leisure; a threat to the vision of economic expansion and development. In effect, deer management became equivalent to wolf eradication (Flader 1974; Leopold 1949: 129–133).

Deer, like livestock before them, came to signify a new type of landscape, embodying in symbolic form the 'tame yet wild' landscape that the Adirondacks themselves had become, in both the literal and the figurative sense. In this new place 'game' ruled supreme and the wolf emerged heretical, an animal without a place. Considered antithetical to successful deer management practices and also an animal whose very presence contradicted the landscape's emerging signification, the wolf's place as 'out of place' was quickly and widely determined. There was no longer room for wolves, cougars or other 'threats' in this urban-defined 'playground' (cf. Gullo *et al.* 1998; Philo 1995); the landscape's new meaning precluded their continued local existence. Aggressive and popular bounty systems emerged, and by the turn of the twentieth century wolves were gone from the Adirondacks and from the state of New York entirely.

Re-placing the Adirondacks

Wolves, however, were not the only victims of these ideological appropriations of landscape. Native Americans had long resided in the Adirondack region. Mohawk, Oneida, Onondaga, Seneca, Cayuga (collectively, the Iroquois), Huron, Mohican and Abenaqui maintained settlements from the St Lawrence River Valley through the mountainous Adirondacks to Lake Erie. Early white explorers and chroniclers of the region employed all manner of tropes to construct Native Americans as 'out of place' and heretical to the emerging working landscape (see below), thereby facilitating and justifying their eventual displacement (i.e. extermination) (Katz and Kirby 1991; cf. Willems-Braun 1997). Like the wolf, the region's Native Americans were posited to belong to a 'wild' nature that was a threat to the domestication of the Adirondack landscape. Sylvester (1877: 21–23) referred to the 'dreadful Iroquois' as 'nature's own wildest child', as a 'league belonging to the old savage wilderness', and as 'savages … [that must follow] the dictates of [their] own wild will': savages in a savage nature. 'Wild' nature and its representatives (Indians and wolves) no longer had a place in the Adirondacks; together, they stood in the way of the experiment of landscape abstraction, redefinition and appropriation. Other descriptions implicated the region's native population as uninterested in and ignorant of the bounty of natural resources (i.e. timber) that the Adirondacks had to offer (Byron-Curtiss 1976: 114; Sylvester 1877: 24). If the natives were not putting the land and its resources to 'good use', then their removal would facilitate the entry into the area of those who would. These descriptions, then, provided the additional incentive and legitimacy necessary to entice the savvy, exploitation-minded settler and entrepreneur into the region.

Perhaps the most insidious and damaging descriptions characterised Adirondack natives in the same terms used to describe the region's wolf population, terms successful in justifying and securing the eventual violent extermination of each. This is well exemplified in the 1976 reprint of Byron-Curtiss's mid-nineteenth-century biography of Nat Foster, an early Adirondack trapper and explorer. In this narrative, Byron-Curtiss and the voices he employs regularly depend upon, and call forth, metaphors and tropes that animalise the native 'other' as predator and as 'vermin':

> [T]he suffering patriots ... *it is needless to say* dealt with the Indians in the same manner as they dealt with wild animals that stole from their flocks. They shot them.
>
> (Byron-Curtiss 1976: 13, my emphasis)

Adirondack Indians were declared 'two-legged varmints' whose raids on and abductions of settlers' livestock were characterised as predator-like 'depredations' (Byron-Curtiss 1976: 13, 68). They were described as just '[a]nother creature that often made things lively' for the 'suffering' white settler (Byron-Curtiss 1976: 181). As such, the killing of an Indian was considered no more grievous than the killing of any other form of 'vermin'. Indeed, their murder was frequently documented and/or described in terms generally reserved for the hunting of the region's larger 'threats' such as bears, cougars or wolves (Byron-Curtiss 1976: 73, 189).

Combined, these representational tropes constructed the Adirondack natives as anathema to landscape meaning, its 'proper' use, and to humanity in general. They were considered to be vermin, weeds, matter 'out of place', and in dire need of removal (Cresswell 1997). These political and racist representations had devastating consequences. It was widely determined that '[the native] and his [sic] forest must perish together' (Francis Parkman in Sylvester 1877: 24), and their disappearance from the area was remarkably swift. Byron-Curtiss (1976: 73) places their 'withdrawal' from New York and New England at the close of the American Revolution. In the Adirondack region, Native Americans were noticeably absent from the mid- to late nineteenth-century accounts of Headley (1849) and Sears (in Brenan 1962).

Ironically, and of particular consequence to wolf restoration, the economic and political marginalisation of the region's early white settlers (i.e. those successful at displacing the region's native population) was achieved through the redefinition of the Adirondack landscape as a place of leisure and sport (cf. Williams 1973). Below, I explore this process of marginalisation through an account of the evolution of landscape description and representation. The similarities which it maintains to that described above are noteworthy. Indeed, local economic activities themselves emerged as 'threatening' to this novel meaning of landscape. In the process, it is my aim to show how these parallel processes of wolf extermination and the marginalisation of early settlers, especially the latter, remain at the core of any resistance to wolf restoration that may exist today.

Landscape definition and exclusion

Early descriptions of the Adirondack region were provided by male journalists from New York City whose travels to and accounts of the area were written up for an anticipatory urban audience eager to learn of this neighbouring wilderness (e.g. Sears in Brenan 1962; Headley 1849; Murray 1869; Sylvester 1877). Their Romantic, contradictory portraits of the region, describing the area as at once 'savage' and 'sublime', enticed the throngs of vacationers that would begin arriving by the latter half of the nineteenth century.

These authors often described a 'wild' and 'untamed wilderness' (see above). Headley (1849), Williams (1871) and Sylvester (1877), writing in the mid-nineteenth century, commonly referred to the Adirondacks using such terms as a 'neglected' and 'unproductive waste', metaphors employed to entice the more productive sensibilities of (urban) society and economy. At the same time, however, these writers found in the landscape a sublimity and grandeur which they considered forged by the hand of God:

> [I]t seems as if I had seen vagueness, terror, sublimity, strength and beauty, all embodied, so that I had a new and more definite knowledge of them. God appears to have wrought in these old mountains with His highest power and designed to leave a symbol of His omnipotence. Man [*sic*] is nothing here, his very shouts die on his lips.
>
> (Headley 1849: 63)

Similarly,

> Pen cannot convey the idea of its sublimity, the pencil fails to even suggest the blended strength and delicacy of the scene. The rude laugh is hushed, the boisterous shout dies out on reverential lips, the body shrinks down, feeling its own littleness, the soul expands and rising above the earth, claims kinship with its Creator, questioning not his existence.
>
> (Stoddard [1878] in Becker 1963: 11)

An evocative blending of the savage and sublime continues with Headley's (1849: v, 145) description of the region's forests, describing them on one occasion as 'a neglected waste' while later referring to them as a primeval and pristine wilderness whose creation 'man has had no hand in'. Contrasting descriptions of the Adirondack landscape and its forests are found throughout early narratives. Williams (1871: 103, 100) referred to parts of the region as a 'desert … repulsive and uninteresting' while describing others as 'shrouded by … primeval forests, [remaining] almost as it came from the hands of its Creator'. Continuing the trend, the Adirondack mountains were frequently likened to the 'glorious' Alps of Switzerland (Headley 1849; Kitchen 1920: 114; Sylvester 1877: 43; Williams 1871: 101; but see Williams 1973: 128), its wilderness to that of the 'darkest' Amazon forest (Sears in Brenan 1962: 16).

These dichotomous descriptions (racial metaphors of light and dark, savage and sublime) had the intended effect of introducing to an urban population of increasing material wealth a landscape of awe and inspiration, a 'neglected' yet wondrous 'playground' in their own backyard there for the taking and the taming. By the 1870s, vacationers by the thousands were arriving in the region via rail and steamboat, staying at any one of a dozen or more of the grand hotels that had sprouted up in the area (Higgins 1931; Williams 1871). They came to hunt and to fish. They came for their health: '[T]here prevails in the atmosphere … a pureness, an elasticity and vitality that impart health and afford an indescribable physical enjoyment in the mechanical process of inspiration' (Williams 1871: 101; see also Headley 1849: 20). It became the seasonal home to the likes of the Vanderbilts, Rockefellers, Morgans and Whitneys. It was lionised in the writings of visiting literati such as Emerson, Cooper, Frost and Longfellow, on the canvases of Homer, Martin and Durand.

Following close behind this deluge of urban visitors and their place-based ideology were frequent and often fervent calls for regional conservation and temperance in exploitation. The genesis of these concerns resided with influential urban vacationers and seasonal residents, among whom there was dwindling tolerance for, among other things, the established natural resource-based economic activities. These were now labelled as environmentally destructive and perverse to the landscape's emerging definition as 'playground'. Writing for *Forest and Stream* magazine in the 1880s, Sears's call for preservation was not atypical:

> [T]here is but one Adirondack Wilderness on the face of the earth. And, if the great state of New York fails to see and preserve its glorious gifts, future generations will have cause to curse and despise the petty, narrow greed that converts into saw-logs and mill-dams the best gifts of wood and water, forest and stream, mountains and crystal springs in deep wooded valleys.
>
> (Sears in Brenan 1962: 16)

The Romantic period in the US was, by this time, in full swing and the artistic products that emerged from the region at this time were of undeniable consequence to the region's eventual protection. By the turn of the century, conservation had prevailed with the establishment of the Adirondack State Park in 1892 and the Adirondack Forest Preserve three years later.

With the legal establishment of the Adirondack State Park, the *de facto* ideological changes in landscape meaning that had been slowly accumulating over the past several years were codified *de jure* by state legislation. The Adirondacks were no longer recognised by the dominant ideology as an 'untamed' landscape of work, one with which the 'locals' had struggled for decades (Fosburgh in Adirondack Museum 1959: 15; Northern Forest Lands Council 1994: 18; White 1967: 37). Instead, as a park, the Adirondacks were now legally a landscape of leisure, health and recreation. The dominant and more powerful urban-based ideology had succeeded in protecting its 'playground' from local 'threats'. The historic economic activities of the local Adirondacker were also defined as heretical to the new meaning of place, their traditional means of livelihood now considered 'out of place'. Unlike their native predecessors, they certainly were not eliminated nor excluded from the area; yet the locals saw their livelihoods, activities, landscape meanings, in effect, their identities, go the way of the wolf and the Native American before them.

Affected were local farmers, miners, loggers and their families who found themselves faced with the unattractive choice between leaving the area entirely in search for work elsewhere or seasonal low-wage labour within the growing tourist industry. Mill and mining jobs quickly gave way to employment in the many hotels, lodges and sporting outfits that had sprung up in the region. Indeed, as Nixon (1992: 28) indicates, to this day the region remains 'a land of [minimum wage] jobs and seasonal employment'. The ideological and legal sequestration of the Adirondack landscape from local economic activities continues to be the source of unending bitterness and distrust by the local communities towards the state and the gaze of its urban vacationers (DuPuis 1996; Heiman 1988: 192–237). As Heiman (1988: 196) posits, '[T]he bitterness felt by local residents at the imposition of extraregional demand for wilderness as leisure space continues to this day. It is particularly intense where preservation is seen as interfering with local livelihood.' This is the context within which wolf restoration to the Adirondacks is currently being considered and debated and, as one writer puts it, the 'battle' between pro- and anti-wolf interests is currently 'simmering' (Thompson 1998).

Wolves in the Adirondacks?

Recently, an outdoor columnist for the *Albany (New York) Times Union* wrote, 'No single animal ... has so consternated a general population as the possibility of wolves in the Adirondacks' (Lebrun 1998). The idea of wolf restoration to the Adirondack region has, in effect, rekindled the age-old yet chronic rupture between New York's rural upstate and urban populations. Bringing the wolf back into the landscape is locally constructed as the latest attempt by an urban-based conservation ideology to dictate local landscape meaning and, subsequently, land-use and property rights (Lebrun 1998; Thompson 1998; also Heiman 1988; Nixon 1992). Indicative of these residual urban–rural hostilities, the wolf's primary advocate, the Washington DC-based environmental non-profit Defenders of Wildlife, finds itself engaged in an uphill battle with an expanding local consortium of anti-wolf interests composed of local farm bureaus, hunting clubs, the Adirondack Association of Towns and Villages, and various county supervisory boards. Indeed, the heavy influence of urban wolf support was critical to the recent prohibition of wolves by New York's Essex County, one of the region's largest and most likely counties for wolf release in the state (McKinstry 1998a). In the words of one local resident, '[I]f [wolf supporters] don't live in the Adirondack Park, they have nothing to lose. The people up here do' (in McKinstry 1998b).

An irony of certain consequence to restoration success now appears. The wolf represents, indeed, is the cultural icon adopted by, the urban-based conservation ideology of which local Adirondackers remain so wary and distrustful. The irony lies in the fact that it was this very same philosophy that was responsible for the wolf's extermination from the region one hundred years ago. Once exclusive of wolves and other 'vermin', conservation discourse today is thoroughly inclusive of these animals and has successfully applied to them remarkable symbolic character. Conservation discourse and ideology have remained, generally speaking, the same, especially in regard to their urban source and its abilities to control landscape meaning. Yet today, two of its earliest casualties find themselves residing on opposing ends of this discourse. Despite remarkably similar histories of displacement at the hands and under the distant gaze of a leisure-seeking urban population, local Adirondackers and wolves today (re-)emerge as discursive antagonists in the chronic battle for landscape meaning and definition.

Conclusion

Cresswell (1996: 60, original emphasis), in his discussion of the place in New York of graffiti artists and their work, portends, 'If the meaning of place changes, the place itself will change. The new meaning will be *their* meaning (the meaning of the 'other'). The place in question will become *their* place (the place of the 'other').' With this in mind, this chapter has historically traced subsequent appropriations of control over the socio-physical Adirondack landscape via the abilities to direct and to control its meaning and definition. Inclusive and exclusive powers emerge from these changes in meaning as ideological adjustments of what constitutes 'in' and 'out of place' become manifest. Exclusion through violence and elimination, as I have attempted to show here, can often be the result. These inclusions and exclusions are time- and space-dependent; what is ideologically 'out of place' at one point in time and in a particular space may, for various cultural, political and social reasons, re-emerge as being suitable to that space at another time. This appears to be the case with the wolf in the Adirondack landscape and, indeed, in the American landscape. Once heretical to landscape meaning, wolves have been culturally redefined and resituated so as to facilitate bringing them back in. Generally speaking, conservation ideology remains the same, yet is today more inclusive of wolves and other 'vermin'. These novel inclusions and attempts at restoration are laudable in their addressing of past evils and injustices. And yet, as we have seen here, understood and addressed acontextually and ahistorically, their potential restoration maintains the abilities to emphasise and expound other exclusions that have to date remained unaddressed.

Furthermore, I have demonstrated how the processes of displacement closely followed by re-placement, each existing in physical (applied) and symbolic (imaginary) forms, facilitate and expedite these changes, providing a virtual one–two punch to 'out of place' heresies. Combined, these geographical processes of 'placement' constituted the ideologically driven means by which wolf eradication and the marginalisation of locals in the Adirondacks was determined and achieved. Restoring animals to the theoretical and physical landscapes (i.e. bringing the animals back in) is an enticing and promising idea. Such acts of restoration maintain the potential to (re)address past exclusions or dismissals, locate and problematise parallel exclusions (e.g. exclusions by gender, race, class or species), and assist in the relocation of our past, our history and ourselves within these animal 'others' and the worlds and the meanings that they occupy. Yet, in each case, we must first identify and problematise the place to which they will be restored and, most importantly, the place from which they originally disappeared. Understood and contextualised together, as parts of a whole, as parts of a process, these places, past and future, offer promise to creativity, thoughtfulness, understanding and restorative success in the present.

Acknowledgements

I am grateful to Jennifer Wolch, Jody Emel, Chris Philo and Valerie Crawford for their helpful and insightful comments and suggestions. Additional thanks to Dianne Rocheleau, Billie Turner, Martyn Bowden, Scott Jiusto and Rachel Slocum for their comments on earlier drafts.

Notes

1 The Yellowstone wolf population has been labelled 'experimental' by the US Fish and Wildlife Service. This designation allows for the removal or 'control' (i.e. extermination) of 'problem' animals (i.e. those that transgress park and/or human boundaries).
2 Red wolf (*Canis rufus*) reintroduction into the Alligator National Wildlife Refuge of coastal North Carolina preceded the Yellowstone release by several years.
3 Philo (1995) provides an excellent example of the 'in/out' of placeness of animals in an urban context.
4 I am indebted to Chris Philo for this term.
5 Entrikin (1991) provides a thoughtful exposition on the role of narrative to the meaning of place.
6 Gray wolves have maintained 'natural' populations in the northern tier of states, most notably Minnesota and Michigan. They regularly cross into the US from Canada via north–south-running mountain chains in the American west (e.g. Cascades, Rockies, Bitterroots), and there is recent sign of their presence in Maine.
7 Brutal and unequal treatment of Native American populations does, of course, still exist in the US, yet in more insidious and 'invisible' ways (Deloria 1985; Matthiessen 1991).
8 These 'visitors from the woods' also included cougars, subjects of a hatred even more intense than that towards the wolves. In the words of an early explorer, cougars were 'the slyest, strongest, bloodiest ranger in the woods' (Murray 1869: 60). Their extermination too was inevitable.
9 See Mau (1944: 238) for an apt description of an early New York 'wolf drive.'
10 The present-day counties of Hamilton, Herkimer, Fulton, Montgomery, Lewis and Oneida make up this interior region.
11 Especially represented here were animals of economic significance (e.g. beaver, marten and mink) and 'predators' such as cougar, lynx and bear.
12 By this time, rail and, especially, steamboat regularly shuttled between New York City and Lake Champlain.
13 This term was used by early settlers and explorers also to describe local Native American populations (e.g. Byron-Curtiss 1976).
14 The Adirondack Museum's 1961 publication, *Journal of a Hunting Excursion to Louis Lake – 1851*, provides an excellent account of an urban hunting expedition to the Adirondacks. It well exemplifies the carnage of which these unexperienced 'sportsmen' were capable.

References

Adirondack Museum. (1961) *Journal of a Hunting Excursion to Louis Lake – 1851*, Blue Mountain Lake, NY: Adirondack Museum Publications.

Anderson, K. (1995) 'Culture and nature at the Adelaide Zoo: at the frontiers of "human" geography', *Transactions of the Institute of British Geographers* 20: 275–294.

Anderson, K. (1998) 'Animals, science and spectacle in the city', in J. Wolch and J. Emel (eds) *Animal Geographies: Place, Politics and Identity in the Nature–Culture Borderlands*, New York: Verso.

Arnold R. (1996) 'Overcoming ideology', in P.D. Brick and R.M. Cawley (eds) *A Wolf in the Garden*, Lanham, MD: Rowman and Littlefield.

Baldwin, A.D., de Luce, J. and Pletsch, C. (eds) (1994) *Beyond Preservation*, Minneapolis: University of Minnesota Press.

Becker, H.I. (1963) *A History of South Pond and Origin of Long Lake Township*, Author's self-publication.

Brenan, D. (1962) *The Letters of George Washington Sears, Whose Pen Name was 'Nessmuk'*, Blue Mountain Lake, NY: Adirondack Museum Publications.

Brick, P.D. (1995) 'Taking back the rural west', in J. Echeverria and R.B. Eby (eds) *Let the People Judge*, Washington, DC: Island Press.

Brick, P.D., and Cawley, R.M. (eds) (1996) *A Wolf in the Garden*, Lanham, MD: Rowman and Little-field.

Byron-Curtiss, A.L. (1976) *Life and Adventures of Nat Foster, Trapper and Hunter of the Adirondacks*, Harrison, NY: Harbor Hill Books.

Cresswell, T. (1996) *In Place / Out of Place: Geography, Ideology and Transgression*, Minneapolis: University of Minnesota Press.

Cresswell, T. (1997) 'Weeds, plagues and bodily secretions: a geographical interpretation of metaphors of displacement', *Annals of the Association of American Geographers* 87: 330–345.

Crosby, A.W. (1986) *Ecological Imperialism*, New York: Cambridge University Press.

Deloria, V. (ed.) (1985) *American Indian Policy in the Twentieth Century*, Norman: University of Oklahoma Press.

DuPuis, E. M. (1996) 'In the name of nature: ecology, marginality and rural land use planning during the New Deal', in E.M. DuPuis and P. Vandergeest (eds) *Creating the Countryside*, Philadelphia: Temple University Press.

Elder, G., Wolch, J. and Emel, J. (1998) '*Le pratique sauvage*: race, place and the human–animal divide', in J. Wolch and J. Emel (eds) *Animal Geographies: Place, Politics and Identity in the Nature–Culture Borderlands*, New York: Verso.

Emel, J. (1995) 'Are you man enough, big and bad enough? Ecofeminism and wolf eradication in the USA', *Environment and Planning D: Society and Space* 13: 707–734.

Entrikin, J.N. (1991) *The Betweenness of Place: Toward a Geography of Modernity*, Baltimore: Johns Hopkins University Press.

Flader, S. (1974) *Thinking Like a Mountain: Aldo Leopold and the Evolution of an Ecological Attitude Toward Deer, Wolves and Forests*, Columbia: University of Missouri Press.

Gullo, A., Lassiter, U. and Wolch, J. (1998) 'The cougar's tale', in J. Wolch and J. Emel (eds) *Animal Geographies: Place, Politics and Identity in the Nature–Culture Borderlands*, New York: Verso.

Haraway, D.J. (1989) *Primate Visions: Gender, Race and Nature in the World of Modern Science*, New York: Routledge.

Headley, J.T. (1849) *The Adirondack: Life in the Woods*, New York: Baker and Scribner.

Heiman, M.K. (1988) *The Quiet Evolution*, New York: Praeger.

Higgins, R L. (1931) *Expansion in New York*, Columbus: The Ohio State University Studies (Graduate School Series in Contributions in History and Political Science No. 14).

Katz, C. (1998) 'Whose nature, whose culture?', in B. Braun and N. Castree (eds) *Remaking Reality: Nature at the Millennium*, New York: Routledge.

Katz, C. and Kirby, A. (1991) 'In the nature of things: the environment and everyday life', *Transactions of the Institute of British Geographers* 16: 259–271.

Kitchen, W.C. (1920) *A Wonderland of the East*, Boston: The Page Co.

Lebrun, F. (1998) 'Adirondackers crying no wolf', *Albany (New York) Times Union*, December.

Leopold, A. (1949) *A Sand County Almanac*, London: Oxford University Press.

Lopez, B.H. (1978) *Of Wolves and Men*, New York: Charles Scribner's Sons.

Matthiessen, P. (1991) *In the Spirit of Crazy Horse*, New York: Viking Press.

Mau, C. (1944) *The Development of Central and Western New York*, Rochester, NY: DuBois Press.

McKinstry, L. (1998a) 'Essex County anti-wolf law easily passes', *Plattsburgh (New York) Press Republican*, December.

McKinstry, L. (1998b) 'Wolf law hearing is feisty event', *Plattsburgh (New York) Press Republican*, November.

Mech, L.D. (1995) 'The challenge and opportunity of recovering wolf populations', *Conservation Biology* 9(2): 270–278.

Miller, W.J. (1917) *The Adirondack Mountains*, Albany: The University of the State of New York (New York State Museum Bulletin No. 193).

Murray, W.H.H. (1869) *Adventures in the Wilderness; Or Camp-Life in the Adirondacks*, Boston: Fields Osgood and Co.

Nixon, W. (1992) 'Fear and loathing in the Adirondacks', *E Magazine* 3(5): 28–35.

Northern Forest Lands Council (1994) *Finding Common Ground: Conserving the Northern Forest*, Concord, NH: Northern Forest Lands Council Publications.

Ohnuki-Tierney, E. (1991) 'Embedding and transforming polytrope: the monkey in Japanese society', in J. Fernandez (ed.) *Beyond Metaphor*, Stanford, CA: Stanford University Press.

Philo C. (1995) 'Animals, geography and the city: notes on inclusions and exclusions', *Environment and Planning D: Society and Space* 13: 655–681.

Scarce, R. (1998) 'What do wolves mean? Conflicting social construction of *Canis lupus* in "Border-town"', *Human Dimensions of Wildlife* 3: 26–45.

Shepard, P. (1996) *The Others: How Animals Made us Human*, Washington, DC: Island Press.

Smith, N. (1984) *Uneven Development*, Cambridge, MA: Blackwell.

Snow, D. (1996) 'The pristine silence of leaving it all alone', in P.D. Brick and R.M. Cawley (eds) *A Wolf in the Garden*, Lanham, MD: Rowman and Littlefield.

Sylvester, N.B. (1877) *Historical Sketches of Northern New York and the Adirondack Wilderness*, Troy, NY: William H. Young Publishers.

Terrie, P.G. (1992) 'The changing wildlife of the Adirondack Park', *The Conservationist* 46(6): 36–40.

Thompson, M. (1998) 'Essex County schedules public hearing on proposed wolf law', *Glens Falls (New York) Post Star*, October.

Whatmore, S. and Thorne, L. (1998) 'Wild(er)ness: reconfiguring the geographies of wildlife', *Transactions of the Institute of British Geographers* 23: 435–454.

White, W.C. (1967) *Adirondack Country*, New York: Knopf.

Willems-Braun, B. (1997) 'Buried epistemologies: the politics of nature in (post)colonial British Columbia', *Annals of the Association of American Geographers* 87: 3–31.

Williams, A. (1871) *A Descriptive and Historical Guide to the Valley of Lake Champlain and the Adirondacks*, Burlington, VT: R.S. Styles' Steam Printing House.

Williams, R. (1973) *The Country and the City*, New York: Oxford University Press.

Wilson, A. (1992) *The Culture of Nature*, Cambridge, MA: Blackwell.

Wishart, D.J. (1994) *An Unspeakable Sadness: The Dispossession of the Nebraska Indians*, Lincoln: University of Nebraska Press.

Wolch, J. (1996) 'Zopolis', *Capitalism Nature Socialism* 7: 2–48.

Wolch, J. and Emel, J. (1995) 'Bringing the animals back in', *Environment and Planning D: Society and Space* 13: 632–636.

Wolch, J. and Emel, J. (eds) (1998) *Animal Geographies: Place, Politics and Identity in the Nature–Culture Borderlands*, New York: Verso.

8   What's a river without fish?

Symbol, space and ecosystem in the
waterways of Japan

Paul Waley

Putting the fish back into rivers

Somewhere out in the eastern reaches of Tokyo, a stream of pure water flows
gently past bamboo groves, pebble beaches, mountain ravines and waterfalls,
artfully placed to suggest the four seasons and to symbolise some of the most
celebrated features of the Japanese landscape, a modern urban equivalent of a
seventeenth-century lord's garden. Children splash in the pond at the foot of a
waterfall and clamber over the rocks placed on either side of the stream. In
summer, festivals are held with the stream as a focus, and groups volunteer to
clean the bed and banks. This is the Komatsugawa Shinsui Kōen literally,
Komatsugawa affection-for-water park), a narrow stream of water together with
the paths and greenery along its banks, about five kilometres in length but on
average only about thirty metres wide. Like so many other waterways in the
suburbs of Japanese cities, the Komatsugawa was once part of an extensive and
intricate network of irrigation channels.

Urbanisation occurred with extraordinary rapidity on the edges of Japanese
cities in the post-war decades. Infrastructure, however, lagged in a number of
important respects, in particular the sewerage system. For example, it was only in
1994 that the whole of Edogawa Ward, through which the Komatsugawa flows,
was connected to the mains system. Years before this, however, smaller waterways
had become polluted, clogged by effluent disgorged not only from factories but
also from houses, most of which had septic tanks that failed to treat kitchen waste
water. The lack of natural slope in so many of the low-lying areas of post-war
urbanisation exacerbated the problem, and many waterways were culverted out
of the landscape. In the case of the Komatsugawa, however, the decision was taken
in 1970 (through the personal initiative of the mayor) to 'bury' the water course
and to create a new landscaped stream above.[1] Because the land here lies below
sea level, the flow of water is controlled by gates and pumps, but in addition there
are a number of treatment plants, treating the water with sterilising agents and
eliminating animal life from the stream.

For some years, the rich textures of the stream's landscaping and the community activities for which it formed a focus made the Komatsugawa a considerable success, with visitors appearing from all parts of Japan and from further afield too. But ideas change, the *zeitgeist* moves on, and criticism mounts. 'What's a river without fish?' was the accusatory question levelled at local government officials. Komatsugawa and similar projects were condemned. 'People have begun to realise', wrote the influential landscape designer Yamamichi Shōzō[2] in 1992, 'that *shinsui* [affection-for-water] and similar projects interfere with rivers and cause considerable dislocation' (Yamamichi 1992: 28). In response to this criticism, the ward office decided to convert another redundant irrigation channel into a new fish-friendly waterway. The result is Ichinoe, a 3.4-kilometre-long stream completed in 1996 at a cost of over ¥300 million (£1.5 million at 1995 exchange rates). Fish are periodically released into the waters, and, although pumps are needed to keep the water flowing, there are no installations for water purification. The stream is cleaned occasionally, when the water is considered too dirty. Signs explain that fish do the cleaning (see Figure 8.1). The landscape remains unnaturally neat and bureaucratically precise, a thoroughly artificial creation of a supposedly natural space (see Figure 8.2). While some residents had been unhappy earlier with the river without fish, others now, according to ward officials, are upset that the water in their stream is too dirty for children to play in. They accuse the ward office of being parsimonious, unwilling to find the extra funds to install a water purifier.

In the eastern suburbs of Tokyo, the local government in Edogawa Ward has responded to criticism of the sterile aquatic environment of the Komatsugawa by introducing fish into a nearby water course, but in doing so it has retained a landscape idiom reliant on a traditional symbolic approach. By contrast, in the western suburbs of Tokyo, re-landscaping work on a number of waterways has been undertaken according to an ideology of concern for ecosystems. In this chapter, these two projects involving waterways and fish within the predominantly urban context of industrial Japan are placed alongside two campaigns to reintroduce (or, more properly, to introduce) animals into an urban environment. The aims of the four projects and the approaches adopted vary widely, from the recreational and aesthetic to the educational, all of them heavily mediated by human agency. Each of these projects involves a greater or lesser degree of manipulation or extension of a 'traditional approach' to animals and river landscapes. This traditional element expresses itself in the use of symbol; not symbol as pictorial device for idea or concept but symbol as synecdoche, the one standing for the totality. Even ideas cast largely within an ecocentric framework are expressed through an essentially anthropocentric use of symbol. In the process new values are attached to traditional symbolic associations, and selected animals are imbued with a new symbolic meaning, that of nature resurgent.

Figure 8.1 Along the Ichinoe waterway in east Tokyo, a sign reassures passers-by that fish clean the river. There is no need to be concerned about the apparently murky state of the water.

Source: Author's photograph

These projects occur within a broader ideological context of long-term state-led capitalist development, based on a consensus, built through adroit social orchestration and ideological manipulation, around the twin notions of a harmonious and homogeneous society. The post-war Japanese economic trajectory –

Figure 8.2   Rocks are strewn across the Ichinoe waterway in east Tokyo with artificial casualness.
Ecologists have criticised this redesigned waterway, and drawn attention to failures to
provide natural habitats.

Source: Author's photograph

whether in its earlier high-growth or later low-growth phase – has several salient
features which have a particular bearing on the state of the Japanese environment.
These include the state's success in co-opting and then diluting opposition to
industrial development; the size and influence of the construction industry; the
reliance on public works to act as a motor for economic growth; and the
continued faith placed by Japan's elites in the redemptive power of technology. In
other words, the state has attempted to ensure political stability through a
rigorous application of a technocratic programme of construction-led growth
(Broadbent 1998; Hein 1993). Even in recent years of capitalist crisis in Japan,
and despite the recent strengthening of the position of forces campaigning for
greater 'ecological justice', the ideological linkage of economic growth with
construction in concrete and the political alliance between politicians and agents
of the construction industry both remain solid. All this has been highly damaging
to human health and the natural environment. It has also resulted in the casting in
concrete of mountain slopes and long stretches of Japan's heavily indented coast-
line. A similar fate has befallen the beds and banks of many waterways as they flow
through Japan's densely populated urban areas, most of which are located on flat
alluvial flood-plains on or near land reclaimed from the sea. Rivers and related
issues of water management have become the principal focus of debate between a
powerful orthodoxy steeped in the traditions and practices of conventional technocratic engineering and a newly emergent school that challenges the old

imperatives of economic growth and insists instead on an ecological equity, a harmonisation of human activity with local ecosystems.

The use of river water for irrigation has long been an essential part of Japanese wet-rice farming, and agriculture remains the main consumer of water, with about two-thirds of the total. Beginning in the early 1900s, however, rapid urbanisation and industrialisation produced intense added demands on water resources. The increased incidence of flooding that resulted, with its greater toll in terms of human casualties and damage to property, was followed by a programme of state-led engineering works within a new framework of legislation that laid out the state's responsibilities (Takeuchi Keiichi 1996: 221). Gradually, an infrastructure was created that aimed at enabling a maximum utilisation of water resources and then a rapid channelling of water into the sea. To this end, multi-purpose dams were built; artificial outlets were constructed for turbulent rivers; and waterways were concealed behind high concrete embankments, cut off from human sight in deep beds. The whole vast programme (conceived though it was in piecemeal fashion) was possible because of a complicated but comprehensive system of state control over the management of rivers. Many of the smaller streams and irrigation channels that criss-crossed the terrain lost their function in the decades of rapid urbanisation after the war. Like Komatsugawa, they became outlets for household and industrial effluent. But as more and more areas were linked to sewage treatment plants, local governments were faced with the question of what to do with the redundant waterways. Malodorous and unsanitary, many were covered up. The pollution of rivers led to an inevitable decline in numbers and diversity of fish and invertebrates and before long to the disappearance of species.

This loss was noted and mourned, but the quality of water in many rivers began to improve again in the 1970s with the introduction of controls on effluent and better treatment facilities. The reappearance of fish in urban rivers was picked up by the media and transformed into a parable of natural rebirth. A broader process of reappraisal was beginning at about this time. A spate of serious flooding in the mid-1970s called into question the premises of the engineering-led approach to river planning (Ōkuma 1988: 232). The seemingly ubiquitous corruption of the 1980s and early 1990s also played its part in bringing a new popular mood to the fore, one in which the earlier post-war sacred cows of economic growth and development were questioned. At the same time, responses to a changing international environment as well as the strength and shape of reaction to the 1995 earthquake in Kobe led to the formation of a large number of NGOs (non-governmental organisations), bringing together people from official and lay worlds to campaign for improvements in the environment. Finally, recent years have witnessed growing anger at engineering projects considered to be unnecessary and harmful to the environment. Most of the high-profile campaigns against the construction of dams and weirs have been unsuccessful, but they have

served to highlight various issues, to radicalise public opinion, and even to reinforce the position of the Environment Agency against other central ministries
(Hesse 1999; Kitagawa and Amano 1994; Morishita Kaoru 1998, 1999).

Both at local and at national level, as well as among planners, academics and
community leaders, there has been a move away from the harsh surfaces of the
technocratic approach to river landscaping and a search for a new idiom. At the
central level, the Ministry of Construction, under whose aegis fall most issues
involving rivers, initiated a profound shift in its policy. It would direct its energy
no longer to the elimination of flooding but rather to the minimisation of damage
from flooding. Principal among the measures called for were the securing of
water retention areas and retarding basins (Kiya *et al.* 1992: 148; Ōkuma 1988:
236). A whole new concept of 'comprehensive river planning' was introduced,
and some level of citizen participation was mandated in the 1997 revision of the
New River Law. Planners have been influenced by developments elsewhere in the
world. New methods of river planning pioneered in Germany and Switzerland,
known as *Naturnaher Wasserbau*, have been studied and, increasingly, implemented
(Larsen 1996). The national five-year river plan stipulates the use of materials
other than concrete in new riparian work. In approximate imitation of its
German counterpart, the whole movement is known as *tashizen-gata kawa-zukuri*
(literally, multi-nature-style river planning).

Two organisations affiliated to the Ministry of Construction have been set up
to propagate best practice through publications and to fund a limited number of
experiments. As a result, several guides have been published in recent years. The
underlying theme throughout these works is the creation of a landscape that helps
animal life and local ecosystems. Before and after photographs show best use of
ox-bow ponds, meanders, inlets and other sinuous and irregular shapes for rivers.
They explain how to build irregularities into the design of river beds to create
eddies and shallows. They demonstrate techniques of restoring natural slopes to
banks, where possible using traditional methods such as *jakago*, or gabions, rock-
filled bamboo baskets. Ecologists and biologists have been active in these projects,
giving advice on a host of issues such as the best types of banks and beds for fish
and invertebrates and the best place for fish ladders. A rich literature exists, much
of it written in an unashamedly pedagogical style, describing the preferred habitats of fish and of insects and water invertebrates and their larvae. The discourse
within which these activities are placed is an essentialising one. 'Multi-nature-
style river planning', according to a recent official publication, 'protects Japanese
culture' (Ribaafuronto 1996: 14).

New ecologies and new orientalisms

Not only in Japan but in other rich countries too, river restoration is now an
important part of planning objectives and a central element in a much broader

discourse involving human responsibilities to nature, human interactions with nature, and their impact on life both human and non-human (Adams 1997; Newson 1997). That rivers should be the centre-piece for a new discourse of the environment is not in itself surprising. Water is the cardinal barometer, the essential indicator of the state of nature, being particularly sensitive to the effects of development and construction. Rivers are powerful symbols, cutting through and across rural–urban divides and bringing people and places together. New practices in Japan are at least to some extent a reflection and an interpretation of new ideas about ecology; ideas that are critical of rationalist science-based thinking, and that some have sought to link to traditions of Eastern thought.

Until recently, an hierarchical and hypothetico-deductive explanatory framework has been dominant, deriving from an essentially Platonic and Cartesian worldview that defined a rational natural world distinct from the ineffable world of the soul. This science-based approach is now subject to a widespread challenge posed by new ways of thought in physics and ecology (Adams 1997; Murphy 1994; Pahl-Wostl 1995). Callicott and Ames (1989: 5) characterise the divergent thought systems as 'the atomistic–mechanistic image of nature inherited from the Greeks, institutionalised in modern classical science, and expressed in modern technology, on the one hand, and the holistic–organic reality disclosed by contemporary ecology and quantum physics … on the other'. Murphy (1994: 23) writes of a 'move from a technological to an ecological social paradigm'. The old order is seen as reductionist and excessively reliant on science-based modalities, as managerial and technocratic, seeking to control and to direct according to hierarchies of expertise, employing the techniques and idioms of large-scale, automotive intervention. Ecological science belongs to this old order, being regarded as 'a technocratic recipe book for directing and controlling change' (Adams 1997: 278). Humanity, as controlling agent, restores nature by re-creating ecosystems as they existed before its intervention. Environmental crises, both actual and potential, as well as new theories concerning the composition of matter have cast a different sort of ecology as flag-bearer of the new order (Pahl-Wostl 1995). This new ecology identifies ecosystems as dynamic and non-linear; it conceives human actions as being intrinsically bound to the natural environment. It sees energy as 'a more fundamental and primitive reality than material objects of discrete entities' (Callicott 1989: 59). At its heart lie ideas about the chaotic structure of the universe.

This affirmation of a non-linear and non-dualistic approach, these new ways of thinking about ecology, bring with them a number of implications, in terms both of human relations with other humans and of human relations with non-human animals. Some of the consequences are eminently political, and have alimented the recent proliferation of writing on political ecology (Blaikie and Brookfield 1987; Escobar 1995; Peet and Watts 1996). This process occurs both in metaphorical and in literal terms. Societies and cultures are no longer so readily seen as discrete, auto-regenerative entities. Social hierarchies are interrogated; systems of knowledge are deconstructed. The integrative power of the global capitalist system is taken to have its referents in the workings of ecosystems. The direct, literal relevance of the new ecology to the political structuring of human society is manifested, for example, in movements attempting to create wider recognition for the benefits of traditional subsistence agricultural techniques. In the words of Parajuli (1998: 187), 'the language of economism is one-dimensional and does not adhere to the language of nature. On the contrary, peasants who practise so-called traditional agriculture try to approximate the multi-dimensionality of nature.' Cartesian dualities (to use a convenient short-hand) are increasingly seen as instrumental in digging a conceptual trench between the realms of rational humans and purely sensate animals (Anderson 1995: 277). New ideas of ecology have led to a readjustment in conceptualisations of landscapes and animal habitats. While the deleterious impact on local fauna of habitat fragmentation is recognised (Wolch *et al.* 1995: 738), a more sophisticated and nuanced reading of the relationship between animal habitats and human moulding and re-moulding of landscapes has been forged. According to this reading, for example, 'suburbs have become ecosystems of their own, though humans are probably unaware of their part in attracting the wildlife with whom they coexist' (Anderson 1995: 275).

Of apparently natural environments, few have been so thoroughly engineered and landscaped as rivers and their banks. Rivers have long been considered the most natural of objects for engineering work. The resource that they represent, for so many purposes so varied and yet essential, as well as the apparent threat posed when they spill their banks, have legitimated a regime of technocratic management by state bureaucracies and parastatal agencies based on scientific principles of calculated containment and hazard aversion. Yet rivers are in a constant state of flux; the forces which shape their flow – precipitation and sedimentation among them – are by their nature unpredictable. For this very reason, rivers can be seen as a key testing ground for new ideas of ecology (Adams 1997: 286). As in Japan, in Germany, Britain and elsewhere, a series of new ideas and techniques have been developed that reject the old managerial, technocratic and engineering prescriptions and instead reflect new thinking about ecosystems and about the best way to counteract damage from flooding. A number of books have been published incorporating this new thinking and setting out new and more ecology-friendly techniques (e.g. Brookes and Shields 1996; Harper and Ferguson 1995; Petts and Calow 1996). Among the new techniques and practices advocated are re-meandering of water courses, landscaping of sloping banks, planting shelves in beds, and the re-creation of flood plains (Patrick 1998: 24). The rehabilitation of ecosystems and natural features is a primary aim of much of this work.

Critical comment on the new culture of ecological concern exists at various levels. In its concentration on restoring specifically local fauna and flora, it is seen by some as encouraging a sort of eco-nationalism. It has been argued, for example, that Japanese ecologism has at times been distorted into an eco-nationalism with a new quasi-official discourse of ecology glorifying a pure past age predating Western influences (Morris-Suzuki 1991). Others see the spaces of ecological action as belonging to a global discourse, no longer pertaining to the particular traditions of local space (Pedersen 1995: 265). A different range of potential contradictions exists in the assumptions commonly made about the unity of thought and practice, intention and action – the gap between ideology, institutional systems, social norms and daily practice (Bruun and Kalland 1995: 16). The shift to an ecological social paradigm, as signalled by Murphy, may be occurring in the realm of ideology, and this may have filtered its way through into social norms and daily practice among some groups, but institutional structures are much slower to adapt, and resistance is deep-rooted and widespread. Furthermore, this representation of the forces involved fails to take into account the role of the market in conditioning the form and nature of human action. Another way of conceptualising these difficulties is through potential contradictions between higher-level knowledge and local-level custom and practice (Bruun and Kalland 1995: 6). This is traditionally illustrated in terms of a modernist ethos posing as higher-level knowledge and swamping particularistic, indigenous techniques. But in the contemporary world in which environmental concern has become the ideology *par excellence* of the rich and powerful, it is easy to see why people in areas of low income should be eager to adopt modernist technological fixes that are cheap and available (Parajuli 1988: 188; Peet and Watts 1996: 10).

Debates about the human relationship with the environment, about science and rationality, about the dualism that lies at the heart of most Western thought, have a tendency to incorporate, implicitly or explicitly, an oppositional referent, and this in many cases is an 'Eastern other'. The classic debate, initiated by White in the 1960s, posited a Western tradition of thought enmeshed in the deep-rooted dualisms of Greek philosophy and Judaist theology as the cause of environmental crisis. Since then, various writers have described an Eastern sphere that is more attuned to an ecological outlook, creating thereby what Bruun and Kalland (1995: 3) have called the 'counter-myth of a prevailing Asian ecology-mindedness'. These arguments draw, for example, on Buddhist ideas such as that of the Net of Indra and the interconnectedness of all living things (Cook 1989). 'Since individual organisms, from an ecological point of view, are less discrete objects than modes of a continuous, albeit differentiated whole, the distinction between self and other is blurred' (Callicott 1989: 61). For a number of reasons – not least of them, perhaps, the frenetic nature of development in much of East Asia in the 1980s and much of the 1990s – the notion of a more benign philosophical outlook in Asia has recently been subjected to greater critical scrutiny (Bruun and Kalland 1995). Nevertheless, nature remains a key element in orientalist discourse (Asquith and Kalland 1997).

Japan has been cast as an enigma in broad terms of the environment. On the one hand, the Japanese have been widely seen as a nature-loving people, a characterisation that they themselves have felt at home with and (not surprisingly) have been keen to perpetuate (Bruun and Kalland 1995: 20). On the other hand, they have been portrayed (again at times by the Japanese themselves) as desecrators and despoliators of their own beautiful landscape, and subsequently as exporters of pollution. This dichotomy has often been treated along chronological lines, with a juxtaposition between a purer past and a modern period sullied by the intrusive panoply of managerial technology and entrepreneurial industrialism. According to this explanatory structure, the Japanese possessed a rich diversity of both life-generating gods appearing in mythical narratives and spirits which gave energy and power to the natural features of the landscape (Eisenstadt 1995: 190). To this was added the fertile legacy of Buddhism, with its idea – central in Japanese traditions – of the immanence of the transcendental in the mundane world, a vision of Buddhahood within the objects of everyday surroundings. Confucianism was a later but still distinctive influence on Japanese social consciousness, with its emphasis on personal improvement and the amelioration of the environment through respectful exploitation and careful stewardship. As a result of Western intervention after the fall of the shogunal regime in 1868, however, it has been argued that a new, more damaging, less respectful approach to development was introduced in Japan. While there is a certain inevitability to this interpretational construct, it presents several dangers, of which the most obvious is a tendency to ascribe all manifestations of environmental degradation to the modern wave of Western influence, neglecting thereby the radical transformations of the environment effected at various stages of Japan's history (Totman 1989).

The contours of the Japanese relationship with nature and the environment bear a number of features that distinguish it from Western traditions of thought. First among these is a lack of abstract ideals. Without abstract ideals, the Japanese ontological world is free of the opposed dualities of moral absolutes. This is not to say that it lacks all dualities. Japanese life is infused with a basic hemispheric duality of purity and impurity, expressed in terms of harmony and chaos, of ritual cleanliness and defilement, of beauty and deformity, to which can be added other and cognate dualities, of inside and outside, of inner feelings and outer posture. This dual conceptual landscape is reflected in the natural world (Kalland 1995: 95). On the one hand, we find a tamed and idealised nature, seen in the carefully cosseted symbolic landscapes of gardens and the minutely crafted and regulated landscape of the paddy fields of narrow valley floors and the densely populated flood plains. On the other hand, there is a wild and original nature, found not only in the forested mountain slopes whose crenellated contours line the valleys and ring the plains but also in symbolised miniature in the thick growth of trees around country shrines to the gods. But these are opposite ends of a spectrum within which we live, not dichotomous dualities between abstract ideal and physical substance.

A second feature of the Japanese (and indeed the broader Asian) ontological landscape is the overriding importance of context, what Eisenstadt (1995: 190) refers to as 'an emphasis on relational and indexical criteria (as against abstract, linear ones rooted in principles transcending existing reality)'. 'This contextualism', write Bruun and Kalland (1995: 11),

> frequently applies to the concept of morality. Nature and morality are closely linked in many Asian cultures, man [*sic*] and environment forming a moral unity. Yet, as there is no absolute good or evil, there is no absolute morality, at least not for commoners. Thus, people's moral obligations towards nature are contextual.

Third, the lack of abstract ideals, of essences abstracted from the phenomena of everyday life, encourages the use of analogy and models (Callicott and Ames 1989: 15). This fondness for analogy is extended and embroidered into a use of symbol, of metaphor, and in particular of synecdoche, as is suggested in several of the cases studies examined here. On a relatively undifferentiated ontological plain, one manifestation of nature, one species of animal, indeed one animal, can be taken without prejudice as representative of the infinitely varied but otherwise undifferentiated totality.

Last, and following on directly, we find a relative lack of prioritisation of the human situation in the natural world. The absence both of abstract ideals and of a vision of a transcendental plain and the emphasis on the immediate context bring an instinctive directness to human contact with the environment (Berque 1982). The human situation is not privileged, thus creating the conditions for what today might be defined as an ecocentric approach to the environment. The hemispheric congruence of dualities (as opposed to Western dichotomic, absolute opposites) encourages the identification of mediatory presences, often on liminal terrain. Thus foxes, appearing at the edge of agricultural land or along the skirts of hills, act as messengers for the rice deity Inari and then as actual embodiments of the god. Monkeys have been cast in a similar role as mediators; they are considered not in terms of a simian, animal Other to the rationalising human, but as 'simultaneously similar, yet distinct from, humans' (Ohnuki-Tierney 1987: 22; see also Asquith 1995). The absence of a privileged position for the human, the lack of a hard line drawn between human and animal, encourages the sort of animal anthropomorphism that has been identified as a particularly strong feature of Japanese attitudes towards animals (Kellert 1993).

In the short accounts that follow, these features of the Japanese ontological landscape appear in various guises, each one that much further stretched to accommodate new thought and practice. In the first of these, salmon are removed both from their original symbolic home and from their normal habitat to be released into Tokyo's main river and treated there as symbols of nature recrudescent. The aim of the exercise is cast not in the terminology and conventions of new ecological thought, however, but rather in a more explicitly anthropocentric idiom. Fireflies play a much more central place in the history of a traditional relationship with animals, but, as is shown in the second of these stories, along with changes in the context of human life in Japan, this role is being transformed to reflect a rapid shift in human priorities and style of living. New ecological thinking involves a reworking of the human relationship with other animals. As the third of the following case studies shows, the setting aside of areas as 'natural' animal habitats represents a clearly articulated attempt to introduce and put into practice ideas drawn from the lexicon of new ecological ideas and practices. Starting with the Komatsugawa stream, each project represents a more substantial re-definition of traditional practice than the last.

Salmon news

Fish have always presented a rich source of symbolism – Christianity is suffused with ichthyological symbols. In Japan, the cleaning of the environment in the wake of rampant industrialisation was marked in popular and scientific discourse by the return of fish to water habitats which they had once occupied, and in particular of specific species of fish whose presence indicated cleaner water. Articles in national newspapers reported sightings of this or that fish in stretches of rivers where they had not been seen for years, and, as part of this mood of rebirth, attempts were made to introduce salmon to the rivers of the Tokyo conurbation.

In his own published account of his promotion of the release of salmon into Tokyo's rivers, Baba Rensei, a journalist with the *Yomiuri* newspaper, describes how his initial inspiration came not from traditional Japanese practice but from the release of salmon into the Thames in 1976 by the local Water Board 'as a symbol of the cleansing of the [river]' (Baba 1985: 18). Seven years later, one of the salmon released there returned to spawn, and, thought Baba (1985: 20), 'Would it not be possible to launch in Tokyo the same campaign as on the Thames – to reintroduce salmon as symbols of the cleansing of the Tama river?' In his book, *Salmon Return to the Tama River: Nature Education Expands*, Baba details, in the terms of a quasi-spiritual quest, his campaign to win support and interest in the release of salmon in the hope that they return to spawn. In terms of numbers of returning fish, the campaign has been an extremely successful one, with the first certified return in 1984 and over one hundred subsequent catches of returning salmon (*Sake shinbun: salmon news* 1996:

13). Artificially incubated salmon had first been released into the Tama river in 1877, but the modern trend was initiated in 1978 in Hokkaido (Baba 1985: 172). The rivers of Hokkaido, like those of the north of the main Japanese island of Honshu, have always been spawning grounds for salmon. There had, however, been no solid evidence of salmon returning to spawn in rivers as far south as those of Tokyo. This has not dissuaded others in the Tokyo region from following Baba's lead. Yanagisawa Hiromichi, a retired glazier in his eighties, launched a similar operation in Tokyo's main river, the Sumida, in 1993.[3]

Baba's project has had a wide-ranging impact. The mayor of Setagaya Ward (an area with a population of 700,000 people in the affluent southwest of Tokyo) has lent official support, including space in a park for the salmon incubation tanks and facilities for the publication of an annual newsletter, bearing the bilingual title *Sake shinbun: salmon news*. An association was formed, and funds raised from prominent individuals and organisations. Above all, a growing number of schools have become involved. Close on one hundred, most of them junior schools, are listed in the 1996 issue of *Sake shinbun: salmon news* as participating in one form or another. Children draw pictures of salmon during incubation and the hatching process; they give speeches to mark the salmon's release. Seminars, even international conferences, have been held for children, organised around the theme of salmon. Exchange visits and international conferences for children have been arranged with Canada and other countries where salmon are an important component of the natural environment (*Sake shinbun: salmon news* 1995: 12). Baba himself signals his intent in the title to one of the chapters in his book, 'Environmental education with salmon as symbol'.

One of the older symbolic ritual acts involving animals is the Buddhist ceremony known as *hōjōe*, in which on the fifteenth day of the eighth month birds and fish were released in a gesture affirming belief in the sanctity of all animal life (Nihon Fūzoku Gakkai 1979: 594). Later the ritual spread from the grounds of temples, and fish, turtles and other animals could be purchased for immediate release and acquisition of instant karma. There is no suggestion that those involved in releasing salmon fry into Tokyo rivers are conscious of any link to the *hōjōe* ritual, but they are insistent on the symbolic nature of the act. Beliefs and myths connected with salmon can be found in many parts of Japan, especially in the north of the country, where salmon were farmed in the eighteenth and nineteenth centuries and where artificial incubation has been practised since the year 1900. Salmon farming in the north of Japan was part of the rhythm of the seasons, an activity that dovetailed comfortably with work in the rice fields (Baba 1985: 51).

While the annual programme to breed and release salmon from Tokyo's rivers has been considered a significant success, from an ecological point of view it raises a number of awkward questions. Indeed, the release of hundreds of thousands of salmon fry into rivers in which there is no reliable historical evidence of their previous presence has invited criticism from ecologists and others more concerned about biological authenticity and the viability of ecosystems than about salmon as symbols of clean water and nature redeemed (Kishi Yūji, interview, 15 November 1996).

Fireflies, dragonflies and the culture of insects

The release of salmon is a fairly crude way to measure environmental health, and salmon themselves are peripheral within Japanese animal folklore. The same cannot be said, however, of fireflies, whose nocturnal twinkling has been celebrated ever since the beginning of written Japanese records. Over recent years, fireflies have become central to a popular discourse concerning community rebirth and the reintroduction of nature into urban areas, to the point almost where they have become a media-enhanced obsession.[4] In urban areas large and small up and down the country, attempts have been made, many of them less than successful, to introduce or reintroduce fireflies and dragonflies to streams and ponds. These activities have been linked to a broad programme of community reintegration and urban regeneration.

In Japan's second most populous city, Yokohama, dragonflies and fireflies have been selected to represent a cleansed and purified nature. They have been selected, according to Mori Seiwa, the originator of the campaign, because they make abstract concepts more directly understandable and because they give both local government and local residents a target in their work to restore wildlife habitats (Mori 1995: 26). The significance of the choice of fireflies and dragonflies lies in their need for clean water. Uniquely among firefly species, the two types most commonly found in Japan, *Luciola cruciata* and *Luciola lateralis*, have larvae that live in water, where they feed on snails, while dragonfly nymphs are completely aquatic (Mori 1988: 80). Not all projects to introduce fireflies have been successful, and, as in Yokohama, dragonflies have sometimes been used in their place. The presence of dragonflies is regarded by naturalists as a good indicator of the conditions and diversity of animal and plant life. In Yokohama, the designation of dragonfly ponds and similar natural habitats began in 1981 and was officially incorporated into municipal policy in 1986. The city now has an extensive network of over thirty ponds and other small nature places, nearly all of which have water as a central feature, using dragonflies and fireflies as symbols of nature resurgent. These include ponds in parks and school yards, rivers, water treatment plants and set-aside fields. Two associations have been formed, principally to interest children, around the city's dragonfly ponds (*tonbo ike*), and annual national dragonfly summits are now held.

The Yokohama project represents an attempt to break down barriers of space and of time: of space, by (re)fashioning natural corridors and nature spaces within the city where all sorts of insects, as well as frogs and other small animals, can live; and of time, by re-creating the experiences of animal life enjoyed by an older generation of people now in their fifties and over, who grew up before or during the early stages of rampant post-war urbanisation. More generally, the spaces being evoked are those agricultural landscapes of wet-rice farming that the urban dweller has left behind, flooded fields and streams and irrigation ditches where fireflies and dragonflies once lived in abundance. The rice fields are but one prominent example of a whole series of features associated with memories of childhood and with an ideal, stereotypical rural landscape. In recent decades, partly because of the growing proportion of the population born in big cities and partly because of the steady transformation of rural landscapes and the decline of rice farming, this landscape has assumed an aesthetic and moral significance, one that has been promoted by the state, as an embodiment of traditional Japanese values (Robertson 1998: 117). City-dwellers have in growing numbers been drawn to these ordinary landscapes of rural areas, and some have become involved in a variety of forms of agro-tourism (Knight 1998). At the same time, a reappraisal is underway of the ecological value of agricultural landscapes, and especially wet-rice farming. Rice, fish and silkworms form a 'self-sustaining ecosystem' that has been exploited for centuries in east China and south Japan (Bray 1986: 132). Waterways can be seen as serving the function of bringing the ecosystems of rural areas, especially those based on wet-rice farming, into cities; they were, and perhaps will become again, ecological corridors.

Over the centuries, insects and other forms of small animal life have been celebrated in Japan. Fireflies appear in some of the earliest collections of Japanese poetry, including the eighth-century *Manyōshū*, as well as in the *Tale of Genji*. They were poetic symbols for the heart burning with longing and desire. Firefly hunting (*hotaru-gari*), an aristocratic pursuit in the Japan of Genji's day, had become a popular pursuit by the time of Tokugawa shogunal rule in the seventeenth, eighteenth and nineteenth centuries (Nihon Fūzoku Gakkai 1979: 599). Around the imperial capital of Kyoto and the shogunal capital of Edo were a number of places famous for the bright lights of their fireflies. The firefly's historical associations are reinforced in the names given to two of the most common species found in Japan, the Genji and Heike firefly (*Genji-botaru* [*Luciola cruciata*] and *Heike-botaru* [*Luciola lateralis*]), named after two feuding families whose battles were celebrated in romances and whose rise to power signalled the decline of the court nobility. Fireflies, however, have not been the only insects to be appreciated in this way. In late summer and early autumn insect-listening parties (*mushi-kiki kai*) were held in the hills around Kyoto and Edo (pre-modern Tokyo), ostensibly to allow the town folk to enjoy the chirruping of the cicadas. In the last decades of the nineteenth and early part of the twentieth century, the capture of fireflies for urban households or for rural tourist centres became more and more commercialised, giving rise to calls for their protection (Laurent 1998). Until quite recently it was common for children to climb trees in search of cicadas or crawl after stag beetles, netting the insects in their hands and placing them in bamboo cages. A whole culture of affection and enjoyment was built around insects.

The Yokohama project, therefore, is built on ancient literary associations but trades on more recent memories. Its leaders are part of a wider movement that manipulates the symbolic value of fireflies and dragonflies in order to advance an agenda related principally to improving urban environments. At the same time, however, they are conscious, as are some other campaigners, of the dangers that excessive attention to fireflies and dragonflies can engender and of the need for a sensitive approach to the creation of proper habitats. In particular, they argue – against the grain of the urban environmental campaigns – that fireflies prefer water which is not too clean (Mori 1988: 84; Morishita Yūko 1995: 147). In doing so, they attempt to apply an ecological harness to a popularised, media-supported campaign to use fireflies as symbols not only of a reflowering of nature but also of a rebirth of community.

Biotopes in the Tokyo suburbs

Of the various schemes to inject a concern for local ecosystems into planning, few are as radical as that implemented in Hino, a town in the far western suburbs of Tokyo, on low-lying ground near the point where the Asagawa flows into the Tama river. The area underwent a period of pell-mell urbanisation that started around 1955, when Hino's population stood at 27,305. By 1995, it had reached 166,429. The predominant activity had changed from agriculture to manufacturing, notably metals and the auto industry (Hino trucks). Hino's rice fields were being rapidly eaten into, and (as has happened in so many other recently urbanised territories in Japan) its extensive system of largely artificial waterways was no longer needed for the purpose for which it had been constructed, irrigation. By 1995, some 180 kilometres of waterway remained. The failure of the sewerage system to keep pace with new construction of housing meant that in Hino, as in Edogawa Ward in the east of Tokyo, the waterways were becoming polluted, in particular by kitchen waste water. This and related problems prompted the local government in Hino to inaugurate administrative measures to facilitate rehabilitation of its waterways. As part of this effort, a number of projects have been initiated whose aim is to create or re-landscape water courses in an ecology-friendly way. As in Yokohama, if not even more so, the projects are very small in scale, but attempts are made to overcome this disadvantage by linking them together into corridors along streams and ditches.

In the pioneering project, a 400-metre stretch of stream known as the Mukōjima waterway has been re-landscaped (see Figure 8.3).[5] Its concrete banks were replaced with packed (but not cemented) stone using traditional techniques; and stakes, mats and baskets were laid along the water edge. The re-landscaping work cost over ¥250 million (about £1 million at the exchange rate of the time, 1993 to 1994). A path runs down either side of the stream. At one end, the waterway opens out into a specially created pond whose banks fall within the precincts of the local primary school, allowing children direct access to the water. At the other end, a mill has been placed, reconstructed in the local style. To this carefully crafted environment a number of species of fish and bird have returned, including both kingfishers and the *aburahaya*, a sub-species of dace (*Leociscus macropus*) particularly sensitive to changes in water quality. Much has been learnt in the short time since the completion of this project, and local techniques have replaced more expensive imported methods. The project is largely the work of one man, Sasaki Nobukichi, a radical advocate of ecology-friendly planning, who has sought to carry forward and realise his ideas despite opposition from colleagues in the municipal government and the initial hostility of local residents.

Other projects in Hino show a similar concern for simplicity and lack of artifice. Several ox-bow formations have been constructed, with carefully designed eddies and pools forming habitats favoured by fish. Many of the smallest of the irrigation channels are being landscaped and protected to ensure that they are not

Figure 8.3   The Mukōjima waterway in the western outskirts of Tokyo opens out into a pond whose banks and bed have been designed to provide homes and breeding grounds for all sorts of aquatic life.

Source: Author's photograph

cast into concrete. At the centre of these various projects is the creation of what the Japanese call 'biotopes'. The term has become frequently used, not only in specialised literature but also in publications for a general readership. It is defined in a Hino local government publication as 'a place where animal life can be sustained'. A term with close affinities to 'habitat', it emphasises not the animals that make a place their home but the place in which animals can find a home. The concept of biotope originated in Germany, where it was initially associated with nature restoration projects in rural areas (Takeuchi Kazuhiko 1994: 54). In Japan, biotopes are often evoked in the linked discourses of ecology, nature, environment and planning of waterways, especially those that once conveyed water to rice fields. The Mukōjima re-landscaping project is described in terms of biotope creation, as are the dragonfly ponds in Yokohama.

In Hino, as in Yokohama, a human concern for ecosystems that have evolved as a response to human actions is expressed in terms that carry a direct human appeal. Frequent references are heard to a traditional Hino landscape, featuring rice fields and irrigation waterways, and frequent appeals are made for its restoration through the destruction of the complicated irrigation system cast in concrete and oiled by pumps. Education and contextualisation within a discourse of traditional landscape dominate the exterior face of the Hino projects, but the landscaping work itself involves a scrupulous regard for techniques of habitat creation that are favourable to fish, invertebrates, birds – all types of animal life. Within the constraints of an urban (or, perhaps more strictly, a suburban) Japanese setting, it is hard to imagine a more painstaking application of the vocabulary of ecological planning.

The moral within the aesthetic within the ecological

Waterways and their ecosystems have dug their way into official discourse on the environment. Ministry of Construction documents routinely begin with a reference to the need to improve riverine habitats. Within the ministry, a determined campaign among a smallish group of officials has brought about some notable successes, in terms of changing perceptions, setting a new agenda for planning methods, and altering some of the legal parameters (Seki 1994). Improving fish habitats is a central preoccupation of the attempts emanating from the ministry's River Bureau to introduce a new approach to the landscaping of waterways. A vocabulary of ecological landscaping has been elaborated and refined in order to introduce and to adapt to the Japanese environmental context a range of techniques and features designed to enhance piscine habitats. Ideas and practices have been absorbed from abroad, especially Germany and Switzerland, and also to a lesser extent Britain.

Their introduction has been far from universally welcome in Japan. To some they appear to privilege animals at the expense of humans. To others they run counter to older, more 'traditional' understandings of the human relationship with animals. In Hino, residents objected strenuously to the Mukōjima project, arguing that it would lead to a proliferation of mosquitoes among the newly planted reeds and bushes along the banks of the waterway. In Edogawa Ward, people living near the Komatsugawa and the Ichinoe waterways objected, at different times, both to a clean and to a dirty waterway. Some farmers in peri-urban areas have been distinctly unsympathetic. There has been disagreement about the significance of animal as symbol. On the one hand, the release of salmon into Tokyo rivers has provoked opposition from professional ecologists and others who see these exercises as misguided, if not actually harmful. On the other, local residents' leaders and officials of local government have on occasion pressed for the adoption of a more overtly aesthetic landscaping idiom in accord with traditional ideas.

These conflicting views have intensified the strength with which experts and officials have attempted to legitimate the ideology of ecology-friendly landscaping and planning. The most ingrained opposition to their efforts has come from both the tenacious hold of the engineering ethos on large swathes of central bureaucracy and the natural alliance that is forged with the construction industry, both general contractors and the horde of small companies who rely on contracts for their survival. A commitment to economic growth through public construction projects has lain at the heart of government policy for much of the post-war period (Woodall 1996). The alliances and habits that it generates, and the vested interests to which it gives rise, lie deeply embedded within Japanese social structures. The bureaucratic—engineering nexus in Japan has conditioned the nature of attempts to introduce a new eco-friendly style of river planning. It has restricted the scale and extent of projects, leading to an 'eco-corner' approach, where ecosystem creation or rehabilitation is seen as one among a series of competing claims on space. By definition, however, this runs counter to the holistic spirit of ecological concern. There is therefore a continual need to compromise, which again sits uncomfortably with some of the premises of ecosystem rehabilitation. Compromises might arise out of an accommodation with those who call for a more ornamental approach to landscaping or from engineering problems caused by ground below sea level, as is the case in the east of Tokyo. Compromises also originate from a perceived need to educate and the consequent casting of the message in a pedagogical idiom ('Our friends on the water's edge', reads the headline of one of Hino's nature education pamphlets).

Education is central in the strong moral drive among those who advocate a departure into a more ecosystem-oriented idiom of river landscaping. The 'moral message' of ecosystem concern is delivered at schools. In Hino, the Mukōjima waterway is brought into the school grounds, and school children are brought out to the fields to participate in the various annual rites of rice farming. Salmon are valued explicitly for their symbolic potential in educating children about the environment. As we have seen, a large number of schools have become involved in the release of salmon into the Tama river, and international conferences have been held using salmon as a theme to bring together children to discuss environmental issues. Human agency is fundamental in the propagation of ecosystem concern.

In the propagation of new ideas and techniques of eco-friendly planning, compromises are struck and certain animals are selected as signs of nature resurgent and as symbols of an otherwise undifferentiated ecosystem. This is the case among scientists, planners and others with a professional involvement, for whom certain animal presences and activities are regarded as significant of greater and more complex changes in natural life. It is also the case that, for the purpose of diffusion to a wider public, recourse is made to symbol as a device to explain, to reduce, and to make manageable the otherwise overwhelming variety of animal life in an ecosystem. The human purview of nature and of animals within nature is inevitably cultural in its essence and in its confines, in the symbols it chooses and the compromises it makes.

I have argued here that the creation of biotopes and animal habitats can be interpreted as an expression of a sophisticated aestheticisation linked to notions of a purer, agricultural past. In Yokohama, for example, the language of ecosystem concern is often a retrospective one that seeks to re-create memories of a less urbanised past, of a time when childhood was more carefree and nature more accessible, when paddling in ponds and streams in pursuit of insect life was one of the pleasures of summer. In some contexts, then, the call is to the past and to a rural, agricultural setting, cast within both a vaguely eco-oriented revisionism designed to promote tourism and the simpler, more spiritual values of an ill-defined 'other' place. Dragonflies, fireflies, even salmon are a link to a past and to ways of life, that of the rice farmer in particular, which are now under threat. The habitats that are being re-created for them are the basis of a new ecologically informed aesthetic.

Ecologically informed aesthetic or aesthetically expressed ecology – the idiom is as yet unclear. Nor is the overall strength of the movement to place non-human animals at the top of the list of priorities in re-landscaping waterways. In the end, it seems likely that, without an enormous effort of the imagination, the inevitable compromises will lead to co-optation, and the entrenched arguments and interests of the bureaucratic and construction nexus will appropriate some of the ideas and practices of eco-friendly planning and keep them well restricted and contained to pocket or peripheral locations.

Acknowledgements

The research on which this chapter is based was assisted by grants from both the Japan Foundation Endowment Committee and the School of Geography at the

University of Leeds, grants that led to two visits to Japan, one in the summer of 1995 and the second in the autumn of 1996. Through the help of Professor Sugiura Yoshio, I was given space and facilities at the Tokyo Metropolitan University. I am indebted to him, as well as to a number of other people for their assistance in tutoring me in some of the background complexities, in sharing their ideas, and in making information available to me: Wanami Kazuo, Tokyo Metropolitan Government; Sasaki Nobukichi, Hino City Government; Mori Seiwa, Yokohama City Government, Yokohama Rivers Group and National Association for Local Water Environment Groups; Takahashi Katsuhiko, environmental activist, AMR; Kaneko Hiroshi, environmental activist, Mizu to Midori Kenkyūkai; Ōsawa Kōichi, environmental activist, TR Net; Yamamichi Shōzō, environmental planner; Yasuda Minoru, Riverfront Maintenance Centre, affiliated to the Ministry of Construction; Miyamura Tadashi, Kantō Gakuin University; Kishi Yūji, ecologist, Keiō University.

Notes

1 Information on the Komatsugawa Shinsui Kōen was obtained during a number of visits and conversations, principally in 1995 with Toyoshima Eiichi, of the Edogawa Ward Office (29 August) and Kōzaki Hiroji, president of a local manufacturing company and chair of the 'Love the Komatsugawa Society' (Komatsugawa shinsui kōen o ai suru kai) (5 September). Additional information was obtained from the ward's publicity literature.

2 All names of Japanese people are given in the Japanese order, with family name preceding given name.

3 I interviewed Yanagisawa Hiromichi on 10 December 1996. Yanagisawa showed me his incubation tanks and supplied me with some of the background information on the salmon campaign.

4 The information on which this passage is drawn was obtained from a number of sources: from Mori Seiwa, a planner and nature enthusiast in Yokohama City Government, during an interview on 4 December 1996, as well as from Mori's own publications; from Laurent (1998); and from Morishita Yūko (1995). The interpretations of information presented here are strictly my own.

5 Information on ecological landscaping in Hino was derived from interviews with Sasaki Nobukichi conducted during visits to the area on 8 September 1995 and 29 November 1996.

References

Adams, W. (1997) 'Rationalisation and conservation: ecology and the management of nature in the United Kingdom', *Transactions of the Institute of British Geographers* 22: 277–291.

Anderson, K. (1995) 'Culture and nature at the Adelaide Zoo: at the frontiers of "human" geography', *Transactions of the Institute of British Geographers* 20: 275–294.

Asquith, P. (1995) 'Of monkeys and men: cultural views in Japan and the West', in R. Corbey and B. Theunissen (eds) *Ape, Man, Apeman: Changing Views since 1600*, Leiden: Leiden University.

Asquith, P. and Kalland, A. (1997) 'Japanese perceptions of nature: ideals and illusions', in P. Asquith and A. Kalland (eds) *Japanese Images of Nature: Cultural Perspectives*, London: Curzon.

Baba, R. (1985) *Sake Tamagawa ni kaeru: hirogaru shizen kyōiku* (Salmon return to the Tama river: nature education expands), Tokyo: Nōsan gyoson bunka kyōkai.

Berque, A. (1982) *Vivre l'espace au Japon*, Paris: Presses Universitaires de France.

Blaikie, P. and Brookfield, H. (1987) *Land Degradation and Society*, London: Methuen.

Bray, F. (1986) *The Rice Economies: Technology and Development in Asian Societies*, Oxford: Blackwell.

Broadbent, J. (1998) *Environmental Politics in Japan: Networks of Power and Protest*, Cambridge: Cambridge University Press.

Brookes, A. and Shields, F. (eds) (1996) *River Channel Restoration: Guiding Principles for Sustainable Projects*, Chichester: Wiley.

Bruun, O and Kalland, A. (1995) 'Images of nature: an introduction to the study of man–environment relations in Asia', in O. Bruun and A. Kalland (eds) *Asian Perceptions of Nature: A Critical Approach*, London: Curzon.

Callicott, J.B. (1989) 'The metaphysical implications of ecology', in J.B. Callicott and R. Ames (eds) *Nature in Asian Traditions of Thought: Essays in Environmental Philosophy*, New York: SUNY.

Callicott, J.B. and Ames, R. (1989) 'Introduction: the Asian traditions as a conceptual resource for environmental philosophy', in J.B. Callicott and R. Ames (eds) *Nature in Asian Traditions of Thought: Essays in Environmental Philosophy*, New York: SUNY.

Cook, F. (1989) 'The jewel net of Indra', in J.B. Callicott and R. Ames (eds) *Nature in Asian Traditions of Thought: Essays in Environmental Philosophy*, New York: SUNY.

Eisenstadt, S. (1995) 'The Japanese attitude to nature: a framework of basic ontological conceptions', in O. Bruun and A. Kalland (eds) *Asian Perceptions of Nature: A Critical Approach*, London: Curzon.

Escobar, A. (1995) *Encountering Development*, Princeton NJ: Princeton University Press.

Harper, D. and Ferguson, A. (1995) *The Ecological Basis for River Management*, Chichester: Wiley.

Hein, L. (1993) 'Growth versus success: Japan's economic policy in historical perspective', in A. Gordon (ed.) *Postwar Japan as History*, Berkeley: University of California Press.

Hesse, S. (1999) 'Squeaky environment issues are going to get the grease', *Japan Times*, 25 January.

Kalland, A. (1995) 'Culture in Japanese nature', in O. Bruun and A. Kalland (eds) *Asian Perceptions of Nature: A Critical Approach*, London: Curzon.

Kellert, S. (1993) 'Attitudes, knowledge and behaviour toward wildlife among the industrial superpowers: United States, Japan and Germany', *Journal of Social Issues* 49: 53–69.

Kitagawa, I. and Amano, R. (eds) (1994) *Kyodai na gukō: Nagaragawa kakō seki* (Giant act of folly: the barrage at the mouth of the Nagaragawa), Nagoya: Fūbaisha.

Kiya, F., Nakamura, Y. and Ishikawa T. (1992) *Toshi o meguru mizu no hanashi* (Talking about water as it relates to cities), Tokyo: Inoue Shoin.

Knight, J. (1998) 'Tourist forests in Japan or re-commoditising the Japanese forest', Paper presented at International Convention of Asian Scholars conference, Noordwijkerhout, Netherlands, 25–8 June.

Larsen, P. (1996) 'Restoration of river corridors: the German experience', in G. Petts and P. Calow (eds) *River Restoration: Selected Extracts from the Rivers Handbook*, Oxford: Blackwell.

Laurent, E. (1998) *Les Mushi dans la culture Japonaise: Approche ethnozoologique à partir d'une étude de terrain*, Lille: Éditions universitaires du Septemtrion.

Mori, S. (1988) 'Hotaru bunka to mizube eko-appu' (Firefly culture and the ecological improvement of river banks), in *Mizu to midori no dokuhon* (A reader for water and nature), Tokyo: Kōgai taisaku gijutsu dōyūkai.

Mori, S. (1995) 'Yokohama de no tonbo-ike-zukuri senryaku' (Strategies in Yokohama for making dragonfly ponds), *Konchū to shizen* (Insects and nature) 30(8): 24–29.

Morishita Kaoru (1998) 'Benefits of public-work outlays questioned', *Nikkei Weekly*, 8 June.

Morishita Kaoru (1999) 'One man's love of the sea sparks drive to save wetlands', *Nikkei Weekly*, 18 January.

Morishita Yūko (1995) 'Hotaru ni kiita kawa to seibutsu to no kakawari' (Learning from fireflies about rivers and their relations with animals), in T. Ōuchi, Y. Takahashi and J. Shinmura (eds)

Ryūiki no jidai: mori to kawa no fukken o mezashite (The era of river basins: aiming for a rehabilitation of forests and rivers), Tokyo: Gyōsei.

Morris-Suzuki, T. (1991) 'Concepts of nature and technology in pre-industrial Japan', *East Asian History* 1: 81–95.

Murphy, R. (1994) *Rationality and Nature: A Sociological Inquiry into a Changing Relationship*, Boulder CO: Westview Press.

Newson, M. (1997) *Land, Water and Development: Sustainable Management of River Basin Systems*, London: Routledge.

Nihon Fūzoku Gakkai (Japan study society for customs) (1979) *Nihon fūzokushi jiten* (Historical dictionary of Japanese customs) Tokyo: Kōbundō.

Ohnuki-Tierney, E. (1987) *The Monkey as Mirror: Symbolic Transformations in Japanese History and Ritual*, Princeton NJ: Princeton University Press.

Ōkuma T. (1988) *Kōzui to chisui no kawa shi: suigai no seiatsu kara juyō e* (A river history of flooding and water control: from suppression to absorption of flood damage), Tokyo: Heibonsha.

Pahl-Wostl, C. (1995) *The Dynamic Nature of Ecosystems: Chaos and Order Entwined*, Chichester: Wiley.

Parajuli, P. (1998) 'Beyond capitalized nature: ecological ethnicity as an arena of conflict in the regime of globalization', *Ecumene* 5: 186–217.

Patrick, K. (1998) 'The battle for the Quaggy', *Landscape Design* 27 (4 October): 21–24.

Pedersen, P. (1995) 'Nature, religion and cultural identity: the religious environmentalist paradigm', in O. Bruun and A. Kalland (eds) *Asian Perceptions of Nature: A Critical Approach*, London: Curzon.

Peet, R. and Watts, M. (eds) (1996) *Liberation Ecology*, London: Routledge.

Petts, G. and Calow, P. (eds) (1996) *River Restoration: Selected Extracts from the Rivers Handbook*, Oxford: Blackwell.

Ribaafuronto seibi centaa (Riverfront Maintenance Centre) (ed.) (1996) *Tashizen-gata kawa-zukuri no torikumi to pointo* (Programmes and aspects of multi-nature-style river planning), Tokyo: Sankaidō.

Robertson, J. (1998) 'It takes a village: internationalization and nostalgia in postwar Japan', in S. Vlastos (ed.) *Mirror of Modernity: Invented Traditions of Modern Japan*, Berkeley: University of California Press.

Sake shinbun: salmon news, Tokyo: *Tamagawa sake no kai* (Tamagawa salmon society), annual publication.

Seki, M. (1994) *Daichi no kawa: yomigaere – Nihon no furusato to kawa* (Rivers of the earth: revive – native places and rivers of Japan), Tokyo: Sōshisha.

Takeuchi Kazuhiko (1994) *Kankyō sōzō no shisō* (Thought and environmental creation), Tokyo: Tokyo University Press.

Takeuchi Keiichi (1996) 'Official and popular approaches to water resources in Japan: the failures of an applied geography', in V. Berdoulay and J.A. van Ginkel (eds) *Geography and Professional Practice*, Utrecht: Universiteit Utrecht.

Totman, C. (1989) *The Green Archipelago: Forestry in Preindustrial Japan*, Berkeley: University of California Press.

Totman, C. (1992) 'Preindustrial river conservancy: causes and consequences', *Monumenta Nipponica* 47(1): 59–76.

Wolch, J., West, K. and Gaines, T. (1995) 'Transspecies urban theory', *Environment and Planning D: Society and Space* 13: 735–760.

Woodall, B. (1996) *Japan under Construction: Corruption Politics and Public Works*, Berkeley: University of California Press.

Yamamichi, K. (1992) '*Nogawa wa yomigaetta ka? Ikita kawa wo torimodosu tame ni*' (Is the Nogawa rehabilitated? Bringing life back to rivers), *Japan Landscape* 21: 28.

9   Fantastic Mr Fox?

Representing animals in the hunting debate

Michael Woods

Introduction

The hunting of wild mammals with hounds is one of the most controversial political issues in contemporary Britain. For supporters, hunting forms an essential
part of rural life; to opponents it is a barbaric and cruel sport, violating the rights
of animals. In November 1997 the Wild Mammals (Hunting with Dogs) Bill was
presented to the British Parliament by the Labour MP Michael Foster, and, if it
had become law, would have made hunting with hounds illegal. As the newly
elected Parliament was calculated for the first time to have a large anti-hunting
majority, the bill was considered to have a realistic chance of success. It eventually
fell victim to parliamentary procedure, having failed to secure enough debating
time to allow its completion, but not before provoking a vociferous public debate
about hunting.

The arguments mobilised in this debate ranged across issues of morality, civil
liberties, economics, rural–urban conflict, class, conservation and animal welfare.
Yet any political discussion of hunting is inescapably an issue about the representation of animals in the political arena. As I have discussed elsewhere (Woods 1998a),
while animals are by nature barred from physical participation in the political
process, their representations are frequently evoked in political discourse. This
includes not only issues of animal rights and welfare where representations are
directly made on behalf of animals, but also policy fields including agriculture,
conservation and environmental health where animals are intrinsically represented
in the definition of problems and solutions (see also Philo 1995). Furthermore,
representations of animals may be mobilised in other contexts as symbols of certain
discourses of place – for example, the hunting debate cannot be separated from the
contesting of rurality, with representations of animals featuring prominently in many
social constructs of the rural (Woods 1998a, 1998b). These in turn draw on ideas of
human–nature relations and of nature and rurality (see Bunce 1994; Burgess 1993;
Cloke *et al.* 1996; Eder 1996; Fitzsimmons 1989; Harrison and Burgess 1994; Short
1991; Wilson 1992; Woods 1998a).

However, as implied above, the meaning of 'representation' is multi-faceted. Burgess (1993) comments how 'representing nature' can refer both to 'speaking on behalf of nature' and symbolising nature in cultural artefacts and processes – or, indeed, through scientific knowledge (Latour 1993). Moreover, Latour describes how modernist discourse attempts to order these differing forms of representation – humans are represented through political representation ('speaking for'); non-humans are represented through scientific representation. Science is hence given the responsibility of constructing representations of non-humans which can inform rational decision-making, with any other form of representation of non-humans being dismissed as irrational.

Yet the process of representation itself – of either type – is complex and therefore open to contestation. Most significantly, representation is not just a *re-presentation* – the reproduction of an object in the same form in another arena – but a *translation*, such that an object cannot be represented without taking on a new form. Politicians translate the agency of citizens into the power of authority; scientists translate the objects they study into matrices of scientific knowledge. In both cases the process of representation gives voice to those who represent, and silences those who are represented (Latour 1993).

The significance of translation is three-fold. First it gives objects a mobility which is both physical and metaphorical. Thus a fox, photographically captured in its natural environment, can be transported in the form of a photographic image into other arenas such as newspaper advertisements. Photographs, cinematic film, video tapes, scientific reports, statistics, anecdotal stories, all become 'immutable mobiles' (Latour 1990) that allow political debate to take place by presenting to decision-makers convincing representations of things which they have not directly experienced. Second, as the translation of an object into an immutable mobile necessarily detaches the subject from its representation, so the subject and representation must be treated as different entities. In translation the subject loses control over its representation, such that the representation must be read as the product of the actor(s) responsible for its construction and mobilisation. Third, as the subject is detached from its representation, so the subject assumes the possibility of dissenting against its representation (see Callon 1986).

This chapter discusses the representation of two animals – the fox and the deer – in the debate about hunting provoked by the Foster Bill in 1997. It outlines three conflicting representations – those of sporting foe, pest and victim – and explores how the animals were translated into each of these representations, and how the process of translation and representation creates opportunities for those representations to be challenged. As claims to scientific knowledge feature prominently in these stories, the chapter then briefly considers the problematic of scientific representation, focusing on the political contestation of the Bateson Report into the effects of hunting on deer.

Take 1 – cunning Reynard: the fox as a sporting foe

One of the traditional representations of the fox mobilised in hunting discourse is that of the fox as an equal and cunning contestant. The elevation of hunting from a mere functional activity to the status of a sport relies on the representation of the quarry as a competitor which is given a 'sporting chance' to outwit its pursuers in the chase. As Serpell (1996: 202) describes, 'its status is elevated to that of a worthy opponent in an amusing game of life and death; an opponent who enjoys nothing more than pitting [its] own strength, speed or cunning against that of the well-armed human.' This representation was important in the construction of fox-hunting as 'the sport of gentlemen' in the late nineteenth century – the pretence of equal competition glorifying the skill of the hunter, while the ultimate victory of the hunt reaffirmed the position of human supremacy over nature in the revered 'natural order'.

In mobilising this representation, hunters allude to the folk figure of Reynard, the fox hero of the satirical medieval French epic *Roman de Renart*. The character of Reynard is wise and cunning, and this representation of the fox became entrenched in European folklore, such that by the late nineteenth century lay discourses in rural England understood the fox through this mythological filter, thus providing a base from which the representation of the fox as a sporting foe could be reproduced through the sporting and political literature of the period:

> [F]oxhunting is one of the most characteristic of our national sports, giving the maximum of pleasure and exercise to those who follow the hounds with a minimum of cruelty to the object of their pursuit, who very often, no doubt, has as good a reason to be satisfied with the result of a run as [its] pursuers.
>
> (Gaskell 1906: no pagination)

Similar representations were still being mobilised nearly a century later in the debate surrounding the Foster Bill. The Countryside Alliance, for example, in a letter to MPs argued that '[a] fox in prime condition is faster and smarter than any foxhound. The odds are in favour of the fox and most that are hunted survive' (quoted in *The Times*, 28 November 1997). This sentiment was echoed by the journalist Anne Perkins (1997: 22), drawing a distinction between hunting and proscribed activities, such as badger-baiting and bull-fighting: 'A strong, brave fox escapes. If it does not, it dies swiftly. Those are the critical distinctions between hunting and baiting.' For Perkins (1997: 22), the pleasure of hunting is that 'it connects me with the country and the creatures that live in it in a way that nothing else can'. The game of hunting generates an admiration for 'the cunning of the fox' (Moore 1998: 2), such that the relationship between the hunter and the hunted is one of respect. Thus, Sir Nicholas Lyell, speaking in the parliamentary debate, argued that the bill would

> destroy what that wise countryman, our present Poet Laureate, Ted Hughes, recently called the pact between the farming community and country people and those animals – the pact whereby in return for the right and opportunity to pursue ancient sports, those animals live in their natural habitat, with closed seasons voluntarily observed, in controlled but thriving numbers.
>
> (*Hansard*, 301, 28 November 1997: col. 1235)

Lyell's comments are interesting in that they suggest that the hunted animals assent to the hunt. Indeed, this idea implicitly underlies the whole representation of the fox as a sporting foe. It attributes to the fox intentions and emotions which cannot be tested – that the fox shows 'cunning', that the fox is 'brave', that the fox 'enjoys' the chase. In this representation the fox is therefore translated not into the form of scientific knowledge, but into the form of anecdote, art, literary description and country knowledge. It is in the form of these immutable mobiles that the fox is displaced in this representation into the political arena. Hence, Moore's admiration of the fox's cunning follows from a description of a day's hunting in which he observes an unusual quick kill, and a fox going to earth after a chase, 'as the huntsman comes slowly towards us through the bushes, weary and hatless and bursting with pride, with tired hounds beside him, it is impossible not to cheer' (Moore 1998: 2). Moore's fox was elusive, but through his cheerful anecdotal account the fox is translated into a cunning victor. Mobilising the same representation of the fox during the parliamentary debate, Peter Brooke drew on Shakespeare for his authority:

> I understand the fox's place in the scheme of things. I derive pleasure from watching the fox in its habitat, and my admiration for its qualities of resource and adaptability are akin to Shakespeare's, who, in 33 references to the fox, is admiring of it in 31.
>
> (*Hansard*, 301, 28 November 1997: col. 1230)

Yet in claiming authority for Shakespeare's representation of the fox, Brooke is inviting scrutiny of the process of representation involved. The translation of the fox into Shakespeare's text or Moore's anecdote requires no scientific or physical enrolment, and the representation here is validated only in that the fox continues to run from its pursuers, more often than not escaping. Through the explanations of fox behaviour offered by country lore, it is effectively legitimised by a mystic sense of 'understanding nature'. Unsurprisingly, therefore, the representation is challenged by the opponents of hunting, who claim to have captured a more authentic representation of the fox. So, for example, the proposer of the Wild Mammals Bill, Michael Foster, employed anecdote to rebut the fox-as-sporting-foe representation:

> I followed the hunt on foot and in a four-by-four vehicle. Three foxes were sighted and chased, but none was caught. They were followed by some 40 dogs hot in pursuit. One fox came within three feet of where I happened to be standing. It was certainly not happy; indeed it looked panicked, and did not know what to do. It is therefore surprising that the [Countryside Alliance director] Robin Hanbury-Tenison should have said: 'If you have ever been hunting, you've seen how foxes laugh at the hounds.' Foxes do not laugh at hounds when they are being chased, and to suggest otherwise is to act with ignorant disregard for the welfare of animals.
>
> (*Hansard*, 301, 28 November 1997: col. 1202)

The representation of the fox as a sporting foe is not, however, the only representation mobilised in support of hunting.

Take 2 – vicious vermin: the fox as pest

The second representation is less romantic and more functional than the first. In it the fox is not esteemed and admired, but vilified as a pest, as 'vermin', which needs to be 'controlled'. The British Field Sports Society (BFSS), in its leaflet *Hunting: the Facts* (1995), is quite clear that 'the fox is a ruthless predator, one capable of inflicting great damage'. In representing the fox in this way, hunting is transformed from a 'sport', with its implications of fun and enjoyment, to a necessary public service – as argued by pro-hunting MPs both before and during the parliamentary debate:

> I do not want to hear any more romantic ideas about pretty little innocent foxes. They are pests and they need to be controlled.
>
> (Kate Hoey, *Hansard*, 301, 28 November 1997: col. 1237)

> People don't hunt because they revel in killing animals but to control the number of foxes.
>
> (Lembit Öpik, quoted in Settle 1997: 1)

> It's clear fox numbers have to be controlled and I don't see this method as any crueller than other methods. It might well be less cruel.
>
> (Cynog Dafis, quoted in Settle 1997: 1)

Not only are foxes represented here as being pests, but they are represented as potentially forming an infestation if not controlled – for example, the *Western Mail* reporting that the Farmers' Union of Wales claimed that a ban on hunting would 'mean some farmland would be overrun with foxes and some farmers would go out of business' (Settle 1997: 1).

To produce this representation of the fox-as-pest, the fox is translated into the forms of scientific knowledge, statistical data and anecdote. As modernist politics dictates that the representation of non-humans is the field of science, MPs mobilised scientific representation of the pest status of the fox to validate their argument. Of particular debate was a report produced by the Ministry of Agriculture, Fisheries and Food (MAFF), part of which was quoted by the bill's proposer, Michael Foster. However, both Peter Brooke and Sir Brian Mawhinney challenged Foster's reading of the report, arguing that subsequent passages did confirm the fox's status as a pest:

> [Foster] quoted the second sentence of this passage, which states: 'On the basis of current evidence, the Ministry does not consider foxes to be a significant factor in lamb mortality.' He stopped there, creating the impression that the phrase ended with a full stop. His hon. Friends murmured their agreement. As [Peter Brooke] said, there is no full stop – it is a semicolon. The sentence continues: 'but it should be stressed that this is against a background of widespread fox control by farmers.' The document goes on: 'Foxes can cause serious local problems for individual farmers and the Ministry therefore considers that foxes do need to be controlled to minimise lamb losses.'
>
> (Sir Brian Mawhinney, *Hansard*, 301, 28 November 1997: col. 1242)

Significantly, though, while summative statements lifted from scientific reports were repeated with a weight of authority, very rarely were any actual figures quoted on the scale of farm-stock losses to foxes. No such references were made during the parliamentary debate, and whilst the BFSS (1995) offers that 'if only 2 [per cent] of a lamb crop is lost to foxes, this means that a typical hill farm with 1,200 ewes will lose 31 lambs, a considerable blow to a marginal hill farmer,' it is rather more vague on losses of pigs and poultry: 'Free-range poultry and piglets are frequent victims.' What is claimed as 'scientific evidence' is therefore often highly anecdotal in its nature. What is reported is not that scientific research shows that foxes are pests to farmers, but that farmers *perceive* them to be pests:

> I immensely enjoyed, both for its scholarship and its narrative, David Macdonald's 'Running with the Fox', published in 1987. ... It is a tribute to Mr Macdonald that his relations with farmers were so good, but it comes

forcefully and vividly through his narrative that the farming community were overall implacably opposed to the fox as a pest.

(Peter Brooke, *Hansard*, 301, 28 November 1997: col. 1231)

The logic of this argument was extended by the Wildlife Network, a group founded by a former director of the League Against Cruel Sports which campaigns for hunting to be reformed and regulated, not banned. The Wildlife Network lays claim to be the true representative of the fox, as the only organisation to put the animal's interests first. Thus, in a leaflet entitled *Putting the Fox First*, it argued that, whilst scientific studies did not necessarily show that foxes were a major pest, if farmers perceived them to be a pest then farmers would hunt foxes by one means or another:

> Most farmers – especially hill farmers – would say yes, the fox is a pest. ... [I]f landowners and farmers *believe* that foxes are vermin – and banning hunts will do nothing to change that perception – they will continue to kill them or have them killed.

(Wildlife Network 1997)

As such, they concluded that the 'Wildlife Network believes that for the fox's sake, hunting with hounds should be reformed, not banned.'

Anecdotal representations have been more directly mobilised in order to court public opinion. Here the fox is translated not into the form of scientific knowledge, but into a narrative form which conveys a sensationalised and emotive representation of a villainous fox. One intervention during the parliamentary debate highlighted a recent case of a child's two pet guinea pigs killed by a fox in central London; while an earlier incident in 1996 where a baby was attacked by an urban fox received considerable press coverage. Although the child received only a small scratch, the attack was reported with headlines such as 'Starving fox savages pram tot' (*Daily Star*) and 'Horror of the fox that savaged my baby boy' (*Daily Mail*), and comments including: 'A starving fox crept into a couple's home and savaged their baby as he slept in his pram' (*Daily Star*) and '[he] was recovering at home yesterday after being mauled by what his parents said was a fox' (*Guardian*) (in Rocks 1997) – which served to translate the fox concerned into a monstrous creature. The urban location of both incidents is significant as the reports were manipulated to produce a representation of the fox as a menace to urban dwellers as well as to farmers.

The pest status of the fox is challenged by opponents of hunting. In contrast to the relative paucity of statistical evidence offered in support of the representation, scientific research has been employed to question the impact of fox attacks. A report by biologists from Bristol University, published shortly before the parliamentary debate, found that only 0.5–3 per cent of lambs were lost to foxes compared to 17 per cent lost to natural causes within a day of birth, and that the estimated 32,500 hens lost to foxes was a tiny proportion compared to other causes of death (Brown 1997: 9). The report's conclusion was quoted by Michael Foster in the debate:

> Foxes do not warrant their reputation as major pests of agriculture. Nationally, losses of lambs, piglets and poultry are insignificant relative to other causes of mortality. Vastly greater improvements in lamb survival can be made by improving husbandry than by fox control. Local problems of fox predation of livestock can be prevented by electric fencing and secure housing. Foxes can be beneficial by consuming rabbits and other pests of agricultural crops.
>
> (*Hansard*, 301, 28 November 1997: col. 1209)

'Scientific' knowledge was also mobilised to refute suggestions that most farmers perceived foxes to be pests. The Bristol report stated that 70 per cent of farmers did not believe that having foxes on their land was harmful, and 36 per cent considered foxes to be useful (Brown 1997). A separate survey in Wiltshire was more equivocal, finding that 60 per cent of non-dairy livestock farmers and 30 per cent of dairy farmers considered the fox to be a pest, but that, overall, farmers ranked the fox as a less significant pest than badgers and rabbits, and that in financial terms the cost of fox damage was less than that caused by badgers, rabbits, pigeons and people (Baines 1997). Baines also concluded that even if the fox is accepted as a pest, hunting is not an effective form of fox population control, an argument echoed by hunt opponents in Parliament.

However, while scientific and anecdotal representations of the fox as a pest have been challenged, another means of translation has proved more durable. On the eve of the parliamentary debate in November 1997, the Countryside Alliance placed an advertisement in the national press reproducing a letter sent to MPs by the Alliance's director. The letter contained the statement that 'the fox is a recognised pest'. The Advertising Standards Authority (ASA) received a number of complaints contesting the veracity of this statement. In its adjudication, the ASA rejected the complaints, ruling:

> The advertisers said [that] the fox was recognised as a pest by farmers and game-keepers, [that] landowners are required by the Agriculture Act 1947 to control foxes, and [that] the Ministry of Agriculture, Fisheries and Food has the power to serve notice on landowners who do not do so. They submitted section 98 of the Agriculture Act 1947, which was headed 'Prevention of damage by pests' and stated that it applied to foxes. The Authority accepted the claims.
>
> (ASA 1998)

In other words, the fox is legally represented as a pest in the form of the Agriculture Act 1947, and this legal definition enables the fox to be represented as such in other arenas without further evidence needing to be produced.

Take 3 – cuddly cub: the fox as victim

The two pro-hunting representations of the fox discussed above are counter-poised by a third, anti-hunting representation of the fox as the victim of the hunt's cruelty. This representation involves two stages of construction, as illustrated by a leaflet produced by the League Against Cruel Sports (LACS). The front panel, with the heading 'Before the hunt …', shows a photograph of a healthy young cub looking directly at the camera. In its size, pose and inquisitive facial expression, the cub mimics the features of a domestic cat or dog. Below the picture, a caption reads: 'Bright-eyed and inquisitive, this young fox is at home in the British countryside where it has every right to live in peace.' The leaflet unfolds to reveal a second panel, entitled '… and after', with a smaller photograph of a disembowelled, dead, fox, and longer caption which describes the image as 'the almost inevitable end to a fox-hunt'.

It is the contrast between the two representations that constructs the fox's status as a victim. The first not only plays on popular sentimentality towards animals, but also positions the fox as a *natural* part of the countryside, where it has a *right* to live free from interference from humans, who, it is implied, are not naturally part of that countryside. Hunting hence becomes a violation of nature, as is conveyed through the second representation, which emphasises the cruelty and the *unnaturalness* of the hunt. Though this representation is mobilised most forcibly through photography, it can also assume a textual form, especially in arenas where the mobility of photographs is limited, such as the debating chamber of the House of Commons. Thus, in proposing the Wild Mammals Bill, Michael Foster firstly verbally constructed the hunting of foxes as being unnatural: 'The fox has no natural predator in this country' (*Hansard*, 301, 28 November 1997: col. 1205). Second, he constructed the fox as being the victim of unnecessary cruelty during the chase and the kill, through reference to quoted anecdote and veterinary reports:

> I am told that the average fox hunt lasts less than 17 minutes before either a kill takes place, or the scent is lost. I shall read out a boast from the Worcestershire hunt which states: 'I cannot let the occasion pass without putting forward the claims of the Worcestershire Hunt to the longest recorded run. … From the place where they found to that where they killed is more than 50 miles in a straight line and with the compass of the ground which must have been covered, the distance could not have been less than 80 to 90 miles.' So says Alan Cure, the former master of the Worcestershire

hunt. That is the equivalent of the circumference of the M25. I wonder when the fox stopped laughing.

> (*Hansard*, 301, 28 November 1997: col. 1206)

A vet who conducted the post mortem of a fox killed by the Cottesmore fox hounds concluded: 'I feel that the most likely cause of death was that of shock (in the pathological sense) brought about by blood loss, organ damage, lack of oxygenation of the blood due to lower respiratory dysfunction and upper airway obstruction. ... In short, the fox died a painful and unpleasant death, which probably was not as quick as evidenced by the areas of haemorrhage seen at many sites.'

> (*Hansard*, 301, 28 November 1997: col. 1207)

Much of the cruelty argument turns on the representation of foxes as sentient beings – as conscious beings with the perception of sense, the ability to feel sensations, and particularly in this case the sensation of pain:

> The debate is about how we treat other species and our relationship with them. The compassion and consideration that we show those species is a mark of our civilisation. We cannot treat them as quarries for our entertainment or pleasure. They are creatures like us, with nervous systems. They feel pain, suffering and fear, just as we do. We cannot treat them as targets or a collection of inert, non-suffering chemicals.
>
> (Paul Flynn, *Hansard*, 301, 28 November 1997: col. 1229)

> I am against the chase, the cruelty involved in the prolonging the terror of a living, sentient being that is running for its life.
>
> (Ann Widdecombe, *Hansard*, 301, 28 November 1997: col. 1251)

The sentience of animals is shared by the representation of the fox as a sporting foe, which attributes feelings and emotions to the fox, and by some of those who represent the fox as a pest and defend hunting as the least cruel form of control. But it is not universally acknowledged. A poll of prospective parliamentary candidates conducted by MORI for the International Fund for Animal Welfare in December 1996 found that while 98 per cent of successful Labour candidates and 92 per cent of successful Liberal Democrat candidates believed that 'animals have feelings, they are sentient beings that can suffer pain, fear and deprivation from behavioural needs', only 60 per cent of successful Conservative candidates concurred. While only 7 per cent of Conservatives believed that animals were not sentient beings, the balance apparently subscribed to a half-way position, acknowledging some perception of sense, but limited in the scope of those sensations.

As with the other representations, the representation of the fox as victim is mobilised through the translation of the fox into a variety of mobile forms. Scientific representations are again important in lending authority to the representation in a political context, but in the absence of detailed scientific studies of the effect of hunting on foxes, this is conveyed through the form of vets' reports of post-mortems conducted on foxes killed by hunts. Michael Foster quoted from three reports in the parliamentary debate, each involving a translation of the fox corpse into an assemblage of biological and medical terms which enabled statements to be made about the manner of death.

However, to appeal to public opinion a clearer, more emotive, form of representation was required. This was provided by the use of photography, involving a fairly straightforward translation of the fox into a photographic image and the mobilisation of that image through its reproduction in press reports, posters, leaflets and, most notably, newspaper advertisements. A series of advertisements were placed in the national press in the lead-up to the parliamentary debate by the Royal Society for the Prevention of Cruelty to Animals (RSPCA) and the International Fund for Animal Welfare (IFAW), both using photographs supplied by the LACS. These advertisements clearly represented the fox as a victim, combining photographs of the corpses of foxes killed by hunts with headlines such as 'How much longer can foxes tolerate this kind of pain?' and 'Whatever you think about foxes, you have to admire their guts.'

Significantly, although the bill covered a range of wild mammals, nearly all of the advertisements carried photographs of foxes. Mink and hares were not portrayed, although one RSPCA advertisement did focus on deer-hunting. Under the headline, 'It's hard to swim when you've been running for three hours,' a photograph shows a stag swimming across a river, pursued by sixteen hounds (see Figure 9.1). The text of the advert supports the representation, mobilising scientific accounts of the suffering endured by a stag during the chase, but it is the photograph which is the most compelling representation. The deer is in the foreground, its head barely above water, its dilated eye staring out at the observer. One dog is nearly upon it from behind, others appear to be flanking it from the front – the stag is clearly cornered. The story of the photograph reveals something of the process of translation. It is a clip from video footage shot by a monitor from the LACS, who have been regularly filming stag- and fox-hunts since the mid-1980s. On this particular occasion in September 1994, the Devon and Somerset Staghounds had pursued a stag into the River Barle near Dulverton. The LACS video evidence resulted in a five-week hunting ban being imposed on the hunt by the Masters of Deerhounds Association for 'inhumane killing', not for the nature of the chase, but because the video went on to show that hunt marksmen had taken ten minutes to destroy the deer once it had been caught. The translation of the stag from an actual animal living on Exmoor to a photographic

It's hard to swim when you've been running for three hours.

On average, a stag hunt lasts 3 hours and covers around 12 miles.

When the stag is finally caught by the hounds it is at the point of total exhaustion.

Scientific analysis of blood samples taken from hunted stags reveals a litany of suffering.

In the early stages of the chase, glycogen and blood sugar levels fall sharply.

As the hunt progresses, fatty acids in the blood rise, indicating high physiological stress levels.

Red pigment in blood plasma increases, caused by ruptured blood cells.

In the later stages of the hunt, high levels of muscle enzymes appear in the blood, indicating life-threatening muscle damage.

Despite its name, stag hunting is not confined to the male of the species.

Hinds are hunted too, sometimes when they are pregnant or with a calf at heel.

Stag or hind, the end is the same. A free wild animal is hunted to death.

The RSPCA has long campaigned against all hunting with dogs.

In areas where deer need culling it is more efficient and more humane for them to be shot by a marksman.

We believe that the hunting of wild animals is cruel and unacceptable in a civilised society.

And the vast majority of the people in this country agree with us.

A Private Member's Bill seeking to ban hunting with dogs comes before Parliament on November 28th.

A MORI poll taken in October 1997 shows that 73% of people support the Bill. We want to turn that overwhelming weight of public opinion into legislation.

The 28th is a Friday when many MPs will be back in their constituencies.

We want you to persuade them to stay in the House and vote to end this cruel 'sport' once and for all.

You can write to your MP direct at the House of Commons, or call the RSPCA on 01403 223284 (9-5 weekdays) and we'll send you a campaign pack.

And if you need further motivation, look again at the stag in the picture.

Look him in the eye.

And tell him you can't be bothered.

 Ban hunting with dogs.

Figure 9.1   'It's hard to swim when you've been running for three hours.'

Source: RSPCA advertisement, 1997; photograph in advertisement © League Against Cruel Sports

image reproduced in the press is hence the product of a carefully managed process, involving the LACS monitors stalking and videoing the hunt (but not intervening to protect the animal), the LACS processing the video tape (selecting this particular image and making it available for wider distribution), and then the RSPCA selecting the image for inclusion in its advertisement (constructing a text around the image and purchasing the space in newspapers).

The process of translation therefore raises a number of questions about the representations being mobilised, and in particular about their 'representative-ness'. These issues were aired with regard to a second advertisement, by the IFAW (see Figure 9.2). Entitled 'The effect of a pack of twenty on an expectant mother,' this contained a photograph of a dead fox with its thorax ripped open – again taken by the LACS monitors – accompanied by text including quotes from veterinary post-mortems. However, the implication of the headline that the photograph showed a pregnant vixen was challenged by the pro-hunting lobby, who claimed, first, that the photograph was of a male fox, and, second, that the photograph had been taken at a time of year when foxes would not be pregnant. Whatever the truth of these allegations, they reveal the dilemma of displacement inherent to representation. For an object to be represented it needs to be trans-lated into a mobile form, yet that translation separates the representation from the subject such that its veracity is always open to question. But if it is the repre-sentations which are important in achieving outcomes, if those representations are always constructed in the image of those who mobilise them, and if our knowledge of any subject can only ever be partial, is the correspondence between a representation and its subject really significant? Or should representations merely be judged by their success in enrolling actors into a particular project? These issues are important in questioning discursive and political representations, but they become fundamental when applied to scientific representations, where the authority of the representation rests on the scientist's perceived ability to construct a knowledge which replicates an objective truth.

Take 4 – the appliance of science

Science plays an enigmatic role in the hunting debate. The discourse of modernism insists that the representation of non-humans is the province of science, and therefore that when questions about non-humans enter the political arena it is on the basis of scientific representations that decisions should be made. Science is presented as a panacea which overcomes the problems of bias inherent to other forms of representation and provides objective, factual, information from which rational judgements can be made. Yet science fails to perform this role in relation to hunting because the appropriate scientific knowledge does not exist. While the reports of veterinary surgeons, biological surveys of animal numbers

THE EFFECTS OF A PACK OF TWENTY ON AN EXPECTANT MOTHER.

It's a disgusting habit. But one that fox hunters seem loath to give up.

In an average year, they get through around 20,000 cubs, dog-foxes and vixens.

Even heavily pregnant vixens are considered fair game.

Selective with the truth, some fox hunters maintain there is no cruelty.

Post-mortem examinations of foxes savaged by hounds prove otherwise.

Typical findings include, "Extensive wounds to abdomen and thorax," "intestines hanging out" and "death caused by pathological shock."

A 'quick nip to the neck' it isn't.

Foxes that manage to go to ground during a hunt face a terrifying and protracted ordeal. Escape routes are blocked, and terriers sent in to corner their prey.

The ensuing underground battle is nasty and brutish. It is not short.

The fox may well die underground, fighting for its life. (The terriers also sustain injuries.)

If it's still alive, the hunters' digging will expose it. The best the poor creature can hope for now is a gunshot.

Those animals that escape the hunt don't necessarily escape the suffering.

The stress and exertion of the chase is traumatic beyond imagination.

This, the fox hunters insist, is sport.

71% of the British people disagree. They think hunting with dogs should be banned (MORI).

Don't be passive.

Please write to your MP. Ensure that on 28th November he or she supports Michael Foster's Private Member's bill to ban hunting with dogs.

Make sure your MP opposes hunting with dogs on November 28th. For more information call IFAW on 0800 10 60 90.

Figure 9.2 'The effects of a pack of twenty on an expectant mother.'

Source: IFAW advertisement, 1997 (© IFAW); photograph in advertisement © League Against Cruel Sports

and the proceedings of the public inquiry on cruelty to wild mammals commissioned after the Second World War (Scott Henderson 1951) were presented as 'scientific knowledge', no direct scientific research of the effect of hunting on foxes has been conducted.

However, scientific knowledge does exist about the effect of hunting on deer, the result of research by Cambridge scientist, Professor Patrick Bateson, funded by the National Trust. Bateson's report was mobilised as evidence by hunt opponents during the parliamentary debate, yet it also became the focus of a sub-play in which the representativeness of its representations was contested by hunt supporters. The Bateson Report was commissioned by the National Trust following a vote by members in 1994 to ban stag-hunting on land owned by the Trust. The Trust's council, which has heavy representation from land-owning interests, was reluctant to implement the decision, and viewed the inquiry as a means of at least delaying, and possibly of shelving, the proposal. In doing so they appealed to modernist principles by insisting that a decision to ban hunting could only legitimately be based on solid scientific proof that the hunting of deer was cruel. Thus the purpose of the inquiry was to challenge the members' representation of the deer with a 'scientific' representation which the council believed would justify their position. Bateson and his research assistant, Elizabeth Bradshaw, spent two years studying the behavioural and physiological effects of hunting on red deer on Exmoor and the Quantock Hills, including analysis of blood samples taken from deer chased and killed by the hunts. Both hunt supporters and opponents co-operated with the research, and both anticipated that it would confirm their opinion – a spokesperson for the Quantock Staghounds claiming that, 'whenever we go to a scientific inquiry, it comes up in the favour of hunting' (interview with author).

The Bateson Report was presented to a panel of scientific experts in October 1996 and to the National Trust in April 1997. Its main findings were unequivocal: that 'hunts with hounds impose extreme stress on red deer and are likely to cause them great suffering', and that hunting with hounds 'can no longer be justified on welfare grounds' (quoted in Barron 1997: 1). These summary phrases were the predominant means by which Bateson's scientific representation of the deer was mobilised through the media and political debate. More detailed findings were also quoted in the parliamentary debate by the Labour MP Dr Ian Gibson, however, with the technical terminology involved serving to heighten the scientific authority of the argument:

> The physical studies showed that cortisol – a hormone that is [a] scientific compass for stress in human beings as well – blood sugar and lactate levels rose as oxygen levels fell, due to overworked muscles. Red blood cells broke down to release haemoglobin into the blood stream, and muscle enzymes

leached into the blood. Cortisol levels were 10 per cent higher in deer chased by hounds than in animals shot by stalkers. Blood lactate levels reached peak levels after 5 km of chase, and dropped as the lactate was used to fuel muscle activity. The consequence was that, after 15 to 30 km, energy stores were completely depleted. ... Haemoglobin spilled into the blood after a few kilometres of chase. Those are the facts.

(*Hansard*, 301, 28 November 1997: col. 1244)

In response to the Bateson Report, the National Trust council imposed an immediate ban on the hunting of deer with hounds on its land. Wider reaction was more varied, and was later described by Bateson and Bradshaw (1997: 51) in a *New Scientist* article:

Those who felt their sport was under threat understandably tried to find fault with the science or implied that experts, poor things, are always squabbling among themselves. On the other hand, those who found our conclusions unsurprising were indifferent or condescending. They didn't need scientists to show them what was already obvious.

Challenges to the Bateson findings from the pro-hunting lobby took two forms. First, the validity of the scientific representations was questioned. As with the politicisation of science in the BSE crisis, arguments emerged about 'good' and 'bad' science (see Woods 1998a). Qualifications and doubt which were acceptable and anticipated in a scientific context were seized upon as failings and grounds for dismissal when that same scientific knowledge was presented in support of political action. Thus the pro-hunting lobby searched for alternative scientific representations which would invalidate Bateson's evidence. These included correspondence from veterinary surgeons, an article by an anonymous physiologist published in the *Daily Telegraph* (critiqued by Bateson and Bradshaw 1997), and, more substantially, a new commissioned scientific study, the Joint Universities Study on Deer Hunting. The report of this second study was no less controversial. A pre-publication leak to *The Times* hinted that it had corroborated Bateson's conclusions, and its findings received little media coverage. However, without directly quoting any detailed results, the pro-hunting lobby mobilised the message that the new report gave 'clear and unequivocal reasons to believe that the Bateson Report is no longer a sound basis for policy' (letter to the *Guardian*, 1 October 1998). The claim provoked an exchange of correspondence between hunt supporters and opponents in the *Guardian*, culminating in an intervention by Bateson:

The recent research only revises some of the physiological details. It does not even attempt to shift Bateson's central point, which was comparative. His team found that the bodies of deer killed by hunting showed a far more advanced stage of exhaustion and debilitation than those dying by any other means.

(Letter to the *Guardian* from Mary Midgley, 8 October 1998)

The Bateson science is doomed for the bin. I have had an interest in the red deer for 23 years. Professor Bateson's science in his report to the National Trust bears no resemblance to the clinical picture. Consequently, I too had blood samples taken from deer. Bateson came to the conclusion that because he found high levels of the muscle enzyme (CK) that the hunted deer have a myopathy (a severe muscle cramp-muscle damage). Unlike Bateson I took a series of samples from each deer. This demonstrated that most of the CK elevation is contamination from damaged muscle caused by the shooting. This demolishes the cornerstone of Bateson's report.

(Letter to the *Guardian* from D.J.B. Denny, 13 October 1998)

Mr Denny's name is not to be found on the worldwide list of scientific papers which have been published since 1981. He has not published any of the details of his so-called study. We are told nothing of his sample sizes or his methods. Even if he had carried out an adequate study, his findings would not impinge on the clear evidence of high levels of stress in hunted red deer. ... If he would care to expose himself to normal peer review, it might be possible to discover how much credibility to attach to his claims.

(Letter to the *Guardian* from Patrick Bateson, 15 October 1998)

Hence the representativeness of the 'deer' mobilised by Bateson in the form of scientific knowledge was questioned through the translation of other deer in other 'scientific' representations which subsequently became open to scrutiny against a criterion of established scientific practice.

The second form of challenge to the Bateson Report involved the discrediting of science as an appropriate means of representation and the privileging of local lay knowledges. As Bateson and Bradshaw (1997: 51) comment, hunt followers 'felt that their impressions and personal anecdotes counted for more than the weight of the quantitative evidence'. In particular they questioned not the scientific detail of the report, but the conclusion that banning hunting with hounds would reduce the suffering caused to deer herds and hence be in the best interests of the deer. The pro-hunting lobby instead argued that banning hunting would result in the large-scale slaughter of deer by farmers – an argument which was swiftly converted into more mobile and graphic representation. Three days before the parliamentary debate in November 1997, a photograph showing the antlers of over thirty stags killed by farmers on the Quantock Hills was released to the press. This was reproduced in national newspapers, along with statements that at least thirty-six of the seventy-six stags on the Quantocks had been shot since the introduction of the hunting ban by the National Trust. The newspapers quoted the joint-master of the Quantock Staghounds, commenting that, 'I feared this would happen. It is nothing to do with revenge. I can only assume that the deer have been doing damage and that farmers can't see any way forward and have taken things into their own hands' (in Gibbs 1997: 8). In other words, scientific representations were dismissed as inappropriate because, unlike local lay knowledge, they failed to anticipate the consequences of their solutions. Thus, while science was presented by many as being the only form of representation of animals that should influence political action, the experience of the Bateson Report proves that in practice scientific representations are open to as much political debate and contestation as other means.

Conclusion

Animals have an enigmatic presence in the hunting debate. Their representations are omnipresent, the arguments of both sides are built around representations of the fox and the deer, and of their interests. Both pro- and anti-hunting lobbies claim to 'represent' the animals in a political sense. Yet the animals themselves are totally absent from the political debate: only their ghostly representations, rendered as immutable mobiles, haunt the parliamentary chambers, television studios and newspaper columns where the human actors 'speak on their behalf'. It could be argued that the animals have no agency to participate; nor, indeed, any agency to challenge or question the way in which they are represented (in both senses of the word). However, this understanding of 'agency' as a possession is itself challenged by 'actor-network theory' (ANT) perspectives which wrestle with the 'agency' of non-humans (Brown and Capdevila 1999; Hetherington 1997). Here, agency is repositioned as an 'effect' which is realised or released through the configurations of entities into actor-networks. Thus there emerges a peculiar agency without intentionality,

> characterised by an ability to generate strange topological effects or 'foldings' upon the ordering of space. This happens when an artefact, by virtue of its apparent 'blankness' – that is, its apparent lack of overt significations – disrupts the space into which it is placed.
>
> (Brown and Capdevila 1999: 40)

It is the disruption of space which holds the key to the empowerment of animals in the hunting debate. The political marginalisation of the fox and deer in our story through the process of representation is intrinsically spatial. The displace-ment of the representation is geographical as well as metaphorical. It is precisely because the political debate does not occur on the hunting field – but rather in the chamber of the House of Commons, or in television studios – that the animals need to be translated into mobile forms of representation in order to be enrolled into arguments on either side of the debate. As a consequence of this spatial displacement, representations of animals can become signifiers for certain geographical spaces. Thus the photograph of the innocent fox cub reproduced in the national press, or the biological picture of the dissected fox translated through post-mortem examination and quoted in debate in Parliament, represent not just the animals concerned but also the British countryside in an urban space.

In this way, the activities of foxes and deer have (unintended) agency-as-effect in the disruption of space at two poles. At one end, the politicisation of hunting is provoked by the actions of animals disrupting the discursive space of the country-side: by the fox attacking a chicken-house and disrupting an agricultural network of production; by the deer swimming a river vainly trying to escape pursuing hounds, disrupting the image of the countryside as a space of nature. At the other end, the arrival of representations of the animals and their actions, in the form of immutable mobiles, in the House of Commons or other debating forums disrupts political space and causes politicians to act in ways they would not otherwise have done (by speaking, debating, voting on the issue).

Moreover, the increasing distanciation of the animal from its representation creates a third means by which animals can achieve unintended agency-as-effect. This is where the animals *in actu* continue to behave and act in ways contrary to their representations, disrupting their own representational space, with the effect of undermining the political legitimacy of those representations in the human realm. In all these ways animals have the capacity to act in a manner which affects how they are legislated for by humans. Animals thus become enrolled into polit-ical participation in a manner which is not wholly fixed in human hands. Humans control the rules of decision-making and the technologies of representation, such that the representations of animals mobilised in support of human-derived polit-ical positions are ones constructed by humans and tainted with human interests. Yet, without the participation of animals in those representational networks, their objectives cannot be achieved. Latour (1993: 144) imagined a 'parliament of things' in which the franchise of humans and non-humans to participate politically becomes entwined, but for the extensive range of issues in which non-humans are fundamentally implicated – animal welfare, conservation, agriculture, the envi-ronment – the realisation of such an assembly may require only recognition of the complex micro-processes of representation through which politics already proceeds.

Acknowledgements

The author wishes to thank the editors for their helpful comments on earlier drafts of this chapter.

References

Advertising Standards Authority (1998) 'Adjudication: Countryside Alliance', ASA Web-Site <http://www.asa.org.uk/adj/adj_2482.htm> (accessed 10 June 1998).

Baines, R. (1997) 'Economic impact of foxes and fox hunting on farming in Wiltshire', Paper presented to Annual Conference of the Rural Economy and Society Study Group, Worcester, September.

Barron, B. (1997) 'Hunting hangs in the balance', *Somerset County Gazette*, 11 April.

Bateson, P. and Bradshaw, E. (1997) 'Cruelty at the hands of the critics', *New Scientist* 154(2080): 51.

British Field Sports Society (1995) *Hunting: The Facts*, London: BFSS.

Brown, P. (1997) 'Foxes "innocent of vermin charge" ', *Guardian*, 28 October.

Brown, S.D. and Capdevila, R. (1999) '*Perpetuum mobile*: substance, force and the sociology of translation', in J. Law and J. Hassard (eds) *Actor Network Theory and After*, Oxford: Blackwell.

Bunce, M. (1994) *The Countryside Ideal: Anglo-American Images of Landscape*, London: Routledge.

Burgess, J. (1993) 'Representing nature: conservation and the mass media', in B. Goldsmith and A. Warren (eds) *Conservation in Progress*, Chichester: John Wiley.

Callon, M. (1986) 'Some elements of a sociology of translation', in J. Law (ed.) *Power, Action, Belief: A New Sociology of Knowledge?*, London: Routledge and Kegan Paul.

Cloke, P., Milbourne, P. and Thomas, C. (1996) 'The English National Forest: local reactions to plans for renegotiated nature–society relations in the countryside', *Transactions of the Institute of British Geographers* 21: 552–571.

Eder, K. (1996) *The Social Construction of Nature*, London: Sage.

Fitzsimmons, M. (1989) 'The matter of nature', *Antipode* 21: 106–120.

Gaskell, E. (1906) *Somersetshire Leaders: Social and Political*, London: Queenhithe Printing and Publishing.

Gibbs, G. (1997) 'Somerset's stags "paying price for Trust hunting ban" ', *Guardian*, 26 November.

Harrison, C. and Burgess, J. (1994) 'Social constructions of nature: a case study of conflicts over the development of Rainham Marshes', *Transactions of the Institute of British Geographers* 19: 291–310.

Hetherington, K. (1997) 'Museum topology and the will to connect', *Journal of Material Culture* 2(2): 199–218.

Latour, B. (1990) 'Drawing things together', in M. Lynch and S. Woolgar (eds) *Representation in Scientific Practice*, Cambridge, MA: MIT Press.

Latour, B. (1993) *We Have Never Been Modern*, Hemel Hempstead: Harvester Wheatsheaf.

Moore, C. (1998) 'The people's hunt', *Daily Telegraph, Weekend Section*, 21 February.

Perkins, A. (1997) 'Hunting for the facts of the matter', *Guardian*, 24 November.

Philo, C. (1995) 'Animals, geography and the city: notes on inclusions and exclusions', *Environment and Planning D: Society and Space* 13: 655–681.

Rocks, S. (1997) 'The fox, the baby and the propaganda machine', Wildlife Network Web-Site <http://hot.virtual-pc.com/wildnet/extra4.html> (accessed 3 November 1997).

Scott Henderson, J. (1951) *Report of the Committee on Cruelty to Wild Animals*, Chairman J. Scott Henderson, Cmd 8266, London: HMSO.

Serpell, J. (1996) *In the Company of Animals: A Study of Human–Animal Relationships*, Cambridge: Cambridge University Press.

Settle, M. (1997) 'Hunters rally to defeat MP's Bill', *Western Mail*, 17 June.

Short, J.R. (1991) *Imagined Country: Society, Culture and Environment*, London: Routledge.

Wildlife Network (1997) 'Putting the fox first!'. Wildlife Network Web-Site <http://hot.virtual-pc.com/wildnet/leaf2.htm> (accessed 3 November 1997).

Wilson, A. (1992) *The Culture of Nature: North American Landscape from Disney to the Exxon Valdez*, Oxford: Blackwell.

Woods, M. (1998a) 'Mad cows and hounded deer: political representations of animals in the British countryside', *Environment and Planning A* 30: 1219–1234.

Woods, M. (1998b) 'Researching rural conflicts: hunting, local politics and actor-networks', *Journal of Rural Studies* 14: 321–340.

10 'Hunting with the camera'

Photography, wildlife and colonialism in Africa

James R. Ryan

Introduction

Visitors to the Royal Albert Memorial Museum and Art Gallery in Exeter, England, are often particularly impressed by the life-size stuffed giraffe which takes a prominent place in the centre of the displays of 'natural history'. What kinds of historical processes and spatial practices were involved in the transformation of this animal from eastern Africa to England and from living, wild animal into a lifeless, domesticated, museum display of 'nature'?[1] As it turns out, this giraffe is one of a number of large animals which were donated to the Museum in 1919 by the big-game hunter Charles Victor Alexander Peel (*c.*1869–1931). A wealthy Englishman, Peel pursued his 'sport' – from elephants to polar bears – as far afield as Somaliland and the Outer Hebrides, recounting his exploits in a series of books whose photographs show the author posed with various hunting trophies (Peel 1900, 1901). Most of Peel's big-game trophies, including the prize giraffe, were shot in Africa between 1890 and 1910, and many found their way into public exhibitions of 'natural history'.[2] The giraffe – shot, preserved, stuffed and given the pet name 'Gerald' – acquired a prominent position within such displays of 'wildlife'. These displays not only fabricated and domesticated the animals of distant colonial territories within new public and private spaces 'back home', but served as monuments to the big-game hunters who seemed to their contemporaries to wield such mastery over the natural world.

This is a useful example with which to begin since the exploits of 'big-game' hunters like Peel and the after-lives of the animals killed and collected are by no means atypical of the Victorian and Edwardian eras. Indeed, from the 1840s trophy collections, along with illustrated books, were essential in the making of profitable reputations by hunters. Roualeyn Gordon Cumming, for example, author of the highly popular and illustrated *Five Years of a Hunter's Life* (1850), returned to England in 1849 with some thirty tons of trophies from Southern Africa. Cumming exhibited the trophies in London for over two years (1850–2) and at the Great Exhibition of 1851, later selling them to the famous showman

Barnum for his American Museum (Brander 1988: 44–48). Hunting trophies, along with live zoological specimens and exotic peoples (MacKenzie 1988: 31; Schneider 1982: 125–151), were part of a thriving popular culture of exhibitions in the mid-nineteenth century; their display was a central element in the imperial project to capture the world 'under one roof' and to transform colonial spaces into exhibition-like landscapes (Altick 1978: 290–92; Greenhalgh 1988; Mitchell 1989). Displays of the remains of 'wildlife' and 'big-game' in particular were used in a range of settings – from scientific institutions to entertainment venues – to convey a variety of meanings, including the colonial prospects of the territory traversed and the manliness of the intrepid hunter.

Colonial hunters cut some of the most striking figures on the Victorian and Edwardian imperial landscape, and remain compelling characters to this day (Brander 1988; Bull 1992). Yet, hunting was not limited to those who made it a full-time career. Many British colonial men (and indeed a few women) participated in the chase and killing of wild animals, both as a form of sport and as a scientific pursuit. Interest in pursuing zoological 'specimens' for private and national collections was fostered by both the dramatic upsurge in the popularity of natural history and the proliferation of popular literature and images of hunting in Britain, which frequently pictured the hunter as a manly adventurer and hero of Empire.

The varied interests and activities of Victorian and Edwardian hunters must therefore be set within the wider contexts of European imperialism and the relations between human groups, animals and the environment. As a number of historians have shown, the hunting, collecting and display of wild animals was intimately associated with the ideology and enterprise of empire (Beinart 1990; MacKenzie 1988, 1990; Ritvo 1990; Storey 1991). This is evidenced in the work of hunters and collectors like Charles Peel, with whom I began. Peel had long collected animal trophies for his own pleasure, and stuffed animals like the giraffe would have at first been housed at Peel's Oxfordshire home. However, in 1906 Peel established his 'Exhibition of Big-Game Trophies and Museum of Natural History and Anthropology' in Oxford in order to show, 'by means of beautiful objects of Natural History, that there are other countries in the world besides over-populated little England' (Peel 1906: 1). The message Peel sought to convey through his hunting trophies was a proudly imperial one. As Frederick Selous, arguably the most famous big-game hunter of the time, reportedly put it in his speech to open Peel's Museum in July 1906:

> [T]he love of adventure, love of a nomadic life, love of hunting – these instincts had descended to the English-speaking race, and it was due to that that this wonderful Empire had been formed. The hunter had always been a pioneer of Empire.
>
> (Peel 1906: 4–6)

The intimate connection between hunting and imperialism was obvious to a man like Selous who had spent some two decades (1872–92) hunting and empire-building in Africa; activities which he publicised through his books and lectures (Selous 1881, 1893a, 1893b; Tidrick 1990: 48–87). Selous collected extensively for natural history museums and gained widespread recognition for his scientific services. Hunting not only provided useful knowledge; it was said to symbolise certain national characteristics. Like Cecil Rhodes, Selous used the supposed superiority of the white, Anglo-Saxon race as justification for colonisation in Africa. For men like Selous and Peel the white hunter and adventurer represented a type of energetic, pioneering Englishman upon which empire depended.

Such sentiments were thus reflected in Peel's Museum, which was laid out to provide visitors with a tour of the geography of empire and its natural history, the latter represented in naturalistic displays and straightforward hunting trophies. Arguing that big-game hunting 'exercises all the faculties which go to make a man most manly', Peel hoped that his trophies would inspire young men to go out and 'do good service to the Empire' (Peel 1906: 4–6). The attempt of such hunters to put displays of animals to the service of empire signals some of the ways in which the natural world was constructed through forms of colonial imagination during the Victorian and Edwardian eras.

The entangled historical geography of just one giraffe, as an animal and colo-nial object with mutable meanings (Thomas 1991), provides a useful entry-point to this chapter, which focuses on aspects of the geography of 'wildlife' in British colonial Africa in the late nineteenth and early twentieth centuries. As part of a larger project on the uses of photography within British colonialism (Ryan 1997), it pays particular attention to how photography was used in parallel with practices of hunting and taxidermy to capture and to reproduce 'wild' animals. It empha-sises how such practices were inter-related and bound up with broader discourses of 'nature', 'colonialism' and 'preservation'. In so doing it draws on recent work elucidating the cultural geographies of nature and spaces of human–animal rela-tions (Anderson 1995; Philo 1995; Whatmore and Thorne 1998; Wilson 1992; Wolch and Emel 1995). Such work points usefully to how human–animal rela-tions are mediated by a range of cultural practices and formed through a spectrum of spatial settings and processes. Thus this chapter seeks to show how particular practices of hunting and photography, enacted in colonial African terri-tories, constructed the 'wildness' of African animals – especially animals known as 'big-game' – and landscape as part of an untouched world of pristine nature. Moreover, such conceptualisations of wild nature as a space outside of human civilisation, forged through discourses of hunting, photography and conservation, both reflected and informed new kinds of colonial relations between Africans and Europeans, humans and animals, and humans and the environment.

Capturing wildlife: photography and taxidermy

A number of writers on photography have noted how, in present-day Western society, where taking photographs consists of 'loading', 'aiming' and 'shooting', the camera has become a 'sublimation of the gun' (Sontag 1979: 14–15). Yet few have noted how this process of sublimation was well underway in the second half of the nineteenth century, most especially within the language and practices of Victorian and Edwardian big-game hunting.

From the late 1850s, explorers, soldiers, administrators and professional hunters began to employ the camera to record images of animals, skins and horns for purposes of scientific documentation and as evidence of their hunting achievements. For many hunters, photography, alongside the related practice of taxidermy, was a convenient means of recording hunting trophies for lectures, books and personal, private collections. Well-to-do Victorian hunters such as Guy C. Dawnay, who undertook hunting and exploring expeditions in East and South Africa in the 1870s and 1880s, compiled photographs of hunting trophies, friends and landscapes into albums as mementoes of colonial adventures (Anon. 1889; Dawnay c.1880). Such photographs of white men with dead animals or antlers, tusks and skins are a common, even clichéd, feature of the repertoire of Victorian and Edwardian colonial photography, and they testify further to the significance of hunting as a ritualistic display of power by white colonial elites over land, subject peoples and nature. Hunters did not only make photographs of themselves with animals they had shot. Another photograph of 1876 in an album of Guy C. Dawnay shows nine preserved lion heads set up on an easel (see Figure 10.1). The photograph presents a disturbing display of dismembered heads, preserved as if part of live, wild animals. Although the dismembered and isolated aspect of the lions' heads quickly shatters any illusion of naturalism, the heads have nevertheless been given expressions of snarling ferocity appropriate to such 'wild' animals. Here photography re-presents the art of the taxidermist; capturing the spirit of the animal as alive and as a hunter might encounter it. Indeed, photography and taxidermy shared the power to fabricate nature in new spaces; to capture animals in their supposedly natural attitudes and display their form to audiences eager for glimpses of exotic wildness and knowledge of natural history.

While photography would eventually supersede taxidermy as an efficient and economical means of arresting time and capturing the surface appearance of animals, as modes of representation the two practices are closely related (Hauser 1998). Early photographers employed taxidermy in order to capture portraits of animals in a seemingly live pose and outdoor setting. In the 1850s J.D. Llewellyn took photographs of stuffed deer, badgers, otters, rabbits and pheasants posed as if photographed in the wild. Just as photographers drew on the skill of the taxidermist to overcome their cameras' technical shortcomings, taxidermists drew in

Figure 10.1 'From the Settite & Royan R.s, N.E. Afr. 1876.'

Source: G.C. Dawnay (1876) in the Royal Geographical Society (with the Institute of British Geographers) Picture Library, C82/006285

turn on the photographer to provide them with an appropriate model of realism for their displays. In his popular *Sportsman's Handbook* (1882), the famous taxidermist and publisher of sporting books Rowland Ward (1848–1912) urged sportsmen to take photographic apparatus on their expeditions. Ward argued that

'photographic pictures of living *ferae naturae*, in their native jungle or forest ...
represent the perfect specimen for our contemplation.' In other words, such
photographs represented the taxidermist's goal of re-creating the life-like forms
of animals within their habitat. To this end, Ward recommended the making of
'instantaneous photographs of animals and objects in motion', going on to
develop his own 'Naturalists' Camera' for the purpose (Ward 1882: vi).

Ward's guides on preserving hunting trophies were followed by many hunters
and taxidermists who employed photography as a means of rendering the
modelling of animal forms more naturalistic. Indeed, by 1911 Ward could
declare:

> [T]he taxidermist could never have reached his present advanced stage
> without the aid of instantaneous photography. ... Previous to the invention
> of the instantaneous camera I used to have to go to the 'Zoo' and model an
> animal in wax before I could mount its skin to my satisfaction.
>
> (Ward 1911: 22)

Using photography, Ward developed his own 'artistic setting-up' of trophies and
specimens whereby, rather than being 'stuffed', the skin was preserved and
remounted on a model constructed from a substance of Ward's own invention,
incorporating limb bones and the skull. In achieving a greater attention to detail,
expression, gestures and habits, this 'Wardian taxidermy' (as it became known)
received considerable praise when examples of Ward's work were exhibited at
various international and colonial exhibitions (Ward 1911: 139). Thus Ward used
photography – already itself an aetheticisation of the 'real' – as the basis for his
supposedly truthful modelling of animal gesture and expression. Yet, as most
products of taxidermy at this time showed, from mammals in cases with dried
grasses and silk leaves to Rowland Ward's rhino-head drinks cabinet, what
mattered was not so much accuracy as the naturalistic appearance within a
domestic, and domesticated, setting.

In this circulating process of the production of nature, then, taxidermists and
photographers drew on each other's codes and conventions in their constant
staging of animal life. Just as taxidermists like Ward modelled animals from
hunters' photographs, hunters like Dawnay took photographs of remodelled lion
heads as a memento of the sight of living wild animals. Hunters like Selous often
shot particular animals and preserved their skins in order to see them subsequently resurrected in realistic displays in, for example, the British Museum
(Selous 1893a: 90–91). In 1907 the big-game hunter and photographer Carl
Schillings (1907, Vol. 1: viii) noted the pleasure to be gained by the European
hunter 'when, making a tour of the museums of various places at home, he sees
awakened to new life the wild creatures he formerly observed and laid low in far-

off lands'. To many hunters, the space of the metropolitan museum thus replicated the space of the colonial frontier. Natural history museums could themselves even become hunting grounds. During his travels 'with gun and camera' in Southern Africa in 1890, the hunter and friend of Selous, H. Anderson Bryden photographed 'several natural specimens' in the Cape Town Museum, including the head of a white rhinoceros shot by Selous in Mashonaland in 1880 (Bryden 1893: 491–492). Significantly, nearly all of the photographs of animals reproduced in Bryden's book were of stuffed ones. Stuffed animals had become the ideal photographic target: a re-creation of nature as apparently authentic, yet utterly docile. Photographs of stuffed animals such as that in Dawnay's album (see Figure 10.1) thus represent a kind of double mimesis, and reinforce the shared ways in which photography and taxidermy are manifestations of a desire to possess and control nature (Hauser 1998).

Photographing wildlife

Photographs of stuffed animals were made in part for the practical reason that until the 1880s photographing animals 'in the wild' remained difficult. The easiest means of securing photographs of living, wild animals, though perhaps the least intrepid, was at London's Zoological Gardens; and a number of commercial photographers followed this route in the 1860s and 1870s (Haes 1865; York 1872). Photographing animals in the wild, as opposed to in captivity in zoos or as stuffed exhibits in museums, was made easier by the technical developments of the 1880s and 1890s: notably, the use of new roll film, the increasing portability of cameras, the reduction in exposure times, and the development of telephotographic lenses. Contemporary advertisements for cameras, telephotographic lenses and telescopes, such as those produced by the London-based firm of Dallmeyer, show the promotion of the use of this technology in practices such as natural history, mountaineering and hunting (see Figure 10.2). As technical developments in photography enabled a reduction in the length of time necessary for an exposure, photographers and naturalists elaborated their experiments with photography to record images of animals (Reid 1882). Eadweard J. Muybridge's photographic studies of animals in motion in the late 1870s became particularly well known (Muybridge 1881). Such improvements in technology were often themselves driven by the contemporary fascination for capturing movement in nature. Thus the French physiologist Étienne Jules Marey (1830–1904) turned to photography in the 1880s in his quest to capture movement of animals in graphic form. In 1882 Marey developed a 'photographic gun' to follow sea gulls in flight, which photographed as rapidly as 1/720th of a second (Didi-Huberman 1987). That Marey should choose to house his rapid exposure camera within the body of a gun is especially interesting, given that such photographic advances also coincided with the improvement in hunting rifles and bullets, chiefly through the development of cordite (1893). Hunting and photography; 'shots' and 'snapshots' thereafter embarked on an ever-closer association (Bennet 1914).

The use of photography in bringing mass audiences face to face with wild animals from exotic places was closely associated with the widespread and

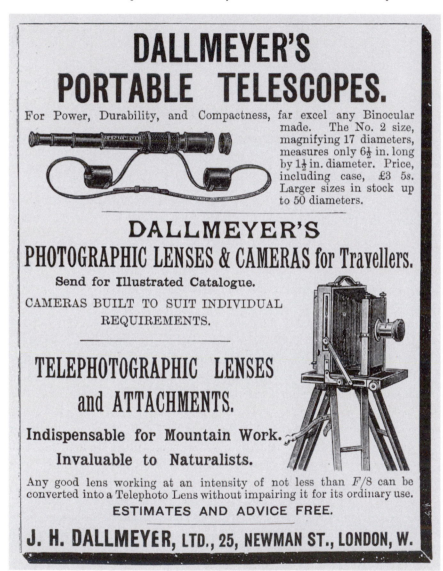

Figure 10.2 'Dallmeyer's portable telescopes, photographic lenses & cameras for travellers.'

Source: Coles, J. (ed.) (1901) *Hints to Travellers: Scientific and General*, London: Royal Geographical Society, loose advertisement

popular interest in natural history. Around the turn of the twentieth century, popular natural history publications such as *The Living Animals of the World*, with contributions by well-known 'specialists' including F.C. Selous, H.A. Bryden, and Sir Harry Johnston, drew on the collections of naturalists and sportsmen to reveal the globe's fauna through numerous photographs (Cornish undated). Stimulated by such publications and aided by the increased portability of cameras and the development of telephoto lenses, 'nature photography' became increasingly popular, spawning a range of guides and handbooks (Bedford 1909; English 1901; Johnson 1912; Snell 1905). Although aimed at amateur naturalists and photographers in Britain, many such guides drew on the wider popular fascination with 'wild life'. Richard Kearton's popular *Wild Life at Home: How to Study and Photograph It* (1898), for example, contained photographs by his brother Cherry which showed the intrepid photographer working 'in the wild' at home. Cherry Kearton subsequently embarked on adventures around the world with his camera photographing 'wildlife' (Kearton 1913). Indeed, the construction of the wildness of wild animals went hand in hand with the representation of wild spaces (and often people) of distant lands. And it was beyond Europe to Africa, particularly British East Africa, that photographer-hunters looked to practise their new sport.

Camera hunting in Africa

Opened up by colonial rule and the Uganda railway in the 1890s, British East Africa became the prime hunting ground for wealthy tourists, settlers and collectors, the latter often working on behalf of natural history museums in Europe and America (Babault 1917). Contemporary guides and publicity literature celebrated the 'Highlands of British East Africa' as a 'Field of Nature's own Making' and as the 'Greatest Natural Game Preserve in the World' (MacKenzie 1988: 200). Big-game hunters and naturalists similarly referred to it in glowing terms: as a 'paradise' and 'Garden of Eden' (Dickinson 1908). As a number of commentators have noted, the representation of East Africa as an edenic landscape was central to the production of African 'nature' as a primeval, untainted domain to be controlled by and for Europeans. The preservation and conservation strategies set up through such a way of seeing were to have serious consequences in terms of the environment and its animal occupants, as well as for the rights of indigenous African populations (Neumann 1995a). The fact that both European sportsmen and students of natural history envisaged this part of Africa in such terms made it the ideal place for the emergence of what became termed 'camera hunting'.

In 1902 the British champion of this new domain of sport and study, Edward North Buxton, a wealthy Englishman with a taste for foreign hunting and climbing adventures, described camera hunting as a field of investigation which promises most interesting results. ... [I]t demands more patience and endurance of heat and other torments, more knowledge of the habits of animals – in a word, better sportsmanship than a mere tube of iron with a trigger; and when a successful picture of wild life is obtained it is a higher achievement, even in the realm of mere sport, than a trophy, however imposing.

(Buxton 1902: v)

Buxton's elevation of camera hunting above ordinary hunting was consistent with his support for early preservation movements. Buxton was a key player in the establishment of the Society for the Preservation of the Fauna of the Empire in 1903. Made up of many well-to-do naturalists, such as Sir Harry Johnston, and hunters, such as F.C. Selous, as well as aristocrats and colonial officials such as Lord Curzon, this powerful society was to prove instrumental in generating information on the hunting and trade of animals. It was also involved in the setting up of reserves and the instigation of game laws and regulations throughout Britain's colonies (MacKenzie 1988: 211–16).

On his hunting trips to the game districts of British East Africa and the Sudan in 1899, Buxton (1902: 54, 90) claimed to have used his telephotographic naturalist camera as 'an alternative weapon to the rifle'. Although he included some photographs of live animals in his published account, notably in the section on game preservation (Buxton 1902: 117, 125), his camera was still clearly a 'weapon' and only after he had killed all of the 'specimens' he required did he devote himself 'whole-heartedly to the absorbing pursuit of camera stalking' (Buxton 1902: 106). Moreover, Buxton's animated accounts of 'camera stalking' serve ultimately to reinforce the continuity between shooting and photographing, with both of them portrayed as adventurous, predatory pursuits (Buxton 1902: 90–93). While he was concerned with the potential effects of the destruction of wildlife, Buxton was certainly not anti-hunting. Indeed, he reassured the 'conservative sportsman' that 'no one proposes to interfere with legitimate sport' (Buxton 1902: 2).

Nevertheless, Buxton received much praise from non-hunters for his advocacy of camera hunting. The naturalist and colonial administrator Harry Johnston, for example, proclaimed Buxton to be

the first sportsman of repute having the courage to stand up before a snobbish public and proclaim that the best sport for a man of cultivated mind is the snap-shotting [sic] with the camera (with or without the telephotographic lens), rather than the pumping of lead into elephants, rhinoceroses, antelopes, zebras, and many other harmless, beautiful, or rare beasts and birds.

(Johnston 1906: xiii)

Johnston (1906: xiii) claimed that camera hunting, along lines pioneered by Buxton and Schillings, represented 'the sportsmanship of the future'. Carl Schillings, a hunter in German East Africa during the late 1890s, was especially keen to capture in photographs the game that he thought would soon disappear forever. In 1903, having experimented with photographic apparatus (especially magnesium flashlight techniques for night-time photography), he undertook a large-scale expedition making photographs and collecting specimens – live and dead – for German museums and zoos. As well as a retinue of some 170 porters to carry his stores and photographic equipment, he travelled with his own taxi-dermist and surgeon. Using live bait, such as an ox, and trip wires, Schillings secured photographs of lions, leopard and hyena hunting at night. Using large metal traps placed nearby, Schillings also caught many of the same animals, most of which were shot. By day Schillings operated his camera like a 'photographic gun' following moving animal targets (Schillings 1906). Schillings described his photographic and specimen hunting expeditions as attempts to shed 'new light on the tragedy of civilisation', in order to help stem the destruction of big game in Africa and to promote conservation along lines drawn up by British authorities. His work was highly influential within British conservation circles and had a lasting influence on animal photographers (Shiras 1906).

Another convert to camera hunting was Arthur Radclyffe Dugmore, an enthu-siastic nature photographer (Dugmore 1903) who undertook a four-month expedition in British East Africa in 1908 as well as several subsequent expeditions in Africa (Dugmore 1910, 1912, 1925). Following his first expedition, Dugmore (1910: xvi) argued passionately for the cause of hunting with the camera, claiming that 'unquestionably the excitement is greater, and a comparison of the difficulties makes shooting in most cases appear as a boy's sport'. Dugmore argued that camera hunting was more exciting, skilful and dangerous than shooting, since the camera hunter had to get his cumbersome apparatus within close range of a wild animal, while often remaining unprotected (Dugmore 1925: 9–10). Part of the appeal of such photographs lay undoubtedly with this effect of close spatial proximity between hunter and animal. However, other photographs reveal Dugmore to be armed with both camera and rifle (Dugmore 1910: oppo-site 204).

Much of this points to an important redefinition of the codes and conventions of 'sportsmanship' as the language and practice of conservation gathered hold. Advocates of conservation thus continuously asserted the 'sporting' and 'manly' nature of camera hunting. Much stress was also put on the quality and genuine nature of the resulting photographs. Schillings (1907, Vol. 1: 88–89) emphasised the accuracy of his photographs, insisting that they were 'authentic, first-hand records revealing secrets which the eye of man [sic] had never before looked upon … the first to show really wild animals in full freedom'. Critical of any attempts to alter or 're-touch' images, Schillings (1907, Vol. 1: 99–100) argued that his photographs were 'true to nature'; absolutely trustworthy 'nature-documents' of African wildlife. Likewise, camera hunters such as Dugmore (1910: xvi–xvii) were equally keen to stress the genuine nature of their photographs and the absence of 'faking' or 're-touching'.

These were important facts to be underlined, given that the value of photographic hunting trophies was contested by some of the keenest hunters. The English hunter Denis Lyell, for example, was sceptical of the 'sportsmanliness' of photographing game, particularly with a telephoto lens:

> I have no wish to decry the pluck of the telephotographers, as they are brave men, but I do wish to dispel an erroneous idea which is prevalent amongst people who know nothing about hunting.
>
> (Lyell 1923: 157)

Lyell claimed in particular that 'charge photographs' (i.e. photographs of charging animals) were only 'true' if they were taken from in front of the animal, and that few animals charged unless they were wounded. It was, he argued, much more dangerous to follow a wounded animal with a rifle than to take photographs of an unwounded one (Lyell 1923).

Such debates thus focus on the 'sportsmanliness' of camera hunting and show how resulting photographs were judged on the genuine – or otherwise – spatial proximity between animal and camera hunter. In common with other camera hunters, Dugmore suggested that to hunt an animal with a camera meant approaching it at close range unarmed. However, as noted, like Buxton and Schillings, he also guaranteed the production of his photographs with a rifle. Indeed, despite his claims to avoid killing animals while in British East Africa, Dugmore put himself in positions to get dramatic photographs which more than occasionally resulted in the animal having to be shot. One photograph shows a rhinoceros apparently 'photographed at a distance of fifteen yards when actually charging the author and his companions' (Dugmore 1910: 22). However, Dugmore (1910: 22) admitted that, '[a]s soon as the exposure was made a well placed shot turned the charging beast,' and he described hunting this rhinoceros thus: '[W]hen he seemed as close as it was wise to let him come I pressed the button, and my companion, as agreed, fired as he heard the shutter drop.' This was by no means an unusual practice for camera hunters: Carl Schillings had similar encounters with rhinoceros where cameras and guns were shot simultaneously (Schillings 1906: 229).

This was also true for others like Marius Maxwell, who, inspired by Schillings and Dugmore, set out for East Africa in 1911, and again in 1921, to 'secure photographic records of incidents in big-game hunting … to obtain an accurate shot with the camera instead of the rifle' (Maxwell 1925: xxi). Although Maxwell claimed to use the rifle only in extreme cases, his attempt to photograph elephants in the wild 'exactly as the hunter would see it' led him to shoot elephants with both the camera and rifle (Maxwell 1925: 13). Photographs of camera hunters posing with cameras beside dead animals or holding both gun and camera highlight the ambiguous place of these individuals. In their very attempt to photograph wild and living nature, they were capturing, and then re-creating in photographs, the experience of hunting and killing.

The contradictions implicit in much camera hunting were later highlighted by the wildlife photographer Cherry Kearton, who criticised the activities of the photographic hunters who were 'guilty of maiming their "sitters" by gun-shot' (Kearton undated: 15). Although Kearton did not name his culprits, both Schillings and Dugmore made photographs of wounded animals (Dugmore 1910: 82–84; Schillings 1906: 321). At issue, according to Kearton, was the violence of

> a certain type of photographic expedition or safari, which, whilst pretending to forward the interests of Natural History, frequently takes as big a toll of animal life as the Big Game hunter proper, who goes out with the sole and frank idea of collecting specimens.
>
> (Kearton undated: 14)

This is not such a paradox if one remembers that early preservationists were rarely anti-hunting, but simply advocates of restricting access to game to proper 'sportsmen' (MacKenzie 1988: 295–311).

Hunting, whether with rifle or camera, remained the preserve of the white man.[3] Indigenous African groups were excluded from this new preservationist hunting code. As a number of historians have shown, the discourses of preservation and conservation of which camera hunting was part were intimately tied to the politics of colonial rule in Africa (Anderson and Grove 1987; Beinart 1989; MacKenzie 1988). Many conservation strategies excluded Africans from their traditional hunting grounds, controlling them through new spaces of 'reserves' and 'national parks'. Indeed, preservationist efforts throughout the twentieth century to establish spaces solely for 'wildlife', such as 'national parks', drew upon and reinforced ideas that Africa was a primeval wilderness. Thus, the 'preservation' of 'nature' within spaces such as the Serengeti National Park witnessed a corresponding exclusion of African indigenous groups and practices deemed 'out of place' (Neumann 1995a). This colonial vision of preserving the pristine wilderness of Africa, coupled with an underlying faith in scientific theories of racial evolution and the 'struggle for existence', also supported the characterisation of Africans as closer to the animal ancestry of the human race than Europeans. Such discourses of social Darwinism and preservation were central to both the hunting ethos and its associate practices such as camera hunting (Mackenzie 1987). In 1902 Edward North Buxton, champion of big-game preservation and camera hunting, suggested:

> We are establishing Reserves in which all kinds of wild beasts are to be left to fight it out. Can we not extend such a measure to some of the human species, to this extent – that they shall govern themselves, and the strong shall prevail?
>
> (Buxton 1902: 40)

Indeed, for many hunters and naturalists Africans were 'savage' peoples seen as a form of 'wildlife' to be managed, controlled and even hunted by Europeans. For military men like Major Hugh Chauncey Stigand, a famous big-game hunter, writer and natural historian, hunting and soldiering were blended together in the 'tracking', 'stalking' and 'marking down' of both animals and human 'savages' (Brander 1988: 144–151; Stigand 1907). Moreover, Stigand's *Hunting the Elephant* contained a chapter on 'Stalking the African', which included his recollections of warfare against various human enemies (Stigand 1913: 309–324). Such associations in practice and language fed into the discourse of 'camera hunting'. The hunter Charles Peel, with whom I began this chapter, thus compared Africans to 'tall monkeys' and described 'stalking them with a view to taking their pictures' (Peel 1927: 61). In this way, the practice of camera hunting was an instrumental part of a colonial discourse in which Africans were excluded from their own environments and resources. Similarly, Europeans claimed that Africans could not be 'sportsmen' since they hunted only for love of meat and for the lust of killing (Stigand and Lyell 1906: 5). Thus, although useful for tracking animals, Africans had no 'sporting instinct' comparable with that of 'white men' (Stigand 1913: 17). Furthermore, as photographic safaris became even more bulky than simple shooting trips, they still commanded the labour of Africans as trackers, guides and porters. Although cameras replaced guns, 'gun bearers' simply became 'camera bearers' (Dugmore 1925: 13). Similarly, gun laws, game laws and administrative policies and game reserves transformed indigenous African hunters into 'poachers'. There is certainly a complex history to such shifts, and, although my central concern here has been with the role of camera hunting in constructing African wildlife as a domain of pristine nature to be preserved by and for Europeans, it is also important to note the extensive resistance offered by different groups of Africans to colonial conservation policies (Beinart 1989: 156–8; Grove 1990). Indeed, local and global struggles over conservation agendas continue in much of Africa today (Neumann 1995b).

Conclusion

I began this chapter by noting how the fabrication of African wildlife (in this case through one giraffe) to audiences in Britain in the early part of the twentieth century involved practices of taxidermy and photography operating within discourses of empire, race and nature. As I have gone on to show, such practices involved particular forms of spatial encounter. Both animal photography and taxidermy depended upon a sense of physical closeness of human and animal, eroding the distance between the wild animal (and its habitats) and the hunter-photographer (and 'his' homelands). Such spatial encounters were the very basis of the thrill for both the hunter-photographers and the viewers of their photographed and stuffed trophies. Despite these effects of proximity, however, the display of wild animals in spaces far removed from their natural habitat served simultaneously to maintain the distance between the 'wild' and the 'non-wild'; the 'civilised' and the 'savage'. There was also considerable slippage in the attitudes and practices of white Europeans between the wild spaces of Africa, on the one hand, and the indigenous inhabitants of such spaces, both animal and human, on the other. The designation of African people as 'wild' or 'savage' tied closely with social Darwinist thinking in legitimating forms of colonial domination, including that enacted through conservation policy (MacKenzie 1987, 1988).

The major difference between photography and taxidermy in representing animals, and indeed the key to the conservationist claims of the camera hunters, was that to photograph an animal need not, unlike taxidermy, involve killing it. As the American taxidermist and naturalist Carl Akeley (1924: 45) noted in commending hunting with cameras, 'when that game is over the animals are still alive to play another day'. The contrast between the activities of individual hunters in the nineteenth and early twentieth centuries and the large-scale, professionalised photographic safaris in Africa of the 1920s and 1930s might therefore be interpreted as signalling a wider transformation in European attitudes towards wild animals: a shift away from indulgent slaughter to enlightened conservation. Yet, as I have shown, the photographic hunting that emerged around the turn of the twentieth century was far from exempt from cruelty or killing. For many of its early practitioners, the practice of 'camera hunting' was not always distinguishable from ordinary hunting. Organised photographic hunting in fact marks a shift in the terms of domination, away from a celebration of brute force over the natural world, to a more subtle though no less powerful mastery of nature through colonial management. This shift was inescapably linked to the broader colonial transformation of Africa itself, from an era of exploration and conquest to one of settlement and administration (Anderson and Grove 1987).

Though the British were the chief architects of 'camera hunting', it was by no means restricted to them. Akeley, for example, used the concept extensively during his visits to East Africa (in 1905, 1909 and 1921) for the American Museum of Natural History and in his associated hunting exploits with Theodore Roosevelt in East Africa (Haraway 1992: 26–58). Indeed, 'camera hunting' became increasingly common within the work of both European and American naturalists and film-makers in the first half of the twentieth century. Hunting with the camera should therefore be seen not solely as part of British colonial discourse, but as implicated in broader movements to create and preserve a vision of African nature as a timeless domain for white European and American 'men'.

Acknowledgements

I should like to thank participants at the 'Animals, Agency and Human Geography' session of the Annual Conference of the Royal Geographical Society (with the Institute of British Geographers) in Exeter in January 1997 for their critical feedback. Participants in the postgraduate seminar at the School of Geography, University of Oxford, in 1998 were also generous in their comments on an earlier version. I also wish to thank David Bolton of the Royal Albert Memorial Museum, Exeter; Joanna Scadden of the Royal Geographical Society Picture Library; Kay Anderson; Don Chapman; Gordon Clark; Patricia Daley; Felix Driver; Kitty Hauser; Ben Page; and Dominic Power. Finally, most importantly, I should like to thank Chris Philo and Chris Wilbert for their invaluable editorial suggestions and patience.

Notes

1 These thoughts were uppermost in my mind when, while participating in a conference session on 'Animals, Agency and Human Geography' at the University of Exeter, I came across this particular giraffe. Much of the substance of this essay is derived from a larger study of photography and British imperialism (Ryan 1997).

2 In 1906 Peel set up an 'Exhibition of Big-Game Trophies and Museum of Natural History and Anthropology' in Oxford, to house his collections. When this closed in 1919, Oxford City Council refused the collection and the Royal Albert Memorial Museum in Exeter agreed to house it in a purpose-built annex, the 'Peel Hut', where various stuffed animals were displayed in supposedly naturalistic settings.

3 White women were not generally considered to be proper hunters by white male hunters. Although hunting expeditions by women were something of a rarity, they did occur and were little different in practice to those of their male counterparts (Ryan 1997: 110–112; Stange 1997).

References

Akeley, C.E. (1924) *In Brightest Africa*, London: Heinemann.

Altick, R.D. (1978) *The Shows of London*, London: Harvard University Press.

Anderson, D. and Grove, R. (1987) 'Introduction: the scramble for Eden: past, present and future in African conservation', in D. Anderson and R. Grove (eds) *Conservation in Africa: People, Policies and Practice*, Cambridge: Cambridge University Press.

Anderson, K. (1995) 'Culture and nature at the Adelaide Zoo: at the frontiers of "human" geography', *Transactions of the Institute of British Geographers* 20: 275–294.

Anon. (1889) 'Hon. Guy C. Dawnay: obituary', *Proceedings of the Royal Geographical Society* 11: 422.

Babault, G. (1917) *Chasses et récherches zoologique en Afrique Orientale anglaise*, Paris: Plon-Nourrit et Cie.

Bedford, E.J. (1909) *Nature Photography for Beginners*, London: J.M. Dent and Sons.

Beinart, W. (1989) 'Introduction: the politics of colonial conservation', *Journal of Southern African Studies* 15(2): 143–162.

Beinart, W (1990) 'Empire, hunting and ecological change in Southern and Central Africa', *Past and Present* 128: 162–186.

Bennet, E. (1914) *Shots and Snapshots in British East Africa*, London: Longmans, Green & Co.

Brander, M. (1988) *The Big Game Hunters*, London: The Sportsman's Press.

Bryden, H.A. (1893) *Gun and Camera in Southern Africa*, London: Edward Stanford.

Bull, B. (1992) *Safari: A Chronicle of Adventure*, London: Penguin.

Buxton, E.N. (1902) *Two African Trips: With Notes and Suggestions on Big Game Preservation in Africa*, London: Edward Stanford.

Cornish, C.J. (ed.) (undated) *The Living Animals of the World: A Popular Natural History*, London: Hutchinson and Co.

Cumming, R.G.G. (1850) *Five Years of A Hunter's Life in the Far Interior of Southern Africa* (2 vols), London: John Murray.

Dawnay, G.C. (*c.*1880) Album of sepia prints, Royal Geographical Society photographs C82/006132–301.

Dickinson, F.A. (1908) *Big Game Shooting on the Equator*, London: John Lane.

Didi-Huberman, G. (1987) 'Photography – scientific and pseudo-scientific', in J. Lemagny and A. Rouillé (eds) *A History of Photography: Social and Cultural Perspectives*, Cambridge: Cambridge University Press.

Dugmore, A.R. (1903) *Nature and the Camera: How to Photograph Live Birds and Their Nests; Animals, Wild and Tame; Reptiles; Insects; Fish and Other Aquatic Forms; Flowers, Trees and Fungi*, London: Heinemann.

Dugmore, A.R. (1910) *Camera Adventures in the African Wilds: Being an Account of a Four Months' Expedition in British East Africa, for the Purpose of Securing Photographs of the Game from Life*, London: Heinemann.

Dugmore, A.R. (1912) *Wild Life and the Camera*, London: Heinemann.

Dugmore, A.R. (1925) *The Wonderland of Big Game*, London: Arrowsmith.

English, D. (1901) *Photography for Naturalists*, London: Iliffe.

Greenhalgh, P. (1988) *Ephemeral Vistas: The Expositions Universelles, Great Exhibitions and World's Fairs, 1851–1939*, Manchester: Manchester University Press.

Grove, R. (1990) 'Colonial conservation, ecological hegemony and popular resistance: towards a global synthesis', in J.M. MacKenzie (ed.) *Imperialism and the Natural World*, Manchester: Manchester University Press.

Haes, F. (1865) 'Photography in the Zoological Gardens', *Photographic News* 10: 78–79, 89–91.

Haraway, D. (1992) *Primate Visions: Gender, Race, and Nature in the World of Modern Science*, London: Verso.

Hauser, K. (1998) 'Coming apart at the seams: taxidermy and contemporary photography', *MAKE: Magazine of Womens' Art* 82: 8–11.

Johnson, S.C. (1912) *Nature Photography: What to Photograph, Where to Search for Objects, How to Photograph Them*, London: Hazell, Watson & Viney.

Johnston, H.H. (1906) 'Introduction', in C.G. Schillings, *With Flashlight and Rifle: A Record of Hunting Adventures and of Studies in Wild Life in Equatorial East Africa*, London: Hutchinson and Co.

Kearton, C. (1913) *Wild Life Across the World*, London: Arrowsmith.

Kearton, C. (undated) *Photographing Wild Life Across The World*, London: Arrowsmith.

Kearton, R. (1898) *Wild Life at Home: How to Study and Photograph It*, London: Cassell and Co.

Lyell, D.D. (1923) *Memories of an African Hunter*, London: T. Fisher Unwin.

MacKenzie, J.M. (1987) 'Chivalry, social Darwinism and ritualised killing: the hunting ethos in Central Africa up to 1914', in D. Anderson and R. Grove (eds) *Conservation in Africa: People, Policies and Practice*, Cambridge: Cambridge University Press.

MacKenzie, J.M. (1988) *The Empire of Nature: Hunting, Conservation and British Imperialism*, Manchester: Manchester University Press.

MacKenzie, J.M. (ed.) (1990) *Imperialism and the Natural World*, Manchester: Manchester University Press.

Maxwell, M. (1925) *Stalking Big Game with a Camera in Equatorial Africa*, London, William Heinemann.

Mitchell, T. (1989) 'The world as exhibition', *Comparative Studies of Society and History* 31: 217–236.

Muybridge, E.J. (1881) *The Attitudes of Animals in Motion: A Series of Photographs Illustrating the Consecutive Positions Assumed by Animals in Performing Various Movements; Executed at Palo Alto, California, in 1878 and 1879*, San Francisco.

Neumann, R.P. (1995a) 'Ways of seeing Africa: colonial recasting of African society and landscape in Serengeti National Park', *Ecumene* 2: 149–169.

Neumann, R.P. (1995b) 'Local challenges to global agendas: conservation, economic liberalization and the pastoralists' rights movement in Tanzania', *Antipode* 27: 363–382.

Peel, C.V.A. (1900) *Somaliland: An Account of Two Expeditions into the Far Interior*, London: F.E. Robinson and Co.

Peel, C.V.A. (1901) *Wild Sport in the Outer Hebrides*, London.

Peel, C.V.A. (1906) *Popular Guide to Mr. C.V.A. Peel's Exhibition of Big-Game Trophies and Museum of Natural History and Anthropology*, Guildford.

Peel, C.V.A. (1927) *Through the Length of Africa*, London: Old Royalty Book Publishers.

Philo, C. (1995) 'Animals, geography and the city: notes on inclusions and exclusions', *Environment and Planning D: Society and Space* 13: 655–681.

Reid, C. (1882) 'Some experiments in animal photography', *British Journal of Photography* 29: 216–218.

Ritvo, H. (1990) *The Animal Estate: The English and Other Creatures in the Victorian Age*, Harmondsworth: Penguin.

Ryan, J.R. (1997) *Picturing Empire: Photography and the Visualisation of the British Empire*, London: Reaktion Books.

Schillings, C.G. (1906) *With Flashlight and Rifle: A Record of Hunting Adventures and of Studies in Wild Life in Equatorial East Africa*, London: Hutchinson and Co.

Schillings, C.G. (1907) *In Wildest Africa* (2 vols), London: Hutchinson and Co.

Schneider, W.H. (1982) *An Empire for the Masses: The French Popular Image of Africa, 1870–1902*, Westport, CT: Greenwood.

Selous, F.C. (1881) *A Hunter's Wanderings in Africa*, London: Bently & Son.

Selous, F.C. (1893a) *Travel and Adventure in South-East Africa*, London: Rowland Ward and Co.

Selous, F.C. (1893b) 'Twenty Years in Zambesia', *Geographical Journal* I: 289–324.

Shiras, G. (1906) 'Photographing wild game with flashlight and camera', *National Geographic Magazine* 27: 367–423.

Snell, F.C. (1905) *A Camera in the Fields: A Practical Guide to Nature Photography*, London: T. Fisher Unwin.

Sontag, S. (1979) *On Photography*, New York: Penguin.

Stange, M.Z. (1997) *Woman the Hunter*, Boston: Beacon Press.

Stigand, C.H. (1907) *Scouting and Reconnaissance in Savage Countries*, London: Hugh Rees.

Stigand, C.H. (1913) *Hunting the Elephant in Africa, and Other Recollections of Thirteen Years' Wanderings*, New York: Macmillan.

Stigand, C.H. and Lyell D.D. (1906) *Central African Game and Its Spoor*, London: Horace Cox.

Storey, W.K. (1991) 'Big cats and imperialism: lion and tiger hunting in Kenya and Northern India, 1898–1930', *Journal of World History* 2: 135–173.

Thomas, N. (1991) *Entangled Objects*, New York: Harvard University Press.

Tidrick, K. (1990) *Empire and the English Character*, London: I.B. Tauris & Co.

Ward, R. (1882) *The Sportsman's Handbook to Practical Collecting, Preserving and Artistic Setting-up of Trophies and Specimens* (Second Edition), London: Rowland Ward.

Ward, R. (1911) *The Sportsman's Handbook to Collecting, Preserving and Setting-up Trophies and Specimens* (Tenth Edition), London: Rowland Ward.

Whatmore, S. and Thorne, L. (1998) 'Wild(er)ness: reconfiguring the geographies of wildlife', *Transactions of the Institute of British Geographers* 23: 435–454.

Wilson, A. (1992) *The Culture of Nature*, Oxford: Blackwell.

Wolch, J. and Emel, J. (eds) (1995) Guest edited issue: 'Bringing the animals back in', *Environment and Planning D: Society and Space*, 13: 631–730.

York, F. (1872) *Animals in the Gardens of the Zoological Society, London, Photographed from Life*, London.

11 Biological cultivation

Lubetkin's modernism at London Zoo in the 1930s

Pyrs Gruffudd

Introduction

The most famous zoo structures in Britain are probably the modernist Penguin Pool and Gorilla House built in London Zoo in the 1930s by the Tecton firm led by Berthold Lubetkin. These buildings were the product, in part, of stylistic experimentation led by a radical intelligentsia – the sort of experimentation that elsewhere led to the construction of modernist villas or apartment dwellings. They therefore signified the absorption of London Zoo into a new aesthetic and cultural context. But arguably, and more importantly, a new fusion of social and political radicalism, on the one hand, and popular science, on the other, was the driving force behind the emergence of zoo modernism. In this sense, London Zoo was a symbol of the contemporary concern for 'planning' and for a reformed relationship between humanity and nature. While the stylistics of modernism appeared to place it at an intellectual distance from nature, closer study of the projects reveal that these buildings emerged from a new theorisation of the natural world and the efficient nurture of life. This chapter considers the structures built at London Zoo, alongside Tecton's other contemporary projects, not as attempts to dominate or to transcend nature but as experiments in harmonising living creatures and their designed spaces through the medium of a scientifically informed modernism. While the creatures in London Zoo were obviously non-human, they were nonetheless considered as indicators of humanity's fate, and the Zoo enclosures were widely read as symbolising deficiencies and new hopes for a humane, and human, urban planning.

Writing about the Adelaide Zoo, Anderson (1995: 276) argues that

> zoos ultimately tell us stories about boundary-making activities on the part of humans. In the most general terms, Western metropolitan zoos are spaces where humans engage in cultural self-definition against a variably constructed and opposed nature. With animals as the medium, they inscribe a cultural sense of distance from that loosely defined realm that has come to be called 'nature'.

In London Zoo in the 1930s that sense of distance was arguably diluted somewhat and animals came to be seen as occupying a pivotal relationship between humanity and nature. To the mind of modernist reformers and applied scientists, they were organisms to be understood, nurtured and housed efficiently, as, indeed, were humans.

'Biotechnic' planning

There was a clear relationship between modernist architecture and concern for environment and health in the 1930s (see Gruffudd 1995). That relationship was multi-faceted. It was expressed in concern for the patterns of ill-health, and specific diseases like TB, generated by dark, insanitary and cramped areas of the Victorian industrial city, and in the concurrent belief in the health-giving benefits of the new architecture – natural light, fresh air and sanitary design. The *Architects' Journal* (*AJ*) in June 1933 graphically illustrated the problem of over-crowding in a series of striking textual and visual images. The smoky, huddled Manchester sky-line, for instance, was set against the rational angular forms of a German housing development:

> The Manchester sky-line ... discloses itself fitfully between mercifully frequent wisps of fog as a devil's cauldron of chimneys, belching soot over long funereal slabs of black-slated roof. ... Here free Britons live, are begotten, born and die. The gleaming white and red tenements at Lichtensberg house an identical type of population – not huddled together literally in stall-like hutches, but in spacious, balconied flats, with abundant light and air, a clear sky overhead, and grass to look out upon.
>
> (*AJ* 1933: 832)

Clearly allied to an architectural discourse was a much wider debate about the virtues of 'planning' – the rational, scientific ordering of human social and spatial life. Most histories of modernism stress the radicalism of the modernist project and its emergence, in Britain at least, from socialist-intellectual circles (e.g. Gold 1997), and this chapter builds, in part, on that reading. But it also offers a challenge to the tendency to view modernist architecture and planning as solely the discourse of the machine. There have recently been some reinterpretations of modernism which stress its ambivalent – or even positive – relationship with tradition, with mysticism and (most importantly here) with the organic and with nature (e.g. Matless 1992, 1998). For Lewis Mumford (1945: 412), the historian of technology, for instance, modernism was all about the emergence of a biotechnic age:

'[L]iving' in architecture means in adequate relation to life. It does not mean
an imitation in stone or metal of the external appearance of organic form:
houses with mushroom roofs or rooms shaped like the corolla of a flower. It
was not flowered wallpaper that was needed in the modern house, but space
and sunlight and temperature conditions under which living plants could
grow: not pliant and moving lines in the furniture, but furniture that responded
adequately to the anatomical form and physiological needs of the body: chairs
that favoured good posture and gave repose, beds that favored sexual inter-
course and permitted deep slumber. In short: a physical environment that
responded sensitively to the vital and personal needs of the occupants.

(Mumford 1945: 412)

In city planning, by extension, biotechnic modernism was about air, sunlight,
gardens, parks, playgrounds and all the various forms of social equipment that are
necessary for the stimulating life of the city. Mumford went further in his reorientation of modernism towards nature, claiming that nearly all of the new methods
of construction, the new materials, and the new means of regulating the air of a
building to adapt it to the living organisms inside came directly from the biotechnics of gardening.

Nature tamed with a smile: Lubetkin at London Zoo

This theoretical interest in the nurturing of living organisms, in the form of
animals, would eventually find expression in the work of the Russian emigré
architect Berthold Lubetkin, one of the central figures in British early
modernism. Lubetkin was born in Tblisi in Georgia in 1901 and lived and studied
in Germany, Poland and Paris before arriving in Britain in 1931. British architectural culture at the time was still dominated by the Edwardians, and civic and
corporate patrons tended to favour variants on a neo-classical style. There were,
however, numerous spaces for experimentation, including private houses, leisure
facilities and industrial and transport buildings. As Allan's (1992) outstanding
biography of Lubetkin points out, however, paralleling this growing architectural
innovation was a burgeoning scientific culture that directed disciplines including
architecture to new social and political ends. One of the key figures in that movement was the evolutionary biologist and later Secretary of the Zoological Society,
Julian Huxley, discussed below. In 1932, Lubetkin, together with five other young
architects, formed the partnership Tecton. This co-operative firm has been seen as
demonstrating all of the stylistic and ideological components of 1930s
modernism. The partners built their practice on that newly radical scientific
commitment to large-scale survey and analysis, with a consequential concern for
function and structure. While many of their early patrons were wealthy, they

would later develop several projects explicitly aimed at improving the living standards of inner-London's working-class population, basing this professional work
on a clearly expressed socialist foundation (see Coe and Reading 1981).

It may appear surprising, then, that Lubetkin and Tecton's first major commissions should have been an apparently marginal series of buildings for the
Zoological Society at London Zoo in Regent's Park. But these buildings, as Allan
(1992) notes, gave the British mass public its first taste of modern architecture,
and they were an instant and unqualified success. They also received critical
acclaim. The commissions originated with Solly Zuckerman, a Research
Anatomist with the Zoological Society from 1928 and Demonstrator at
University College London. He recommended to Dr Geoffrey Vevers, the Zoo's
superintendent, and Sir Peter Chalmers-Mitchell, Secretary of the Zoological
Society, both socialists (like Lubetkin), that the Tecton firm might be offered the
commission for designing a home for two young gorillas newly acquired from the
Belgian Congo. The proposal was accepted, but Chalmers-Mitchell imposed a
deadline for the design stage of merely four days. Recalling his thoughts on the
design process, Lubetkin wrote:

> There are two possible methods of approach to the problem of zoo design;
> the first, which may be called the 'naturalistic' method, is typified in the
> Hamburg and Paris zoos, where an attempt is made, as far as possible, to
> reproduce the natural habitat of each animal; the second approach, which,
> for want of a better word, we may call the 'geometric', consists of designing
> architectural settings for the animals in such a way as to present them
> dramatically to the public, in an atmosphere comparable to that of a circus.
>
> (In Allan 1992: 199)

The 'naturalistic' approach, in which animals were presented in cages or enclosures that made some gesture to their native habitat, predominated in Europe at
this time, accounting for the last major building project in London Zoo – the
Mappin Terraces, completed some twenty years earlier. Lubetkin rejected this
fake naturalism or tokenism (see Berger 1980), however, and adopted the
geometrical approach in his designs. This was, in part, pragmatic. One of the
zoo's functions was, he claimed, entertainment; a naturalistic enclosure allowed
shy animals to hide but also failed to provide the more extrovert the stage on
which to perform. The architect's task was to understand the animal in its
essence, and then to coax it through design to display its distinctive characteristics. (Form follows function, as it were.) But the designs were also ideological.
Lubetkin's work was based on a philosophical understanding in which humans
were seen as highly evolved biological organisms in some senses set apart from
nature. So, while Lubetkin demonstrated a real fondness for animals, it was

within a clearly expressed hierarchy where humans were the rational superior. When Lubetkin later became a Cotswold pig-farmer, for instance, he accepted giraffe and elephant 'evacuees' onto his farm during wartime, but equally gave his pigs numbers rather than names (Gardiner 1987). Lubetkin's zoo enclosures are unequivocally modernist in that they affirm the power of the human mind to alter environment through geometry; but they are also honest in that they place that power within a nexus of rights and responsibilities. As Allan puts it, Lubetkin's zoo approach rejected both camouflage and conquest as metaphors of human intervention in nature. Human instrumentality was to be both explicit *and* benign: 'nature tamed – not with a fist, but with a smile' (Allan 1992: 201).

Lubetkin and Tecton's buildings were also significant in that they affirmed the importance of architecture and planning in the nurturing process. They were biological cultivation in essential form, stripping away layers of culture and treating the animal organism as a series of characteristics and desires. The 'scientific' nature of the zoo projects allowed Tecton to embark on the detailed analysis and survey that would become their hallmark. During the four-day design period for the Gorilla House Lubetkin bombarded Zuckerman, an expert on primates, with a series of questions about the habits and environmental needs of gorillas. Zuckerman had already published *The Sexual Life of Primates* and published *The Social Life of Monkeys and Apes* in 1932 (see Haraway 1992). His research methods were unorthodox: he placed a baboon named Betsy in the Hampstead home of the architect Clough Williams-Ellis for several months as he wanted to see how it would react to 'bright, intelligent young society'. The Williams-Ellises were noted for their progressive child-rearing, it seems, and it was even thought that the baboon might baby-sit for them. But, as Clough sadly reminisced, this was '[n]ot the social success that we had hoped, unresponsive and dirty, we bade our disappointing little lodger farewell without regrets' (Williams-Ellis 1980: 157).

Although there was surprisingly little agreement on ideal conditions for enclosed gorillas, Lubetkin generated a consensus based on existing research. This showed that they were particularly prone to catching human diseases like flu and colds. They were also sensitive to bad weather, despite coming from the Belgian Congo with its variable climate (*The Times* 1932). Lubetkin solved this problem in the Gorilla House, opened in April 1933. Rather than attempting to reproduce their native habitat in the Congo, he produced the pure, constructivist form of a concrete cylinder building with a cage frontage that could be closed off by sliding screens during inclement summer days. It could also be permanently closed during the winter, creating an interior viewing space for visitors. This – together with a raised stage and recreational apparatus – created a balance between care for animals and human curiosity. There were also powerful electric lamps to simulate sunshine. In addition, a circulation system was devised that would purify and humidify the air. As one professional journal put it: 'The air [the gorilla] breathes is washed free from dust and fog, heated, humidified and served up as hygienically as possible. If the anticipations of the architects are fulfilled it is obvious that the best place in London for a breath of fresh air will be the Gorilla House' (*Architectural Design and Construction [ADC]* 1933: 317).

Considerable press attention focused on the apparent autonomy granted to the animals. A push-button water fountain in the centre of the cage, for instance, allowed the gorillas to drink and to bathe when they chose. This, doubtless in concert with the apparently 'human' character of the primates (they were often referred to as 'star comedians', 'variety performers' or 'entertainers'), stimulated an immense amount of anthropomorphism in the press. For instance, the revolving walls were built,

> of course, as a direct result of a petition from the higher apes who complained bitterly of the vagaries of our climate. Then, no doubt, the chief gorilla having heard the weather forecast will instruct the head keeper to take the necessary steps to preserve an equable climate in his commodious cage. He – the gorilla, not the keeper – is a delicate subject, liable to catch cold and particularly prone to influenza.
>
> (*ADC* 1933: 317)

Aspects of fashionable, modern urban dwelling were alluded to by referring to the gorillas' new 'flat' into which the animals had moved and where they staged a 'house-warming party' for Fellows of the Zoological Society (*Daily Telegraph* 1933). The *Morning Post* (1933) laid a layer of domestic bliss on the scene:

> Mok soon realised that they had a sunshine roof and a constant stream of purified air. He beat his chest, and tore round the cage. Madame Moina, as the house-wife, was more interested in the revolving walls and dust-proof screens, and shot up to the ceiling on a length of rope, the better to examine them.

While the other animals were allegedly 'jealous' of Mok and Moina's 'luxury home', critical acclaim followed. Zuckerman (1933), writing delightedly from his new post at Yale University, said: 'I shall see to it, Mr Tecton, that you are appointed official architect to all gorillas, chimpanzees, monkeys, human beings, and inhuman beings.' In more measured tones, Mumford (1937), in his column in *The New Yorker*, hailed the Gorilla House as the only example of a truly flexible architecture – a building adapted perfectly for a healthy and evolving function – in the Museum of Modern Art's exhibition on English modernism.

Even greater public and professional acclaim came in 1934 with the completion of the Penguin Pool (see Figure 11.1), possibly the best-known modernist artefact in Britain. It was named as one of the most significant buildings of its time by the *AJ*'s 'vigilance committee', with votes from the likes of poet John Betjeman, artist John Piper and London Transport's Frank Pick (*AJ* 1939). It entailed a strictly and elegantly mathematical concrete ellipse enclosing a shallow pond. In the centre was a pair of concrete spiral ramps forming an apparently interlocking double helix shape, which, without external support, sloped down gradually from a platform into the water. The form was, apparently, inspired by an egg. As Lubetkin put it:

> An egg is one of nature's perfect shapes. It has a purity that I would compare to a Brancusi. In the mathematical interpretation, the setting out of the structure had to retain the eloquence of an egg, for I wanted the cool, intellectual geometry of the form to be intensified by the organic growth of a nearby tree. ... It's the necessary contrast, you know, ... between the wildness of nature and the order of architecture, that's the underlying theme of all Palladio's work.
>
> (In Gardiner 1987)

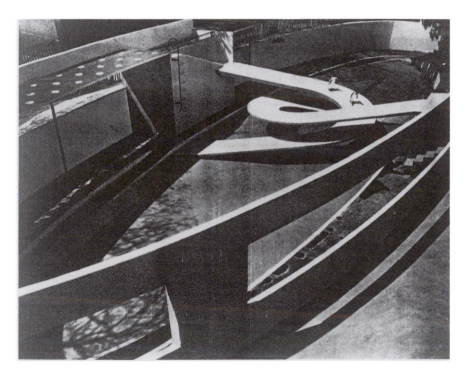

Figure 11.1   The Penguin Pool at London Zoo.

Source: Architects' Journal (1934) 79, 14 June, p. 858

As with the Gorilla House, Lubetkin avoided the 'naturalistic' style, although the abstraction gestured to it. According to Allan (1987), the snow-white concrete structure and azure underwater floor did not intend a representation of Antarctica but offered 'a visual metaphor for it'. The aim was the construction of a biologically efficient space that would also fulfil the Zoo's entertainment and education functions. The Penguin Pool provided flat, concrete areas and shade at various levels in addition to a shallow pool some ten metres long. Nesting boxes were set into the external walls, and viewing areas for the public were provided around the outside. The pool's design also created an underwater view refracted above water level. Again, the building was biologically functional: it was easy to clean, provided shelter for the birds and easy systems for access to nesting boxes, which were, however, away from the public. Several contemporary reports stressed the physiological benefits for the penguins. The variety of surfaces (rubber, slate, concrete) and levels were part of 'a genuine attempt ... to preserve the birds from the boredom which generally overtakes all zoo inhabitants' (*AJ* 1934: 857). Initial studies showed that the penguins were more active than in other enclosures, and spent more time in the water. They seemed to take to the pond immediately and to breed successfully (see Allan 1992). Later evidence was, however, contradictory. When a resident of one of Tecton's renowned Highpoint flats wrote in praise of the pool in 1950, the Superintendent, G.S. Cansdale, replied that he was less interested in aesthetics than in accommodation for the birds, and that '[i]n this respect it has failed dismally' because it was a heat-trap in summer: 'in fact the only point in its favour is that it allows a good view of the birds' (Cansdale 1950; on Cansdale see Philo 1995: 661–662).

That sense of theatricality implied by human intervention in the natural world through zoos was also heavily to the fore here. This was a building in which 'penguinness' was produced. In the water the penguins could demonstrate their natural grace and speed. The ramps and walkways, however, provided the comical contrast. As the *AJ* (1934: 857) put it:

> Penguins have an attractive and faintly ridiculous quality: their shape and black and white colouring produce an almost human effect, enhancing the grotesque awkwardness of their movements on land. ... The two cantilevered ramps – have a theatrical quality and provide a suitable stage for the waddling gait of the penguins, who are shown to be able to hop as well as waddle by the stepped ramps. – The pond, in fact, explains the characteristics of the penguins and 'produces' them effectively to the public.

Again, the penguins 'produced' by the pool were continually anthropomorphised, one cartoon in *Punch* showing a row of snooty penguins impatient at being held up at a 'pedestrian crossing' between the two ramps. The pool was also said to offer 'a suitable setting for any latent publicity talent' on the birds' part (*AJ* 1934: 857). Nonetheless, in transmitting the penguins' character to the watching public, the pool also transmitted lessons for human society. The architecture critic C.H. Reilly (1935: unpaginated) argued that the pool was:

> a complete and unique conception. It has unity and elegance and lightness all in a very high degree. It uses our beloved concrete in a way which opens up new vistas for us all. Who would have thought a few years back that those broad, curved sloping ways, graceful and light as the spring of a watch and associated in the mind at once with the wing of a bird, could be made in it? One can see [the] new world of light and air and glass and silver, which is to come about when our noses are more sensitive, already approaching. I hope I live long enough to have a small town house, I suppose with one ramp for my wife and another for myself as circumscribed, and complete for my needs and with no possible addition or alteration – indeed the perfect unity. No doubt I shall have to satisfy my habits before I am worthy to live in such a thing of beauty, but that would be very good for me as for most of us.

Lubetkin's close scrutiny of animal character and behaviour was further exemplified in the unexecuted designs for a Gibbon House for the Zoological Society's other site at Whipsnade. It was a thin, curved concrete shell acoustically designed to amplify the gibbons' characteristic cries and to provide an artificial environment where they could be viewed both in the 'open' and 'indoors'. Another completed, though now demolished, project was a Studio of Animal Art for Regent's Park. It was designed to facilitate the close study and drawing of animals in a hygienically controlled and calm environment. The main studio accommodated twenty-five art students on three separate platforms – dispersing the observers so as to minimise the animals' nervousness – and the room as a whole was parabolic in shape because

> all animals prefer to have walls on three sides as this gives a sheltered feeling, and they are less apt to move. The parabola was found to be the shape which gave the maximum of protection to the animal, together with the best possible visibility.
> (*Architect & Building News* 1937: 155)

Several other commissions were obtained by Lubetkin and Tecton for the Zoological Society's premises both in London and in Whipsnade. The semi-rural location of Whipsnade allowed Lubetkin to indulge his interest in the contrast between his modernist constructions (and, by extension, human rationality) and nature. Both the Giraffe House and the Elephant House were set against dense foliage, the former being constructed of wood and brick. There was also a separate, and in many ways more comprehensive, commission at Dudley Zoo where a series of buildings echoing the themes pioneered in London were set around a medieval castle (see Allan 1992).

The science of life: Huxley, animals and evolution

The London commissions gained the fervent support of Julian Huxley, the new Secretary to the Zoological Society appointed in 1935. Huxley was the grandson of Thomas Henry Huxley and the brother of the novelist Aldous Huxley (author of *Brave New World*). At the time of his appointment to the Zoo, Julian Huxley was Professor of Zoology at King's College London, where his scientific reputation was built mainly on his influential writings on the theory of evolution (e.g. Huxley 1942). A paper on the courtship rituals of the great crested grebe published in 1914 was regarded as one of the pioneer academic studies in animal behaviour; his works on the effects of genetic changes and on differential growth were turning-points in the study of these subjects. But Huxley was also a progressive, socialist intellectual committed to popularising science (e.g. Wells *et al.* 1931) and making it relevant to citizens' everyday lives. He brought this zeal to his new post at London Zoo, establishing the popular new magazine *Zoo*, and setting up 'Pets Corner' (which later became the Children's Zoo). The Studio of Animal Art was also one of Huxley's innovations. It was intended to be part of a complex to include a cinema where scientific and zoological films could be shown to schoolchildren and visitors.

On his arrival at the Zoo, however, Huxley discovered considerable opposition to Lubetkin and Tecton among Society members. In a lengthy memorandum (Huxley 1937a) in support of the architects, he quoted leading architectural commentators by way of arguing for Tecton's suitability as architects for the Zoo. Huxley framed the matter as an important one of principle. Embracing the innovations of modernist style was essential because it signified a return to 'first principles' in building. Lubetkin – 'an excellent planner of animal accommodation' (Huxley 1937a: 5) – had gone back to these first principles in his studies of animal needs and behaviour: '[H]e is always considering the health of the animals and their qualities as objects to be exhibited to the public' (Huxley 1937a: 4). The innovations of modernism then allowed those characteristics to be reflected in the designs. As Huxley (1937a: 1) put it:

> Modern construction is … valuable in facilitating the use of many unusual forms. … This is of particular importance in zoo architecture since the exhibition of animals presents the architect with many new problems, which can often be best solved by departing from conventional or traditional forms (e.g. the Penguin Pool with its elliptical plan).

He believed that zoo culture as a whole could only gain from continued collaboration with the architects:

> [T]he Society will be earning the gratitude of later generations by giving a lead in the matter of modern construction and by doing so will also play its part in the field of zoo architecture enabling many new types of zoo building to be constructed.
>
> (Huxley 1937a: 4)

Furthermore – and importantly for Huxley's notion of popular science – the embracing of modernism marked the Zoo out as being a pioneering and accessible institution. As he argued, 'in adopting some type of modern construction, the Society is giving a valuable lead in the inevitable confusion of a transitional period, and not merely indulging a passing whim of artistic fashion' (Huxley 1937a: 1).

Huxley's advocacy of Lubetkin and Tecton was ultimately constrained by the Society's Council. In 1937 Lubetkin and Huxley appeared in a television broadcast to show the new plans for an Elephant House. This was a controversial project based on a £10,000 donation from the Maharajah of Bhavnagar. Again, Huxley solicited support from leading figures in the art, architectural and scientific worlds. Kenneth Clark, the Director of the National Gallery, praised the building's concern for the elephants' 'domestic economy' and noted how 'severe concentration' on this and other factors 'has produced a powerful and dignified piece of architecture which has a quality of work from which all superfluities have been eliminated by intellectual labour'. Furthermore, the combination of lighting and the massing of the building 'has led unconsciously to an elevation suggestive of the character of an Elephant' (quoted in Huxley 1937b). However, the commission was later granted to another firm amidst much acrimony as Huxley's relationship with the Zoological Society deteriorated. Huxley was even moved to write to the Maharaja's Prime Minister urging that the gift be withheld in the event of Tecton losing the commission (Huxley 1944a). The more conservative members of the Council were undoubtedly unappreciative of Tecton's modernity, but personal dislike of Huxley seemed equally significant. The original drawings for the Gibbon House, for instance, are annotated with the following note attributed to Lubetkin: 'Unexecuted; backed by Julian Huxley, but (as tended to happen) this very fact put the Zoo directors against it, since Huxley regarded by them as apt to be overbearing and too little willing to discuss with them' (Drawing N40/28. London Zoo, Gibbon House, *c.*1933, Royal Institute of British Architects Drawing Collection).

Nonetheless, this collaboration between Lubetkin and Huxley reveals some of the broader relevance of the zoo buildings in the culture of modernism. This can be further exemplified by the work of the Architects and Technicians Organisation

(ATO), established in 1935 by Lubetkin and other Tecton members. One of its main aims was the improvement of working-class housing, and, as Gold (1997) demonstrates, the members' socialism informed their campaigning and their prescriptions. Among the other members of ATO were scientists like J.B.S. Haldane, Solly Zuckerman and Julian Huxley (Coe 1982). According to Allan (1992: 314), groups like the ATO 'must be seen within [a] wider spectrum of analytical and agitational activity and be identified with the then radical consensus that regarded centralised planning, with its assumed benign paternalism, as the key instrument of social progress'. Science, of course, played a key role in that vision of centralised planning, with the figure of the 'expert' to the fore. Huxley dedicated himself to the social role of science, primarily through his interventions in eugenic debates on nature versus nurture, but increasingly through a belief in the need for a scientifically organised and planned society. In 1934 he published *Scientific Research and Social Needs*, a powerful survey of scientific application to society in such wide-ranging realms as food, building and health, as well as industry and war. Science, he argued, was 'not the disembodied sort of activity that some people would make out, engaged on the abstract task of pursuing universal truth, but a social function intimately linked up with human history and human destiny' (Huxley 1934: 279). He also served on the 'think-tank' Political and Economic Planning in the 1930s and the Hobhouse Committee on National Parks in the 1940s. The finest examples of centralised planning were not, however, to be found in Britain. In *A Scientist Among the Soviets*, published in 1932, Huxley recalled one vision of co-ordinated planning – the Five-Year Plan – based on radical social and scientific principles: '[P]roper planning is itself the application of scientific method to human affairs; and also it demands for pure science a very large and special position in society' (Huxley 1932: 52). Such vision was not limited to totalitarian states, though, and in *TVA: Adventure in Planning* Huxley (1943) outlined his favourable impressions of the Tennessee Valley Authority in the United States, noting positively how it tackled social and environmental problems on a grand scale.

Perhaps the significance of Lubetkin and Huxley's collaboration at London Zoo was the fact that scientific research was being applied in the solution of environmental and domestic problems within a coherent institutional setting. Huxley (1937a) was anxious that the Zoo site at Regent's Park be treated as a coherent architectural whole. In this sense, the Zoo was an experiment in social engineering or planning. Coe (1982: 55) argues that 'the buildings were seen not just as zoo buildings in isolation but actually as a microcosm of a programme of building for a reformed society and that message was not lost on an astute public'. Certainly the press at the time frequently pointed to the irony that humans lived in relative deprivation compared to the animals of London Zoo. Of the Penguin Pool, the journal *Mother and Child* (1938: 297) asked:

How many citizens of London have brooded over the railings of that pool, envying the penguins as they streak through the blue water or plod up the exquisite incline of the ramp – and have wondered sadly why human beings cannot be provided, like the penguins, with an environment so adapted to their needs?

Huxley himself, in *Scientific Research and Social Needs* (1934), illustrated the contrast in housing between human habitations in the slums – dark, narrow and listlessly set against a grey sky – with the Gorilla House at London Zoo, geometrically gleaming against a sunlit sky and framed on one side by trees (see Figure 11.2).

But there was more to this relationship between animal and human than mere jealousy or irony. According to Reading (1992: 29), the animal world at the Zoo 'was presented as an opportunity to observe the working of society in a micro-cosm'. In the culture of modernism, as Mumford (1945) would later note, architecture and planning were about the healthy and efficient nurture and evolu-tion of life. Lubetkin and Huxley's work at London Zoo, viewed from this intellectual standpoint, was a series of central experiments in – rather than marginal diversions from – the pursuit of the healthy and well-ordered modernist society. The animals were read as living beings whose nurture was subject to the same rules of biology, psychology and even sociology as humans. Urging the architectural profession to 'build for today not yesterday', the critic Paul Reilly (1937) noted that Lubetkin began by studying the wants of the animals which his buildings were to house, and then proceeded by inquiring what materials could best satisfy these wants. Human society, he continued, also had to develop in this nurturing fashion, and he – like Huxley and Mumford – assertively championed modernism as the most vital way of satisfying the biological needs of tomorrow.

In an article called 'The descent of man', however, The *Architects' Journal* (1933) pointed to the evolutionary irony of the comparison between animal and human habitations. While evolution had advanced humans from the status of mere beasts, the conditions of urban life in the early twentieth century threatened to return humans down the evolutionary ladder. A pair of photographs (see Figure 11.3) contrasted a squalid, dark and infested tenement bedroom with the cleanliness and functionalism of the new Gorilla House at London Zoo, the photograph showing a gorilla silhouetted against a shaft of bright sunshine. Quotations drawn from descriptions of the Gorilla House and from Medical Officers of Health reports respectively further pointed to the standards established in the Zoo's experimentation in planning:

> The new Gorilla House in the Zoological Gardens, London, shows the thoughtful care devoted to this housing problem. Gorillas are liable to the diseases of man, and are, therefore, protected in winter by movable glass screens.

Figure 11.2   'A contrast in housing. Compare these human habitations – in the slums – with this
home for apes – at the London Zoo.'

Source: Huxley, J. (1934) *Scientific Research and Social Needs*, London: Watts & co, between p. 65 and p. 66;
lower photograph © The Royal Institute of British Architects

Here the place is infected by rats and cockroaches, family in bad health –
suffering with chests. Two children have died of diptheria.

*The windows are provided with blinds which can darken the cage, so that the tropical
conditions, natural to gorillas, of twelve hours darkness and twelve hours light may be
reproduced.*

Bedroom in which the family sleep is practically pitch dark even in the
daytime.

*In the centre of the floor of the winter cage there is a small basin and fountain, oper-
ated by a brass knob in the floor, manipulated by the gorillas.*

A very bad feature of this property is the water and lavatory accommodation
– one tap between six houses and one w.c. between three.

*Inside, the colour scheme was chosen to eliminate the oppressive sense of prison that is
apparent in many animal cages.*

The walls of the room are covered with a pink distemper, the distemper
flaking off with the damp.

(*AJ* 1933: 834)

'In a home fit for gorillas to live in,' the journal maintained, 'the country
possesses an asset which may be regarded as a standard of comparison for future
building – for human beings as well as animals' (*AJ* 1933: 834).

The implication of the irony here is clear. While human society, with the benefits of social progress, had failed to provide its working classes with decent,
healthful homes, 'mere' animals were being advanced and nurtured by enlightened applications of science. But while the zoo buildings appeared to point to a
perverse divide between buildings for humans and for non-humans, there was, in
fact, far more connection than might at first appear between Lubetkin's zoo
buildings and the socially committed science of Huxley or the biotechnics of
Mumford. Many of the analytical techniques and practical solutions utilised by
Lubetkin and Tecton in their zoo buildings were developed in parallel in buildings
for human use and habitation. Lubetkin had, for instance, prepared plans in 1932
for a TB clinic at East Ham that set out elements (common to the Gorilla House)
that would feature in many of Tecton's health-related designs (Coe and Reading
1981). The design was streamlined, hygienic and deliberately promoted the
virtues of modernism in environmental and social reform. It was based 'biotech-
nically' on maximum penetration of sunlight and fresh air. The building's

THE   DESCENT   OF   MAN

Figure 11.3   The *Architects' Journal* on hygiene and overcrowding.

Source: *Architects' Journal* (1933) 77, 22 June, p. 834

modernist concern with zonation and functional efficiency was essential in the treatment of communicable disease. Not only was air circulation modelled in order to provide currents of fresh air and to expel foul air, as in the animals' enclosures, but the circulation of the public (both well and unwell) and staff (both medical and clerical) was modelled to provide clear and hygienic routeways. These design elements were eventually expressed in the Finsbury Health Centre, commissioned by a progressive Labour-controlled borough that also commissioned Tecton to build working-class housing. Lubetkin produced a two-storey central axis with walls of glass tiles, flanked by two wings that splayed out as if projecting open arms into the borough. The Centre drew together the borough's previously scattered health facilities into one functional building that was, in its essence, symbolic of good health (see Gruffudd in press).

Many of Lubetkin's design elements were clearly exhibited in both the health and zoo building. Both types were founded on detailed scientific and practical/functional research. Both were committed to naturalism, in terms of fresh air and sunlight. The same air-circulation technology was utilised, based on the same critical understanding of the importance of hygiene. Some details were carried through also: an innovative system of interlocking locker doors in the TB clinic and Finsbury changing rooms was echoed in the design of the Gorilla House feeding cubicles, for instance. Arguably, the biological cultivation for which the likes of Mumford called was exhibited in pure form in the buildings designed by Tecton for London Zoo. The cultivation of the perfect animal body in the enclosures stood not so much as a metaphor *for*, as an experiment *in*, the cultivation of the perfect human body with which so many modernists were concerned, and to which so much modernist aesthetics alluded (see, e.g., Wigley 1995). The author of the *Mother and Child* article, doubtless alive to the need for effective nurturing, pointed simply but powerfully to architectural modernism's new role in biology: "'Functionalism' is an unattractive word, but it takes on a new meaning as applied to the penguin pool – and to the Finsbury Health Centre' (*Mother and Child* 1938: 300).

Similarly, Huxley's evolutionary concerns can be traced in many of his other planning- and architecture-related activities. 'Man' (*sic*), Huxley (1944b: x) argued,

> must become consciously evolutionary, in his individual thinking, in his collective outlook, and in his social machinery. ... The implementation in practice of even our existing knowledge concerning diet, disease, and positive health will make sweeping alterations in effective human nature, the result of which cannot be foretold: and the results of future discoveries in glandular control, sex-determination and eugenics are still more unpredictable.

The techniques of large-scale over-all planning offer quite new possibilities of controlling man's physical and social environment.

The TVA experiment represented one such attempt at integrating scientific understanding with social improvement in an holistic fashion. In Huxley's narrative of the TVA (1943), the principles of biological cultivation worked out in London Zoo were reflected on a grander scale. He typically wove together the same close understandings of biological systems (soil, flora and fauna, water-bodies, human health), of modernist design and technology, and of the importance of community and of social and regional planning through science.

On a more local scale, the application of evolutionary theory and modernist design to health was also a feature of Huxley's activities. He was a member of the steering committee of the Pioneer Health Centre in Peckham. This was an experiment in family, community and health that was also, significantly, housed – from 1935 onwards – in a landmark modernist building (this time designed by Owen Williams), and was explicit in its adoption of biological and evolutionary thinking on the human body, the family and urban space (Gruffudd 1995). Drawing on the example of the Mexican *axylotl* (the tadpole stage of the salamander), the Centre's directors stressed the importance of environment in nurturing the full potential of any living organism. The Centre's aim was, through a building of steel and glass, to provide a healthful environment involving well-designed physical spaces as well as contact with nature's benefits through natural foods, alongside a healthy social environment – what they called the 'social soil' (Pearse and Crocker 1943). Huxley was active in his support of the Centre, making several widely reported speeches on its behalf and using it as an example of his ideal, scientifically-led social science for campaigns on health-funding and on dietary improvement. In the arena of dietary improvement – as in that of environment and design – the treatment of animals could nonetheless still be viewed as an ironic statement on the neglect of humans. An article in the *Daily Herald* (1937) on the need for national nutritional standards showed a composite photograph of a chimpanzee and two children standing behind a montage of the healthy fresh foods that constituted a typical week's intake for the chimp. Yet, it quoted Julian Huxley as saying that, '[i]f we gave the gorillas and chimpanzees a diet like that actually eaten by millions of human beings in this country we should be rightly blamed for not keeping them in proper condition.'

Conclusion

There is an irony in the *Architects' Journal*'s illustration of 'The descent of man' as a return to the animal stage of hygiene while the animals achieve a new cleanliness. It is clear that, although Lubetkin certainly believed in the power of the rational human mind to modify nature, he believed that it should do so on the basis of a theoretical and biological understanding of the connections and duties between the human and non-human world. There was a significant degree of environmental humility in his treatment of animals, despite appearances. Furthermore, that understanding of nature–culture relationships, of organisms evolving in their respective environments, was – for the likes of Huxley – part of the key to a larger-scale reorientation of human society through planning. More than mere analogy or metaphor, evolution or 'biotechnics', as Mumford might have called it, was a technique to be employed in theory, analysis and, ultimately, design and planning. It is therefore unsurprising that Mumford, in *The Culture of Cities*, should illustrate the new forms of the biotechnic civilisation by mentioning: a Frank Lloyd Wright house ('an organic union of garden and house'); the 'exquisite purity' and functionalism of a Japanese house; a TB clinic by Tony Garnier stressing the healing qualities of 'light, color, order, visual repose' (Mumford 1945: 356); the modern American kitchen, which, together with the bathroom, was one of the great biotechnic utilities of the modern dwelling; and, finally, the responsiveness to environment and to residents of Lubetkin and Tecton's Gorilla House for London Zoo.

Acknowledgements

I would like to thank the Nuffield Foundation for the grant that funded the research on which this chapter is based, and, for their comments, those who attended the original session at the Institute of British Geographers conference and a research seminar at Portsmouth University. Most of all, thanks to the two editors for their advice and patience.

References

Allan, J. (1987) 'A song of summer: the restoration of the Penguin Pool', Ms in Lubetkin Papers [LUB 1/10], Royal Institute of British Architects.

Allan, J. (1992) *Berthold Lubetkin: Architecture and the Tradition of Progress,* London: RIBA Publications.

Anderson, K. (1995) 'Culture and nature at the Adelaide Zoo: at the frontiers of "human" geography', *Transactions of the Institute of British Geographers* 20: 275–294.

Architect & Building News (1937) 'Studio of Animal Art, Regent's Park', 7 May: 155–160.

Architects' Journal (1933) 'Special issue on slum housing', 77, 22 June.

Architects' Journal (1934) 'Penguin Pool Zoological Gardens: Lubetkin, Drake and Tecton', 79, 14 June: 856–858.

Architects' Journal (1939) 'Notes and topics: agrees with Vigilance Committee …', 89, 25 May: 846.

Architectural Design and Construction (1933) 'New Gorilla House for the London Zoological Society', 3(8): 316–318.

Berger, J. (1980) 'Why look at animals?' in *About Looking*, London: Writers and Readers.

Cansdale, G.S. (1950) 'Letter to S. Neuberger dated 19 December 1950', in Lubetkin Papers [LUB 1/7], Royal Institute of British Architects.

Coe, P. (1982) 'Lubetkin and Tecton', *Royal Institute of British Architects Transactions* 2: 51–56.

Coe, P. and Reading, M. (1981) *Lubetkin and Tecton: Architecture and Social Commitment*, London: Arts Council of Great Britain/University of Bristol: Department of Architecture.

Daily Herald (1937) 'Physical training or food', 22 January.

Daily Telegraph (1933) 'New Home for Mok and Moina', 29 April.

Gardiner, P. (1987) 'Pick up a penguin', *The Observer,* 15 March.

Gold, J.R. (1997) *The Experience of Modernism: Modern Architects and the Future City*, London: E & F.N. Spon.

Gruffudd, P. (1995) ' "A crusade against consumption": environment, health and social reform in Wales 1900–1939', *Journal of Historical Geography* 21: 39–54.

Gruffudd, P. (in press) 'Science and the stuff of life: modernism and health in inter-war London', *Journal of Historical Geography*.

Haraway, D. (1992) *Primate Visions: Gender, Race and Nature in the World of Modern Science*, London: Verso.

Huxley, J. (1932) *A Scientist Among the Soviets*, London: Chatto & Windus.

Huxley, J. (1934) *Scientific Research and Social Needs*, London: Watts & Co.

Huxley, J. (1937a) 'Memorandum on Zoo architecture to the Zoological Society of London, 14 May 1937', in Lubetkin Papers [LUB 1/4], Royal Institute of British Architects.

Huxley, J. (1937b) 'Draft letter to the press dated 8 November 1937', in Lubetkin Papers [LUB 1/14], Royal Institute of British Architects.

Huxley, J. (1942) *Evolution: the Modern Synthesis*, London: Allen & Unwin.

Huxley, J. (1943) *TVA: Adventure in Planning*, London: Scientific Book Club.

Huxley, J. (1944a) 'Letter to Berthold Lubetkin dated 8 June 1944', Lubetkin Papers [LUB 1/14], Royal Institute of British Architects.

Huxley, J. (1944b) 'Preface', in *On Living in a Revolution*, London: Chatto & Windus.

Matless, D. (1992) 'A modern stream: water, landscape, modernism and geography', *Environment and Planning D: Society and Space* 10: 569–588.

Matless, D. (1998) *Landscape and Englishness*, London: Reaktion.

Morning Post (1933) 'Mok and Moina's luxury suite', 29 April.

Mother and Child (1938) 'A London health centre – employment of a new architectural idiom', November: 297–300.

Mumford, L. (1937) 'The sky line: penguins and architects', *The New Yorker,* 6 March: 59–60.

Mumford, L. (1945) *The Culture of Cities*, London: Martin Secker & Warburg Ltd.

Pearse, I.H. and Crocker, L.H. (1943) *The Peckham Experiment: A Study in the Living Structure of Society*, London: George Allen & Unwin Ltd.

Philo, C. (1995) 'Animals, geography and the city: notes on inclusions and exclusions', *Environment and Planning D: Society and Space* 13: 655–681.

Reading, M. (1992) 'Animals and men', *Architects' Journal* 195: 28–37.

Reilly, C.H. (1935) 'Ms. of article to *Architects' Journal,* January 1935', in Lubetkin Papers [LUB 1/7], Royal Institute of British Architects.

Reilly, P. (1937) 'Build for today not yesterday', *News Chronicle*, 15 October.

The Times (1932) 'New Gorilla House: novel building at the Zoo', 31 December.

Wells, H.G., Huxley, J. and Wells, G.P. (1931) *The Science of Life*, London: Cassell.

Wigley, M. (1995) *White Walls, Designer Dresses: The Fashioning of Modern Architecture*, Cambridge MA: MIT Press.

Williams-Ellis, C. (1980) *Architect Errant: The Autobiography of Clough Williams-Ellis*, Penrhyndeu-draeth: Golden Dragon Books.

Zuckerman, S. (1933) 'Letter to Berthold Lubetkin dated 20 July 1933', in Lubetkin Papers [LUB1/5], Royal Institute of British Architects.

12 Virtual animals in electronic zoos

The changing geographies of animal capture and display

Gail Davies

In a world despoiled by overdevelopment, overpopulation, and time-release environmental poisons it is comforting to think that physical forms can recover their pristine purity by being reconstituted as informational patterns in a multidimensional computer space. A cyberspace body, like a cyberspace landscape, is immune to blight and corruption.

(Hayles 1993: 81)

To the same degree as man [*sic*] has raised himself above the state of nature, animals have fallen below it: conquered and turned into slaves, or treated as rebels and scattered by force, their societies have faded away, their industry has become unproductive, their tentative arts have disappeared. ... What visions and plans can these soulless slaves have, these relics of the past without power?

(Buffon, quoted in Berger 1980: 10)

Introduction

The creation and exhibition of knowledge about plants and animals in the practices of natural history are intrinsically geographical enterprises. The processes of constructing and representing natural history are those of identifying and defining the spaces that non-human organisms occupy in human culture. This is evident in the localised, place-based, observations and descriptions of amateur practitioners in the wake of Gilbert White, as much as in the wide-ranging collecting and theorising which emerged in European taxonomy following Carl von Linné. Our knowledge of plants and animals is created by moving between worlds: the inner and outer spaces of habit, form and function; and moving across space between field, laboratory and zoological institute (Keller 1996; Outram 1996; Pratt 1992). Animals are fixed in representational spaces connecting these realms as botanical and zoological illustrations in encyclopaedias and notebooks (Blum 1993); as preserved and dried specimens in herbaria and natural history museums (Stemerding 1993); as living representatives in zoological and botanical gardens (Anderson 1995); and, increasingly, as informational patterns in multi-dimensional electronic spaces. In occupying these spaces, plants and animals play important roles in practices which define the complex relationships between human and non-human worlds; processes of domesticating, commodifying, aestheticising, as well as constructing knowledge about the natural world.

Natural history has lost ground as an academic discipline (Vernon 1993), but a number of issues ensure that it has not lost importance as a complex set of sociospatial practices through which relationships between nature and society are defined. The attentions of many biological scientists have moved away from taxonomic description, yet there are still huge gaps in our knowledge about the natural world, with an estimated 90 per cent of species still undescribed (May 1988). Working at these absences on the local scale, natural history is still an area of endeavour for large numbers of amateur practitioners (Harrison 1993); while taxonomy has attracted renewed academic interest through the Darwin Programme instigated after the Rio Conference in 1992. More pertinently for this chapter, natural history has achieved immense significance as a popularisation of biological knowledge, with an explosion of natural history publications and television programmes featuring the natural world, and animals in particular (e.g. Franklin 1999; Marren 1995; Silverstone 1986; Wilson 1992). In the last forty years the various ideas and images of natural history have become more widely available than at any time in the past. These popular presentations of natural history in the media are contested (Crowther 1997; Mills 1997), however, and there are other popularisations of natural history which have recently declined in status. Zoos, menageries and other forms of animal display have closed following attack from an increasingly active animal rights lobby, and in response to perceived public sensibility about the enclosure of live animals (Montgomery 1995). Natural history as a complex, contested and changing network of practices, associated with defining and structuring the borders between the human and non-human worlds, continues to be of material, social and spatial significance.

This chapter is concerned with reviewing and developing the growing area of research on human–animal relations and geography (Philo and Wolch 1998; Wolch and Emel 1995, 1998) through attention to the changing institution of the zoo within the networks of natural history. First, the chapter pulls together literature on the spatial practices associated with natural history and traditional zoos. Second, I focus on the construction and characteristics of a relatively new space within the networks through which animals are woven into human culture, exploring an emerging form of animal display that I term the 'electronic zoo'. In the complex milieu of natural history, with its existing representational strategies that range from notebook to camera, there are now further opportunities for the construction and presentation of natural history knowledges. The latest electronic spaces of digital imaging, the internet and virtual reality, take their place alongside more established technologies such as film, photography and television. These offer new ways of conceiving of and portraying natural history, and introduce the possibility of different relationships between human and animal experiences. Recent developments in image technology mean that images can be digitally reproduced and manipulated for use on video, CD-ROM and the internet. Striking likenesses of animals can be produced on a vast scale in IMAX® cinemas, or perform in three dimensions as animatronic models. New digital procedures for cataloguing and reissuing video and celluloid material offer renewed glimpses of now changed habitats and extinct animals. All of these forms of virtual animals transform the ontological and epistemological status of animals in representation, as well as presenting new opportunities for animal display. Initiatives for replacing living animals with forms of electronic representation are now being developed in various forms of the electronic zoo which are the focus of this chapter.[1]

Traditional zoological institutions and parks are increasingly incorporating new technologies of animal simulation and display to add interpretation, interest and educational value to their established forms of animal exhibition (Veltre 1996). Most recently, Animal Kingdom, a 500-acre theme park which opened in Florida in 1998, claims to fulfil the Disney mission with a tribute to the animals that have formed an important part of its own history. Animal Kingdom, centred on a signature icon of the Tree of Life, offers a mix of high-technology animatronics, featuring dinosaurs and mythical creatures, alongside the display of animals in large safari enclosures. Yet further centres for the display of animals are under way in which the virtual is set almost completely to replace the real. As of this writing (January 2000), Wildscreen-at-Bristol (see Figure 12.1) is planned as the world's first purpose-built wildlife and environment media attraction. Funded as one of the UK Landmark Millennium Projects, and opening in spring 2000, the new centre promises to change our perspectives on the natural world with interactive exhibitions of wildlife films, an IMAX® cinema and ARKive:

> Wildscreen will showcase the wonder of our planet, helping visitors observe and explore life in a way that has never been done before. Live exhibits will be combined with state of the art technology to help visitors get the most from their visit. A stunning botanical house and giant IMAX® cinema – the first in the South West – will further bring the experience to life. Visitors will learn about their own environmental impact, and in the news gallery discover the latest developments in natural history from around the planet. Wildscreen will also be the headquarters for ARKive – an exciting, globally accessible library of wildlife and scientific films, photographs and sound recordings.
>
> (Wildscreen-at-Bristol press pack 1999)

Taking inspiration from a growing body of work on the changing nature of zoological display within different cultures of natural history (e.g. Jardine *et al.* 1996), this chapter explores the motivations for and implications of these sites for the geographies of animal capture and display. A direct comparison reveals many similarities between the spatial practices constitutive of and constituted by traditional zoos and the new spaces of the electronic zoo. Both forms of animal exhibition involve social and spatial practices of collection, accumulating resources from widely dispersed locations, concentrating and ordering these within specific (commonly) metropolitan areas. Both are concerned with defining and describing the kinds of spaces that animals are to occupy, and delimiting the sorts of relationships that people have with these animals. These similarities are drawn out in the following two sections which explore and reflect on the relationship between the networks of the traditional and electronic zoos, developing Latour's (1987) ideas of zoos as 'centres of calculation' which control cycles of accumulation in natural history. Latour (1987) characterises the process of natural history as a practical cycle of knowledge production, whereby things which are far away, invisible and unknown are brought back to a centre where, through mapping, tabulation and other representational strategies, they are made known, well ordered and predictable. The 'centre of calculation' metaphor has been a productive one in the history of natural history (Miller 1996; Stemerding 1993), with obvious appeal to geographers (Parry 2000; Thrift *et al.* 1995), and it is used here to draw attention to the similar spatial dynamics in the accumulation of animals associated with traditional and electronic zoos. The key transformations from the zoological institute to the wildlife media centre revolve around the use of different visual technologies for the purposes of enrolling animals into networks of collection and exhibition. Yet even here there are continuities which can be followed through to interrogate the relationship between technology and forms of knowing about nature. Recent literature on the geographies of cyberspace is challenging the technological determinism which accompanies discussion of the impacts of new technology, seeking a more relational understanding of changing socio-technical relationships (Bingham 1996; Hillis 1994; Kitchin 1998). Here again a network approach can be revealing of what is new and what pervades in these networks through which the technical, natural and cultural are assembled (Bingham 1996; Wise 1997).

In following a Latourian approach, I seek to explore the challenges facing traditional zoos in relation to the success of the electronic zoo. The comparison rejects notions of a radical shift in the historical development of the zoo, exploring the continuities and discontinuities in how these organisations mobilise, stabilise and combine animals in the spaces of their displays. My development of this argument uses an 'actor-network'-inspired approach[2] to understand both forms of zoo as complex places, embedded in differing socio-spatial assemblages of people, devices and documents, seeking through different means of representation to make themselves spokespersons for nature. Through developments in image technology, I argue that the electronic zoo has simultaneously been able to accelerate the circulation of animals through its networks, concentrating value from their images, while dispersing responsibility for their embodied form. The chapter concludes with reflections on how these differing networks construct and constrain human experience of animals, and on the opportunities that they present for expressions of animal agency.

Placing animals in the traditional zoo

Zoos exist in a variety of forms in different societies and cultures. They have been shaped by, and given shape to, a wide range of human responses to animals. Moreover, they are complicated institutions whose form and function reveal overlapping purposes and occupy different places in diverse networks which commodify, control and care for animals, generating and popularising knowledge about animals and accruing value from them. In their evolution zoos have been places of collection, colonisation, agricultural experimentation, education and exhibition. Zoo buildings are therefore rarely found in isolation, and surrounding an urban zoo are usually found a *mélange* of other buildings and institutions such as botanical gardens, lecture theatres, display galleries, libraries, living quarters and dissecting rooms; not to mention more recently cinemas, food halls and souvenir shops. This multiplicity of functions means that there are continual contestations over the distribution and form of space required for the different functions of the zoo. The changing layout of zoological and botanical gardens reflects wider changes to the status and popularisation of certain forms of science (Outram 1996).[3] The scientific, cultural and technological transitions which are revealed in the establishment of what can be called 'classical zoos' in the nineteenth and the early twentieth century[4] provide a useful point from which to develop analysis and understanding of the new space of the electronic zoo. They reveal the symbolic spaces occupied by animals within zoos and the very practical means through which these organisations are created, maintained and changed. I pursue this analysis through a focus on the way that animals are mobilised, stabilised and circulated in the networks surrounding the forms of animal display found in the classical zoo (Latour 1987).

The networks of the traditional zoo

Common to classical zoos is their distinctive array of living animal species. These institutions characteristically bring into urban centres collections of geographically dispersed 'wild' animal species. These are arranged in displays which taxonomically distribute animals into genera, families and species – the big cats, great apes, and so forth – furnishing their cages with maps showing the location of each animal's natural habitat. This geographical displacement is part of what constitutes their attraction, and the process of decontextualisation alone can serve to give exotic value to certain animal species, the majority of which are never seen by people in nature. In this, the genealogy of the classical zoo can be traced to the private collections of caged animals or menageries and cabinets of curiosities that were important status symbols for the wealthy or the royal in the eighteenth century (Findlen 1996). These valorised certain ways of knowing about nature that emphasised adventure, exploration and the development of knowledge by way of moving through space (Outram 1996; Pratt 1992). The animals thus collected stood in metonymically for the regions from which they were derived, providing a visual display of imperial reach, while the bars and cages constraining wild beasts demonstrated human power over the domain of nature. The flowering of zoo building in the nineteenth century was supported by vast networks of animal trading that spanned from Western Europe and the United States to those parts of the world divided into imperial territories. Huge numbers of living animals were mobilised through the formal collecting trips of individuals[5] or organisations such as the Dutch East India Company (Flint 1996), public donations, royal gifts from colonial territories, and more informal mechanisms.[6] This huge influx of animals was readily absorbed into Europe, where in the 1860s zoological gardens were being established at the rate of one a year (Reichenbach 1996).[7] Even when zoo exhibits were derived from pre-existing animal collections,[8] or breeding programmes at other zoos, the value of the animals was premised on this imagined geographical movement from the wild to the disciplining spaces of the zoo.

The movements of animals associated with zoos are significant not only for this displacement, but more precisely in the centralisation of animals in particular locations, notably cities, alongside other institutions such as natural history museums and scientific societies. As Anderson (1995: 279) points out, bringing wild animals into the city was especially meaningful in the nineteenth century as representing 'the ultimate triumph of modern man [*sic*] over nature, of city over country, of reason over nature's apparent wildness and chaos'. While the exhibition and ordering of caged animals within metropolitan zoos demonstrated human power over nature, other forms of animal life such as livestock, seen as dirty, degrading and polluting, were systematically being removed from the urban sphere (Philo 1995). The city was increasingly defined as a civilised space removed from nature. However, animals *within* the zoo were celebrated by city officials and scientific institutes as symbols that lent pre-eminence to established cities such as London (Ritvo 1996); or order, authority and identity to newly established colonial cities such as Adelaide (Anderson 1995). The symbolic, as well as financial, value of these collections of animals derived from their systematic organisation and accumulation according to evolving discourses of natural history and animal breeding.

The presence of such animals in the city was valuable for communicating the moral order of nature and humankind, as revealed through the taxonomic classifications of natural history (Anderson 1995; Haraway 1989; Ritvo 1987). The exhibition of animals acted as a form of living library where this order could be studied, displayed and communicated; and zoological societies strove to realise a collection that could 'furnish every possible link in the grand procession of organised life' (Ritvo 1996: 46). Animals were valuable not merely corporeally, but also for the information about the order of nature manifested in their forms and functions, stabilised and made visible within the zoo (Stemerding 1993; Parry 2000). Individual animals were made to stand in for whole species and families: a single lion signified all lions (Berger 1980). As indicated, the spatial arrangement of animal cages reflected this ordering of nature. Anderson (1995: 283) documents that within Adelaide Zoo 'the exhibits were set out in conformity with prevailing classifications based on visible characteristics – reptiles, birds, mammals and fish – each exhibit was made to stand as a taxonomic specimen of a broader category'. The importance of the visual in this form of knowledge was further underlined as animals not normally visible to humans were put on display for human benefit (Marvin 1994). The seven lions and two tigers at Adelaide Zoo were displayed in cages lined with white tiles 'to furnish an excellent background for visitors including natural history students and writers' (Anderson 1995: 284).

Not all animals within the zoo have been organised in this way. Among the taxonomic exhibits of many zoos were also found the zoo pets: animals named and set apart from the rest of the collection. In Britain zoo pets tended to represent not British scientific culture in the form of developments in natural history, but the native colonies, especially those within significant European populations such as Africa and Asia (Ritvo 1996). Rather than occupying stark cages arranged for study, they were frequently housed in ornate enclosures which offered racialised representations of their native environments (Anderson 1995). These zoo pets were also set aside for special treatment, interacting with people to give rides and frequently featuring in the mass media reports of the zoo. One notably famous example was Jumbo, the African elephant who lived at London Zoo from 1865 to 1882 before becoming the star of the travelling Barnum circus in the States (Flint 1996).

This final destination of Jumbo draws attention to important processes following the mobilisation and stabilisation of animals within the zoo, which is the ability to accrue value from the accumulation and circulation of animals. The consolidation of animals in the zoo not only aided and popularised developments in taxonomy and natural history, but also offered financial returns for its application and display. In utilising the knowledge and animals in their captivity, zoos became entrepôts for developments in stock-breeding and the acclimatisation of animals within overseas colonies. The South Australian Society campaign for the establishment of a zoo in Adelaide included the intention to 'introduce, acclimatise, domesticate and liberate select animal, insect and bird species from England' (Anderson 1995: 281). Even at the heart of empire in London Zoo, stock-breeders who wanted to distinguish their livestock with 'exotic' blood could pay stud fees that ranged from five shillings for a zebu, one pound for a Brahman Bull, and two pounds for a zebra (Ritvo 1996: 45). Commercial concerns played an important role in the acquisition of animals for public display. Ritvo records financial anxiety at London Zoo regarding the short life-spans of its large collection of big cats during its early years. Their average life-span of around two years, with one mortality a month, meant that such deaths diminished the general appeal of the zoo to visitors who found the 'carnivora ... one of the most attractive portions of the collection' (Ritvo 1996: 47).[9]

This ability to derive value from the public display of animals is the final part of my argument before reflecting on the human experience of animals constructed by the zoo. As the drive and funding for biological science shifted from taxonomy and natural history to more mechanical models of biology throughout the late nineteenth and early twentieth century, zoos gradually evolved into organisations whose primary purpose was to exhibit animals to the public not to experts. Public presentation relied on esoteric research only for factual authority, and the actual process of creating public exhibitions bore no necessary resemblance to the process of generating the knowledge that they portray (Allison 1995). Such popularisation was essentially a communication process, linked to the histories and technological networks of communicative production as much as to those of science (Cooter and Pumfrey 1994). The importance of the visual in the zoo has thus never been wholly associated only with the construction of ocular knowledge, but has also entailed an important way of attracting or enrolling visitors into the zoo.

Animals in zoos, however, are attempting to live their own lives that do not necessarily coincide with being properly visible. The partial invisibility of animals within the zoo, alongside the application of new visual technologies such as photography, film and museum dioramas to reveal and to present animal behaviour, ends up challenging the classical zoo's ability to rely on increasing visitor receipts. This can be explored by examining one of the first so-called 'modern' zoos: Carl Hagenbeck's Tierpark in Hamburg (Reichenbach 1996). Hagenbeck's acclaimed animal displays are credited with ushering in a new visual regime at the zoo, one that has been heralded as the birthplace of modern animal displays which are less oppressive to animals, and more ecologically accurate. In these zoos animals were introduced to parkland enclosures, with mixed animal displays and invisible forms of enclosure separating the animals and visitors. Marvin (1994: 197) suggests that 'the mechanism of captivity such as moats ... does not form a central feature of the enclosure, [and] expresses a very different set of attitudes to the natural world.'

The origins of Hagenbeck's zoo suggest a more complex motivation than animal welfare or education for the shift to bar-less moated enclosures. The absence of visual control of animals cannot be used simply to read off a more sympathetic, less dominating or exploitative relationship to the animal world. Hagenbeck was an entrepreneur who started his career as an animal dealer and zookeeper in Hamburg in the 1850s with a wholesale seafood business and pet store. By 1907 he had established a hugely successful animal trading empire and 'built a private zoological garden so spectacular and attractive that it made the old Hamburg Zoo look obsolete and uninteresting' (Reichenbach 1996: 61). Hagenbeck drew inspiration from contemporary revolutions in visual technologies to respond to a public with increased opportunity for travel and a growing familiarity with photography, patenting his panorama in 1896. This was a series of enclosures embellished by artificial rock and hedges and laid out as stages. Each enclosure was slightly higher than the previous and they were separated from each other and from the public by concealed moats. This form of display, which owed much to the film or theatre set, formed the main type of animal enclosure. Other zoos responded quickly to such developments with innovations of their own. Some built new enclosures, such as the Mappin terraces at London Zoo in 1913 (Montgomery 1995: 580), while others introduced events such as feeding times and chimps' tea parties, and worked closely with local media to showcase their new exhibits and to stress the zoo's function as spectacle (Anderson 1995: 287). As scientific support and funding for zoos declined, enlisting the public into the organisations became as important as enrolling animals into the displays.

This is the somewhat ambiguous legacy and architecture inherited from the traditional zoo. Most zoos are now furnished with a range of displays that reflect these changing methods of domesticating, mythologising and visualising the natural world. Subsequently, of course, a further layer of interpretation has been added in the form of conservation and biodiversity displays which give further shape to the processes through which zoos seek to enrol, mobilise and position nature, to which I return later. However, actor-network theory reminds us that the achievements of these networks to speak for nature are never secure (e.g. Callon 1986; Wynne 1993). As one zoo director writes:

> [W]hen zoo curators exhibit highly endangered snow leopards in order to discuss their desperate battle with humans for dwindling resources in the high mountains of Tibet, they must always contend with anthropomorphic visitors who will bring their children to 'see the big pussycats.'
>
> (Veltre 1996: 28)

Human identity and animal agency

The visual classifications of natural history and the visual technology of the zoo not only define a way of being for animals, but also attempt to delimit a particular form of human interaction with animals. Anderson (1995: 278) elaborates that the form of exhibition at Adelaide Zoo produced 'a particular form of human in relation to nature. This is a historically specific type of (white) masculine that is unseen, that is not the spectacle but rather the privileged eye (I), the bearer of reason, the author, the knower.' The visual organisation of nature in the built spaces of the zoo further manifests and imposes a structure of authority between expert and public knowledges, giving importance to some ways of knowing about nature, but not to others (Outram 1996). While the zoo does give primacy to visual knowledge and spectacle, from the viewpoint of the zoo visitor this can be overstated. The visitor to the zoo is presented with an order of nature arranged as a form of living encyclopaedia, but this does not define the limit of the zoo experience. Visiting a zoo is a multi-sensual experience where the sight of animals is combined with an awareness of their whole bodies, their sounds, smells, touch, even taste. The most striking experiences for the visitor at the zoo are not necessarily just the visual, for the zoo also offers the opportunity for direct human–animal interaction, such as animal rides, demonstrations, children's zoos and public feeding. Visitor interest in these experiences is often so strong that people are willing to pay additional fees to participate in them (Kreger and Mench 1995). Additionally, visitors to traditional zoos are almost always having a social day out; with family outings and school parties contributing to a collective engagement with animals. One of the most potent parts of the zoo experience is that, although curtailed by boundaries and unequal separations, you are in a place that is shared between people and animals. In the words of Montgomery (1995: 575): '[T]he real event, the only event, is the physical presence of the animals.' This physical presence can be disturbing, threatening, appealing or amusing, but it is usually challenging.

Animals are enrolled into these networks of animal display through the enclosures of the zoo; the zoo speaks through them and for them, yet they are still actors within this network (e.g. Callon 1986). Their enrolment is always uncertain, and the animals still have the power to surprise us with how they act. Their corporeality reminds us that they are mortal as well as immortal; they are embodied animals as well as information. Their behaviour can challenge our understanding of them as species and as individuals, and it is in this experiential space that we are confronted by the agency of animals. The bear that rocks incessantly despite being moved to what is viewed as a superior enclosure contests our understanding of its behaviour and challenges the sign beneath which it stands (Wilbert 1998). Animals' bodies can attract and repel us at the same time. They can avoid our gaze or they can court it. They can also return it:

> The eyes of an animal when they consider a man [*sic*] are attentive and wary.
> … The animal scrutinises him across a narrow abyss of non-comprehension.
> This is why the man can surprise the animal. Yet the animal – even if domes-

ticated – can also surprise the man. The man too is looking across a similar, but not identical abyss of non-comprehension. ... A power is ascribed to the animal, comparable with human power, but never coinciding with it.

(Berger 1980: 2–3)

Despite their subjugated position within the networks of the zoo, animals are nevertheless active subjects embodying a form of agency in their ability to continue to challenge, disturb and provoke us. This agency is one of the reasons why visiting a zoo can be an ambiguous experience, with pleasure and excitement mixed with fear and nostalgia. It is also one of the driving forces behind the developments of the electronic zoo. Unlike the experiences promised at the electronic zoo, however, the traditional zoo is ultimately an embodied encounter between humans and animals within a shared, albeit unequally constructed, space.

(Dis)placing animals in the electronic zoo

The developments of the electronic zoo promise a new cycle of accumulation for natural history, with once-living animals, now immortalised through various technologies, being mobilised, stabilised and combined into existing assemblages of natural history. The electronic zoo creates new geographies of animal capture and display in the process of bringing nature from far afield to the centre of the city. The developments in film technology, digital imaging and visual presentation offer the opportunity to revive the networks of natural history; new telecommunications networks intersecting with existing ones to offer new spaces and times, and new forms of human interaction, control and organisation which are continually constructed (Latour 1987). These technologies do not simply act to introduce new ways of seeing nature into pre-existing culture, but are formed and re-formed by the existing networks of the traditional zoo. The most compelling work emerging on the geographies of cyberspace challenges the technological determinism implicit in much comment about the advent of virtual reality by rendering us sensitive to how the reality with which we are dealing is already 'virtualised' (Bingham 1996; Robertson *et al.* 1996). The traditional zoo does not construct a more natural image of the animal world than the electronic zoo, and visual technologies have always played a central role in the aesthetic and epistemological regime of the zoo. However, in its extended networks for enrolling and representing animals through the use of virtual technologies, the electronic zoo *does* remake forms of human subjectivity and animal agency. In the networks developing here there are new dimensions, bound up in the ways of enrolling animals through film and digital technology, which have implications for the sorts of spatial practices, the kinds of places and the forms of subjectivity encouraged by the electronic zoo.

At-Bristol is planned as part of the redevelopment of the old Bristol docks, and it will bring a new multi-media science experience as well as the world's first purpose-built wildlife and environment media attraction to the heart of the city of Bristol. Funding for the at-Bristol development will be met by a partnership including the Millennium Commission, English Partnerships, Bristol City Council and the Harbourside Sponsors Group, with other private sector funding. The elements making up the wildlife media centre include a botanical house, an IMAX® theatre, a video theatre and space for talks, access to the ARKive species database and a venue for the biannual Wildscreen wildlife film-making festival. Part of this exhibit can already be accessed on-line: the ARKive or 'digital Noah's ark', which consists of a storehouse of knowledge about the world's endangered species, is available on the internet (http://www.arkive.org.uk/). Pre-publicity assures us that throughout the displays 'models, graphics, live and pre-recorded electronic images, lighting effects, sound and interactive systems will be used' to 'bring visitors face to face with the extraordinary diversity of the world – in vivid close-up' (Wildscreen-at-Bristol press pack 1999). This multiplicity of spaces and functions in the new electronic zoo mirrors the heterogeneity of buildings in the traditional zoo, and gives some indication of the different kinds of socio-spatial practices that constitute and are constituted through these new developments.

The networks of the electronic zoo

The spatial processes of network-building around the electronic zoo bear many similarities to the spatial processes of collection and trade supporting the traditional zoo. Similar to the accumulation of animals in the zoo, the images of animals gain value through their geographical displacement from the sites of filming to the sights of the electronic zoo. This occurs in a number of ways, most evident through the Wildscreen festival to be hosted in the Bristol Centre. This has become the leading film festival for wildlife film-makers, attracting film and television professionals across the globe to watch and to reward each other's films, engage in dialogue, and most of all sell their products. The 'wild' and unpeopled nature presented in the spaces of television wildlife programmes form the raw material for the new collections of the electronic zoo (Davies 1989, 1999). The most valuable images of animals in these markets are ones that demonstrate high production values, achieved through international funding, and feature the most popular animals with international audiences. The animals highly valued in the networks of the traditional zoo remain the most sought after by film-makers, with animals such as the big cats, polar bears, elephants and great apes forming the main guarantors of the high costs of producing natural history films.[10] Although popular with British audiences, there is little money to be made from filming British wildlife as these are not images that will subsequently be widely marketable.

Figure 12.1 A computer-generated projection of Wildscreen@Bristol, New World Square.

Source: *At-Bristol Images*

Additionally, while 'local' film-makers are active in those parts of India and Africa where highly valued indigenous animals can be filmed on location, their relative lack of resources means they are rarely able to produce images of sufficient quality for broadcast in the West. These same animals have also been used to pioneer the development of IMAX® cinema. Projecting images up to 70 feet across and 70 feet high, IMAX® screens offer an all-encompassing vision of wildlife that has been used to convey new experiences of landscapes and animals. The first IMAX® films feature penguins in the Antarctic, East African safaris and images of mountain gorillas. The electronic zoo is at the centre of networks of wildlife film-making companies who facilitate the collection of images of 'exotic' animals filmed at research sites throughout the globe for sale to television networks, cable channels and other forms of display in the West. The unequal spatial movements in these new media empires, which mirror the trading empires of the zoo, remind us that technologies are implicated in, rather than offering solutions to, the realities of uneven geographical development (Gillespie and Robins 1989).

The use of film as an intermediary in the networks of the electronic zoo also enables further dislocation between places and scales, which adds value and authority to the images of wildlife. Film has emerged as an important method-ological tool within scientific disciplines that rely on field practices of observation and description (Cartwright 1992; Haraway 1989; Winston 1993), facilitating attempts to mirror more closely the idea of mechanical objectivity that has consti-tuted the highly mediated world of present-day life science (Mitman 1993; Vernon 1993). Film can also accommodate the conventions of realism central to the traditional representational practices of the museum diorama or zoo panorama (Mitman 1993: 640), thus functioning as a border object in the worlds of both science and entertainment. The positioning of film as a scientific method of inscription enables film-makers to claim authority for their images as trans-parent mediators between field observation and exhibition. The realist coding of film eliminates the evidence of its own creation such that these intense images of animals bear few traces of the material objects of human and animal labour that formed them. There is a long history of removing humans from the images of natural history, as well as removing traces of the human and animal interaction in the creation of these representations.[11] This legacy of authority and trust in natural history film-making looks set to continue in the computer-generated images of virtual animals in the electronic zoo. However, the images created of pristine animals and natural habitats exist in tension with the discourses of a nature under threat. The animals in ARKive, the world's known endangered or extinct species, are being gathered from a growing number of natural history film-making companies (existing images on the ARKive showing provenance from the BBC Natural History Unit, Planet Earth Pictures and the Natural History

Photographic Agency, among others). Despite strong criticism of the conservation claims of traditional zoos, natural history film-makers have been able to distance themselves from responsibility for protecting those animals that they film (Mills 1989, 1997). The collection of these images within the ARKive of the electronic zoo appears to continue the appropriation of the ark as a millennium sign, evident also in the Earth Centre at Doncaster, for representing the natural world without any positive action towards it.

The form of communication space designed for the electronic zoo is not radically different from the traditional zoo. Within the spatial rhetorics of cyberspace are subsumed a large number of interlocking forms of technology, which suggest different kinds of engagement between the subject and object in any communication (Adams 1998). In its design for an exhibition space, supported by an internet presence, Wildscreen fits in closely with established forms of communication involving the dissemination of information by one or a few to the many, familiar from the lecture hall or exhibition; whereas the ARKive follows a well-established encyclopaedia approach, classifying animals by taxonomy or natural habitat, albeit available to all via the internet. Despite the opportunities to create new forms of animal exhibition or experience, value is still placed on the visibility of animal morphological features, with the images of individual animals on-line framed against restricted backdrops or even on white backgrounds. Other framings or means of communication such as bulletin boards, which might have included non-scientific interpretations of animal behaviour or footage of animals filmed by non-experts, do not seem to have been included. Indeed, the stated aim of Wildscreen-at-Bristol has been precisely to popularise an elite interpretation of life, with the opportunity to 'enjoy views and experiences often only accessible by film-makers and scientists' (Wildscreen-at-Bristol press pack 1999).

Perhaps the most striking comparison with the traditional zoo emerges from the ability to derive values from the circulation of animal images in the networks of natural history film-making and exhibition surrounding the electronic zoo. In divorcing the spectacle of animals from their bodily presence, they have been able to speed up the spatial dynamics of collecting and accumulating of value from recirculating images of animals. The images of species displayed in Wildscreen can be multiplied and reused with no diminution of the original, and they can be altered, amended and redisplayed for use across different parts of the Wildscreen complex. More value can be accrued from these kinds of images than was possible from the animals in the zoo, where the bodily needs and ultimate mortality of animals restricted the ways in which they could be exploited. Images of animals on film can be endlessly redeployed for use in multi-media advertising, as well as in natural history film-making. This value is of course protected and controlled through new mechanisms operating within the network. The first page on the ARKive is a copyright notice that has to be agreed before entering the site. The spaces of the electronic zoo are structured through ownership as well as exper-tise. The ditches and walls of international copyright agreements form the invisible means of control over the images of animals in the electronic zoo.

Human identity and animal images

With any new technology or form of communication, it is important to ask what kind of human subjects it suggests (Poster 1996). These are difficult questions to address, and ones beset by all manner of determinisms. However, in the same way that the traditional zoo inscribes a dominant position for viewers, so I would suggest does the electronic zoo. First, the electronic zoo offers a different form of social space which may replace the collective experience of visiting the zoo with the more isolated and individualistic encounter of the cinema. Many exhibits in the electronic zoo are to be viewed in a dark theatre where visitors are disen-gaged from both humans and animals. In the electronic zoo the camera has already fixed the animals in a domain which, although visible to the camera, will never be entered by the spectator. The devices used to obtain ever more arresting images – hidden cameras, telescopic lenses, flashlights, remote control cameras – combine to produce pictures which carry with them numerous indications of their normal invisibility (Berger 1980). In the electronic zoo these animals and their lives will be ever more visible, but the fact that the animals who are the subjects of these images could once observe us has lost all significance in this form of encounter. As Berger (1980: 14) reflects, 'what we know about them is an index of our power and thus an index of what separates us from them. The more we know, the further away they are.' The electronic zoo does apparently offer the potential to present more information about more animals to more people, without the manifest cruelty of animal incarceration, and yet I would argue that there is something particularly poignant about the *loss* of embodied animals in the electronic displays of natural richness and biodiversity.

The processes of image-making stabilise the animals in these electronic spaces. This has distinct advantages for the construction of animal displays: the images of animals are guaranteed as intense and eye-catching; their behaviour will be char-acteristic and suitable for the interpretations given it; and the animals will always be visible in the electronic zoo. The knowledge we have of animal behaviour and ecology appears more secure in this zoo, since the film inscribes evidence that supports scientific interpretations of animal behaviour. Indeed, the process of natural history film-making is a collaboration between scientists and film-makers, with the film-makers constructing their images and narratives from the estab-lished body of ethological knowledge (Davies 1998). The filmed animals in the electronic zoo can subsequently be reproduced with the certainty of asexual reproduction, and the images can be recirculated endlessly. Concern for animal welfare and preservation can shift instead to the creation and maintenance of the bioparks and reserves in which many of these animals are 'naturally' located (e.g. Whatmore and Thorne 1998).

This certainty comes at a cost, however, for knowledge always involves some loss of being (Žižek 1996: 280). The uncertainties evident at the traditional zoo are spaces where competing conceptions of animals and alternative animal behaviours could be explored. The animals in the electronic zoo do not misbehave, but are committed to repeating endlessly one interpretation of their complex behaviour. For the visitor to the electronic zoo there is no ambiguity about what the animal might mean, for the kinds of perspective we can take and interaction we can have with these images are curtailed at the site of film-making. The spaces where ordinary people can see themselves interacting with animals are excluded from these admittedly stunning, but experientially empty, images of wildlife. As a consequence we risk losing the opportunity to learn from these animals; not only about them, but also about ourselves. Our interactions with animals are lessened by their coding as merely information. As animal identities are increasingly defined through genetics, with species reduced to strands of DNA whose existence is separate from their corporeality, so their existence as pixellated icons in digital images removes these animals from everyday spaces into the realms of a purified nature where they become visible, knowable and immortal, but untouchable. The embodied animal does remain woven into these networks of the electronic zoo, it is true, but the terms and effects of power – and the ethical issues associated with natural history film-making – become more difficult to trace through the elongated networks around this new medium of animal display.

Of course, the achievement of the electronic zoo will never be secure or certain. Turkle (1995), for one, reminds us that people adopt multiple identities in relationship to different kinds of technology, and I look forward to the new kinds of creativity which are sure to emerge around these images of animals. For the animals themselves, however, the situation is perhaps more problematic. Their bodily presence is central to their expressions of agency and identity, for it is through the places that they inhabit and how they act within these spaces that they are able to contest our understandings of them. As Philo (1995: 656) suggests:

> [M]any animals (domesticated and wild) are on occasion transgressive of the socio-spatial order which is created and policed around them by human beings. ... [A]nimals often squeeze out of the places – or out of the roles that they are supposed to play in certain places – which have been allotted to them by human beings.

I would suggest that the further marginalisation of many animals, as they end up inhabiting either diminishing areas of wilderness or electronic spaces within the city, removes their capacity to interact and to act in ways evident in everyday worlds. Animals are rendered both fully visible and fully controllable within the spaces of the electronic zoo, encoded and transformed into information on ethology and genetic biodiversity. The ability of the electronic zoo to speak for animals through the socio-spatial networks constructed around it is more secure than in the traditional zoo, one of the reasons for its success. The capacity of animals as active subjects within these geographies is correspondingly diminished. The important constituents of their lives are made visible to us in the exhibition spaces of the zoo in the centre of the city, remaining separate from their bodies. This was arguably also the case in the traditional zoo where the body of an animal scientifically stood in for the large species. However, for the viewing public it was the personalities, performances and physical presence of the animal which gave local expression and meaning to these scientific understandings, and allowed them to engage with these animals. Furthermore, the corporeal presence of the animals acted as a potent reminder of their subjectivity and agency. The removal of living animals from the spaces of our cities is to be celebrated in some respects, since animals have suffered cruelty, indignity and death in the networks of the traditional zoo, but their fate as endlessly circulated images – and the further disengagement of humankind from animals – in the electronic zoo gives cause for further concern.

Conclusion

This chapter has examined the networks of popular natural history as they define and delimit the places occupied by animals in human culture. It has re-examined the spatial practices of the traditional zoo, using this to explore the developments of the electronic zoo. In the two empirical examples given there are important overlaps. The electronic media do not determine a new way of looking at nature, but enable the continued expansion and culmination of a subject–object relationship forged in earlier modes of natural history. The geographical relationships are not radically altered, but rather accelerated and extended in these new network assemblages. What is different is the way that the animals are stabilised through different forms of representation, in the shift from animal capture and display through cages, bars and enclosures to the film technology and digital spaces of the electronic zoo. The networks of the traditional zoo remind us that our constructions of nature have always been mediated in various ways. However, I have argued that the electronic zoo offers a different kind of place through which boundaries and interactions between humans and non-human animals are re-created. In this conclusion I speculate briefly on the nature of this new kind of place.

There is of course an extensive geographical literature on the meaning of place. This has been recently enriched by reflection on the contribution of animals to constructions of place (Philo 1995; Philo and Wolch 1998; Wolch and Emel 1998), as well as speculations on the developments of cyberspace for our experience of place (Adams 1998; Graham 1998). The first serves to remind us how people, plants and animals are already interwoven into our everyday experiences of cities, suburbs, country and wilderness. They recapture the ordinariness of the wild (Anderson 1997), seeking to relocate animals from their position within pristine utopias of wilderness and to reclaim the heterogeneity of social spaces as ones shared between humans and non-humans (Whatmore and Thorne 1998; Wolch *et al.* 1995). The second set of literatures on cyberspace contemplates whether these developing socio-technical enterprises offer new spaces for exploring and communicating identity, community and experience; what and who stands to benefit from access to the spaces of virtual worlds, and conversely what is risked. Through conceiving of zoos as places which are points in the branching networks and spatial processes of natural history disciplines, I have been concerned with both of these reworkings of place, and of course with the tensions between them.

In drawing some preliminary and speculative conclusions about the position of animals within the networks of the electronic zoo, issues of animal agency, human experience and human responsibility emerge as a consequence of the dematerialisation of animal forms. For Tuan (1977: 18), places achieve concrete reality when our experience of them is total; that is, 'through all the senses as with the active and reflective mind'. I would suggest that this is equally so for encounters with animals. The visual has arguably dominated expert understandings of nature for several centuries. The visual is the defining sense in both the traditional and the electronic zoo, and its pre-eminence over other senses in science and entertainment drives the development of the electronic zoo. In the emphasis on the visual consumption of animals within the electronic zoo, though, both human experiences and animal agency are curtailed. Limited to visual experiences, the human visitor misses out on the multi-sensory engagement and interaction with animals possible within the embodied spaces of older zoos. Reduced to visual information, the animals are fixed within image spaces which position animals as inanimate inhabitants of wild spaces outside of the human realm. The electronic zoo does not overcome the unequal boundaries created between humans and animals evident in the traditional zoo, and in fact through further developments of visual technology may reinforce the 'apartheid' between human and animal kind (Kiley-Worthington 1990). Consigning animals to the purified spaces of an imagined nature, while constructing new cultural and technological artefacts in their likeness, has implications for understanding how animals are already fully bound into the many networks of scientific understanding and environmental management as well as everyday life.

For Giddens (1984), place is a setting for seeing the consequences of one's actions, for reflexive thought. Perhaps it can be argued that the traditional zoo – as a place shared with animals, experienced with all the senses, and presenting us in concrete form (often literally) with the boundaries and relationships forged between human and non-human animals – offers a better location for such reflections than the more controlled environment of the electronic zoo. The form of natural history made flesh through the networks of the traditional zoo is full of contradictions for the zoo visitor: I have argued that it reduces the freedom of animals, while also giving expression to their agency. The electronic zoo offers a different kind of space, one where we are relieved of coming face to face with the subjugation of animals, and one perhaps where we are relieved of considering them at all. In important ways most animals today are living their lives by consequence of human actions. Rather than further removing living animals from everyday life, there is a need for new spaces where we can explore the different means of taking responsibility for our roles in deciding what these lives will be and how we might be able to share in them.

Acknowledgements

I would like to thank Chris Philo, Chris Wilbert, Jacquie Burgess and Luke Desforges for their criticisms and encouragement. I am indebted to members of the BBC's Natural History Unit for introducing me to the Wildscreen project, and to Wildscreen for use of the image. The support of the Economic and Social Research Council (ESRC) is gratefully acknowledged.

Notes

1 I use the term 'virtual animal' to signal the growing distance between the body of the animal and its image, as created through the increased ability to manipulate and to recontextualise images of animals. There is of course no clear-cut boundary between older forms of visual capture such as photography, film or taxidermy and newer technologies of digital imagery or animatronics. However, through these more dynamic technologies, representations of animals can be made to perform, inhabit and even reproduce within digital spaces. What I have labelled the 'electronic zoo' is a space at the centre of the processes through which such virtual animals are created, managed and displayed. Many of these animals are not truly virtual, in that they ultimately do have an 'original', although this is perhaps not the case for extinct animals only visually known to living humans through their capture on film or re-creation in computer graphics. This issue is increasingly problematic for many of the representational strategies used to create images of wildlife. This use of the term 'virtual animals' in this sense is perhaps provocative, but I hope that it is productive.

2 The adoption of ideas from 'actor-network theory' (ANT) and science studies can be seen as part of an ongoing geographical project to challenge the dualisms of Western experience and intellectual thought. Increasing numbers of geographical writers are using the ideas of ANT (as contained in, e.g., Callon 1986; Latour 1987 1988, 1993; Law 1991, 1994) to explore the assemblages of nature, technology and society with which the discipline is concerned (Bingham 1996; Demeritt 1996; Hinchliffe 1996; Murdoch and Clark 1994; Murdoch and Marsden

1995). There are now a number of thoughtful introductions to the scope and limitations of this work in the geographical literature (Amin and Thrift 1995; Murdoch 1997a, 1997b; Whatmore 1999).

3 This is demonstrated in the work of Outram (1996), who suggests that the Musée National d'Histoire Naturelle in Paris, founded in 1793, acted as a microcosm where debates on the use of space in the museum mirrored a period of transition in natural history. Cuvier, trying to pioneer an approach to natural history which related living beings through their internal structures rather than external characteristics, clashed with his brother Frédéric, director of the zoo, over the appropriate allocation of space between the living animals and the anatomy galleries.

4 There is a burgeoning academic literature on the institution of the zoo, which particularly focuses on the establishment of European and colonial zoos from the mid-nineteenth to the twentieth century, and on the parallel developments of American zoological gardens (Anderson 1995; Hoage and Deiss 1996; Jardine *et al.* 1996). These histories and geographies of zoological institutions usefully contextualise and denaturalise different forms of animal display, revealing rich portraits of individual organisations while reflecting more broadly on the range of processes through which knowledge about nature is (re-)created (Allison 1995).

5 The German animal collector Carl Hagenbeck was perhaps the last of the great animal traders (Montgomery 1995). He built an enormous capture–transport–delivery system that reached across Africa, Asia, Australia, South America and the Philippines, and he supplied nearly all of the zoos of Europe, most of its circuses, and a large number of private and commercial menageries. He is reputed to have shipped as many as 15,000 animals in one single year, and in 1883 he took back no fewer than eighty-three elephants from Sri Lanka alone (Montgomery 1995).

6 Zoo proprietors apparently 'raced one another down the Thames estuary to board ships returning from Africa, the Americas, and especially the East Indies. Such vessels often carried a few living animals along with their more conventional cargo' (Ritvo 1996: 44).

7 By the 1870s Hagenbeck had swamped the market, and in order to diversify he started importing whole ethnographic communities, creating displays that showed Laplanders, Eskimos and Sudanese peoples sharing exhibition spaces with animals, ethnographic displays and dioramas (Montgomery 1995).

8 London Zoo and the collections which moved to Paris after the revolution in France can both trace their histories, and their animals, to the royal collections of animals held at Windsor Park and the Tower of London and at Versailles respectively (Osborne 1996; Ritvo 1987).

9 More recently, the Chinese Association of Zoological Gardens has been able to exploit the rarity value of its pandas, which it leases to about four zoos a year. There is a waiting list for such visits, with a three-month stay for the pandas costing between $500,000 and $1,000 000 (Wilson 1992).

10 These animals are well known by natural history film-producers both in the UK and in the US. A limited assemblage of animals such as the great white shark, panda, wolf, gorilla, polar bear, leopard and other big cats constitute the staple of big-budget co-produced natural history films. As one BBC producer says, 'I call them the National Geographic animals' (Richard Brock, interview, spring 1995).

11 Schmitt (1969) suggests that this has always been a goal of natural history photographers as well as film-makers, and stresses the importance of technology for achieving it. In early wildlife photography,

[m]an's [*sic*] physical presence posed a barrier to his communication with creation's dawn. The wild creatures which met his eye were usually nervous and looked unnatural. Only rarely could he surprise then unaware. Serious hobbyists learned to fasten simple triggers to their cameras so that birds could take their own portraits. Day and night, such cameras stood watch over bird's nests and woodland trails – patient proxies for impatient men. Thus nature lovers found a way to enjoy second-hand the realities of nature which they might have otherwise not seen.

(Schmitt 1969: 140)

References

Adams, P. (1988) 'Network topologies and virtual place', *Annals of the Association of American Geographers* 88: 88–106.

Allison, S. (1995) 'Making nature "real again" ', *Science as Culture* 5: 57–84.

Amin, A. and Thrift, N. (1995) 'Living in the global', in A. Amin and N. Thrift (eds) *Globalisation, Institutions and Regional Development*, Oxford: Oxford University Press.

Anderson, K. (1995) 'Culture and nature at the Adelaide Zoo: at the frontiers of "human" geography', *Transactions of the Institute of British Geographers* 20: 275–294.

Anderson, K. (1997) 'A walk on the wild side: a critical geography of domestication', *Progress in Human Geography* 21: 463–485.

Berger, J. (1980) 'Why look at animals?', in *About Looking*, London: London Writers and Readers.

Bingham, N. (1996) 'Object-ions: from technological determinism towards geographies of relations', *Environment and Planning D: Society and Space* 14: 635–657.

Blum, A. (1993) *Picturing Nature: American Nineteenth-Century Zoological Illustration*, Chichester: Princetown University Press.

Callon, M. (1986) 'Some elements of a sociology of translation: domestication of the scallops and the fisherman of St. Brieuc Bay', in J. Law (ed.) *Power, Action and Belief: A New Sociology of Knowledge?*, London: Routledge and Kegan Paul.

Cartwright, L. (1992) 'Experiments of destruction: cinematic inscriptions of physiology', *Representations* 40: 134–150.

Cooter, R. and Pumfrey, S. (1994) 'Separate spheres and public places: reflections on the history of science popularisation and science in popular culture', *History of Science* 32: 237–267.

Crowther, B. (1997) 'Viewing what comes naturally: a feminist approach to television natural history', *Women's Studies Internal Forum* 20: 289–300.

Davies, G. (1998) *Networks of Nature: Stories of Natural History Film-Making from the BBC*, Unpublished PhD thesis, London: University of London.

Davies, G. (1999) 'Exploiting the archive: and the animals came in two by two, 16mm, CD-ROM and BetaSP', *Area* 31: 49–58.

Demeritt, D. (1996) 'Social theory and the reconstruction of science and geography', *Transactions of the Institute of British Geographers* 21: 484–503.

Findlen, P. (1996) 'Courting nature', in N. Jardine, J.A. Secord and E.C. Spary (eds) *Cultures of Natural History*, Cambridge: Cambridge University Press.

Flint, R.W. (1996) 'American showmen and European dealers: commerce in wild animals in nineteenth-century America', in R.J. Hoage and W.M. Deiss (eds) *New Worlds, New Animals: From Menagerie to Zoological Park in the Nineteenth Century*, London: Johns Hopkins University Press.

Franklin, A. (1999) *Animals and Modern Cultures: A Sociology of Human–Animal Relations in Modernity*, London: Sage.

Giddens, A. (1984) *The Constitution of Society*, Cambridge: Polity.

Gillespie, A. and Robins, K. (1989) 'Geographical inequalities: the spatial basis of the new communication technologies', *Journal of Communication* 39(3): 7–18.

Graham, S. (1998) 'The end of geography or the explosion of place? Conceptualising space, place and information technology', *Progress in Human Geography* 22: 165–185.

Haraway, D. (1989) *Primate Visions: Gender, Race and Nature in the World of Modern Science*, London: Routledge.

Harrison, C. (1993) 'Nature conservation, science and popular values', in A. Warren and B. Goldsmith (eds) *Conservation in Progress*, Chichester: John Wiley.

Hayles, K. (1993) 'Virtual bodies and flickering signifiers', *October* 66: 69–91.

Hillis, K. (1994) 'The virtue of becoming a no-body', *Ecumene* 1: 177–196.

Hinchliffe, S (1996) 'Technology, power and space – the means and ends of geographies of tech-nology', *Environment and Planning D: Society and Space* 14: 659–682.

Hoage, R.J. and Deiss, W.A. (eds) (1996) *New Worlds, New Animals: From Menagerie to Zoological Park in the Nineteenth Century*, London: Johns Hopkins University Press.

Jardine, N., Secord, J.A. and Spary E.C. (eds) (1996) *Cultures of Natural History*, Cambridge: Cambridge University Press.

Keller, E. (1996) 'The biological gaze', in G. Robertson, M. Mash, L. Tickner, B. Curtis and T. Putnam (eds) *FutureNatural: Nature, Science, Culture*, London: Routledge.

Kiley-Worthington, M. (1990) *Animals in Circuses and Zoos: Chiron's World?*, Essex: Little Eco-Farms Publishing.

Kitchin, R. (1998) 'Towards geographies of cyberspace', *Progress in Human Geography* 22: 385–406.

Kreger, M. and Mench, J. (1995) 'Visitor animal interactions at the zoo', *Anthrozoos* 8(3): 143–158.

Latour, B. (1987) *Science in Action: How to Follow Scientists and Engineers through Society*, Cambridge, MA: Harvard University Press.

Latour, B. (1988) *The Pasteurisation of France and Irreductions*, Cambridge, MA: Harvard University Press.

Latour, B. (1993) *We Have Never Been Modern*, Hemel Hempstead: Harvester Wheatsheaf.

Law, J. (ed.) (1991) *A Sociology of Monsters: Essays on Power, Technology and Domination*, London: Rout-ledge.

Law, J. (1994) *Organising Modernity*, Oxford: Blackwell.

Marren, P. (1995) *The New Naturalists*, London: HarperCollins.

Marvin, G (1994) 'Review essay: Maple and Archibald's (1993) *Zooman: Inside the Zoo Revolution* and Bostock's (1993) *Zoos and Animals Rights: The Ethics of Keeping Animals*', *Society and Animals* 2: 191–199.

May, R. (1988) 'How many species are there on earth?', *Science* 241: 1441–1449.

Miller, D. (1996) 'Joseph Banks and the "centres of calculation" in Hanoverian London', in D. Miller and P. Reill (eds) *Visions of Empire: Voyages, Botany and Representations of Empire*, Cambridge: Cambridge University Press.

Mills, S. (1989) 'The entertainment imperative: wildlife films and conservation', *Ecos* 10(1): 3–7.

Mills, S. (1997) 'Pocket tigers: the sad unseen reality behind the wildlife film', *Times Literary Supple-ment*, 21 February.

Mitman, G. (1993) 'Hollywood technology, popular culture and the American Natural History Museum', *Isis* 84: 637–661.

Montgomery, S. (1995) 'The zoo: theatre of the animals', *Science as Culture* 4: 565–600.

Murdoch, J. (1997a) 'Inhuman/nonhuman/human: actor-network theory and the prospects for a nondualistic and symmetrical perspective on nature and society', *Environment and Planning D: Society and Space* 15: 731–756.

Murdoch, J. (1997b) 'Towards a geography of heterogeneous associations', *Progress in Human Geog-raphy* 21: 321–337.

Murdoch, J. and Clark, J. (1994) 'Sustainable knowledge', *Geoforum* 25: 115–132.

Murdoch, J. and Marsden, T. (1995) 'The spatialisation of politics: local and national actor spaces in environmental conflict', *Transactions of the Institute of British Geographers* 20: 368–380.

Osborne, M.A. (1996) 'Zoos in the family: the Geoffrey Saint Hilaire clan and the three zoos of Paris', in R.J. Hoage and W.A. Deiss (eds) *New Worlds, New Animals: From Menagerie to Zoological Park in the Nineteenth Century*, London: Johns Hopkins University Press.

Outram, D. (1996) 'New spaces in natural history', in N. Jardine, J.A. Secord and E.C. Spary (eds) *Cultures of Natural History*, Cambridge: Cambridge University Press.

Parry B.C. (2000) 'The fate of the collections: social justice and the annexation of plant genetic resources', in C. Zerner (ed.) *People, Plants, and Justice: The Politics of Nature Conservation*, New York: Columbia University Press.

Philo, C. (1995) 'Animals, geography and the city: notes on inclusions and exclusions', *Environment and Planning D: Society and Space* 13: 655–681.

Philo, C. and Wolch, J. (1998) 'Through the geographical looking glass: space, place and human–animal relations', *Society and Animals* 6: 103–118.

Poster, M. (1996) 'Postmodern virtualities', in G. Robertson, M. Mash, L. Tickner, J. Bird, B. Curtis and T. Putnam (eds) *FutureNatural: Nature, Science, Culture*, London: Routledge.

Pratt, M.L. (1992) *Imperial Eyes: Travel Writing and Transculturation*, London: Routledge.

Reichenbach, H. (1996) 'A tale of two zoos: the Hamburg Zoological Garden and Carl Hagenbeck's Tierpark', in R.J. Hoage and W.M. Deiss (eds) *New Worlds, New Animals: From Menagerie to Zoological Park in the Nineteenth Century*, London: Johns Hopkins University Press.

Ritvo, H. (1987) *The Animal Estate: The English and Other Creatures in the Victorian Age*, Harmondsworth: Penguin.

Ritvo, H. (1996) 'The order of nature: constructing the collections of Victorian zoos', in R.J. Hoage and W.M. Deiss (eds) *New Worlds, New Animals: From Menagerie to Zoological Park in the Nineteenth Century*, London: Johns Hopkins University Press.

Robertson, G., Mash, M., Tickner, L., Bird, J., Curtis, B. and Putnam, T. (eds) (1996) *FutureNatural: Nature, Science, Culture*, London: Routledge.

Schmitt, P. (1969) *Back to Nature: The Arcadian Myth in Urban America*, New York: Oxford University Press.

Silverstone, R. (1986) 'Putting the natural back into natural history', *Journal of Geography in Higher Education* 10(1): 89–92.

Stemerding, D. (1993) 'How to make oneself nature's spokesman? A Latourian account of classification in the eighteenth- and early nineteenth-century natural history', *Biology and Philosophy* 8: 193–223.

Thrift, N., Driver, F. and Livingstone, D. (1995) 'Editorial: the geography of truth', *Environment and Planning D: Society and Space* 13: 1–3.

Tuan, Y.-F. (1977) *Space and Place*, Minneapolis: University of Minnesota Press.

Turkle, S. (1995) *Life on the Screen: Identity in the Age of the Internet*, New York: Simon and Schuster.

Veltre, T. (1996) 'Menageries, metaphors and meanings', in R.J. Hoage and W.M. Deiss (eds) *New Worlds, New Animals: From Menagerie to Zoological Park in the Nineteenth Century*, London: Johns Hopkins University Press.

Vernon, K. (1993) 'Desperately seeking status: evolutionary systematics and the taxonomists' search for respectability 1940–60', *British Journal for the History Science* 26: 207–227.

Whatmore, S. (1999) 'Hybrid geographies: rethinking the "human" in human geography', in D. Massey, J. Allen and P. Sarre (eds) *Human Geography Today*, Oxford: Polity.

Whatmore S. and Thorne L. (1998) 'Wild(er)ness: reconfiguring the geographies of wildlife', *Transactions of the Institute of British Geographers* 23: 435–454.

Wilbert, C. (1998) 'Dysfunctional animals in functional places: scientific knowledge through the mangle', Paper presented at the Royal Geographical Society/Institute of British Geographers Conference, University of Surrey, Guildford, UK, January.

Wilson, A. (1992) *The Culture of Nature: North American Landscape from Disney to the Exxon Valdez*, Oxford: Blackwell.

Winston, B. (1993) 'The documentary film as scientific inscription', in M. Renov (ed.) *Theorizing Documentary*, London: Routledge.

Wise, J. (1997) *Exploring Technology and Social Space*, London: Sage.

Wolch, J. and Emel, J. (1995) 'Bringing the animals back in', *Environment and Planning D: Society and Space* 13: 632–636.

Wolch, J. and Emel, J. (eds) (1998) *Animal Geographies: Place, Politics and Identity in the Nature–Culture Borderlands*, London: Verso.

Wolch, J., West, K. and Gaines, T. (1995) 'Transspecies urban theory', *Environment and Planning D: Society and Space* 13: 735–760.

Wynne, B. (1993) 'Public uptake of science: a case for institutional reflexivity', *Public Understanding of Science* 2: 321–327.

Žižek, S. (1996) 'Lacan with quantum physics', in G. Robertson, M. Mash, L. Tickner, J. Bird, B. Curtis and T. Putnam (eds) *FutureNatural: Nature, Science, Culture*, London: Routledge.

13 (Un)ethical geographies of human–non-human relations

Encounters, collectives and spaces

Owain Jones

Introduction

The aim of this chapter is to draw out considerations on how ethical relations between humans and non-humans (and in this instance, animals) are deeply uneven, and on how this unevenness has distinct spatial dimensions which require a geographical sensibility within any study addressing them. As Casey (1998: ix) asserts, 'nothing we do is unplaced. … How could we fail to recognise this primal fact?' Human–non-human relations are inevitably embedded in the complex spatialities of the world. The myriad encounters which make up human–non-human relations shape and are shaped by this spatiality in an incredibly rich (in ontological terms) series of 'spatial formations'. Any consideration of human–non-human relations has to confront this geography of the spaces and places of encounter. In particular, my focus is on the ethical implications which may follow from looking at the world in this way.

My starting-point is that all these (emplaced) encounters have an ethical 'resonance' or 'freight', even if they do not fall within the compass of ethics in terms of either formalised systems of thought or more messy emotional/moral 'systems' of judgement. There is hence a spatiality of ethics which shadows the spatiality of encounter, forming a terrain of ethical events which is as variable as the terrain of the earth itself. In normative ethical terms, or from moral/emotional positions, this presents a field of encounters which range from the ethical to the unethical, or from the cruel to the compassionate, but also include whole sequences of encounters which are lost to these particular gazes. I suggest that by trying to take seriously this geography of (un)ethical encounters, we deal with the world as practice in a way which might be more inclusive, incisive and embedded than are abstracted, universalised systems of thought. Such an approach opens up a vast array of questions, some of which revolve around notions of moving towards the irreducible ontology of 'non-representational theory' (Thrift 1999), while still being able to describe, even to prescribe upon, the world in ways which (might) make a difference. Here my aim is to set out in more detail the notion of a geography of (un)ethical relations, and to develop a number of themes which seem to be important within it.

The deeply complex pattern which is human–non-human relations has a number of implications which need to be addressed. First, it presents a 'starting-point' or 'ground' from where, or upon which, it is extremely difficult to build or to impose abstracted, universalised prescriptive or descriptive accounts of the ethics of human–animal relations. Thus approaches which deal with this geography of being, such as Lynn (1998a, 1998b), seem necessary, and perhaps within the richness of encounter there are forms of human–animal relations which can be narrated as fresh perspectives on issues of animal ethics, rights and welfare. Second, the incredible ethical unevenness embedded in the complex patterning of human–animal relations needs much more frank narration. Third, in particular it is the spaces which are customarily closed off from conventional ethical gaze which need urgent consideration. For example, Weston (1999: 189) states, '[as] far as I know, there are no worked-out ethical defences of factory farming; it is hard to escape the conclusion that it is a practice sustained by silent collusion, by the "wish not to know".' The development of a geography of human–non-human relations, and exploring the ethical dimensions of such a geography, is an extensive multi-dimensional project, which, given current concerns for environmental integrity, environmental ethics and reconnecting the natural and social realms within social theory, is likely to feature prominently across a range of disciplines.

As well as making a case for the development of a geography of human–animal relations, this chapter offers three more specific contributions to this endeavour. The first is a view of ethics, which, extracted rather perfunctorily from the work of Levinas, argues for seeing all encounters, and the space(s) of all encounters, as ethically charged, even if they fall outside the realm of established ethical consideration. The second considers how humans often relate to animals not as individual others, but as part of some form of collective. This is a key mechanism underpinning the spatialisation of ethical encounters between humans and non-humans which needs thinking through. Third, with the notion of a geography of (un)ethical encounters established, three types of spatiality are briefly 'opened up', with the aim of raising some key questions and showing how such geographies may look and why their examination is needed.

Non-human geographies and ethics

It hardly needs saying that human relationships with animals have been and remain deeply complex and shifting. They range from the sublime to the obscene, and are twisted and folded into all kinds of paradoxical, ironic, tragic and also cathartic forms. That the differing ways in which humans and animals interact are often spatially constructed is evident in that many typologies of animals have a spatial basis: for example, *farm* animals, *zoo* animals, *laboratory* animals, *wild* animals, *domestic* pets, and so forth. Differing cultural spaces, wherein animals can be conceived of as either 'in place' or 'out of place', also vary widely in their relations with non-humans, and thus a compound differential of spatialised ethical relations is formed. The complexity of human–non-human relations, coupled with the general exclusion of non-humans from normative ethical considerations, has led not only to this fragmentation of the ethical nature of these encounters, but has also rendered many of these spaces of (un)ethical encounter closed from view. These hidden spaces demand consideration because of the deeply questionable nature of some relationships, and because of the huge range of inconsistency between them. Within the processes implicated here, it will be argued, the ethical invisibility of the *individual* non-human other is a key factor in the spatialisation of ethical relations and in how such issues should be addressed. From this position, it can be seen that all of the many differing material and imagined spaces of the world carry differing (un)ethical freights which need to be drawn out and addressed when questioning human–animal relations. This is quintessentially a geographical issue.

Fitzsimmons and Goodman (1998: 194) suggest that across the social sciences and humanities there has emerged a broad if somewhat disjointed front, the concern of which is 'to bring "nature" "back in" to social theory by contesting its abstraction from "society"'. Work such as that by Whatmore and Thorne (1998), Wolch and Emel (1995a, 1995b, 1998), Whatmore (1997), Anderson (1995, 1997) and Philo (1995) indicates that 'human' geography is playing a important role within this new front, particularly in relation to animal geographies of one kind or another. But, despite the work highlighted above, there still remain large gaps in the focus of 'non-human geography'. For example, Whatmore and Thorne (1998: 436) point out that 'geographers have paid remarkably little attention to wildlife in recent times'. Addressing the other half of the animal world *not* seen as 'wild', Anderson (1997: 464) feels that the human sciences have mostly failed to engage critically with the processes of animal domestication, and that this failure 'continues to constrain the imagining of alternative ethical and practical relations between humans, animals and environments'. Similar concerns to these above have been expressed by Philo (1995: 656–657, emphasis added) about animal geographies in general:

> The tendency has been to consider animals as marginal 'thing-like' beings devoid of inner lives, apprehensions or sensibilities, with little attempt being made to probe the often take-for-granted assumptions underlying the different uses to which human communities have put animals *in different times and places*.

My chapter supports this position by trying to show that within all of these geographies of human–animal relations (as perhaps in all spatial practices), there are accompanying issues of ethics which are always shadowing these spatialities. As such, my aim follows Proctor's (1998: 15) call for 'understanding ethics as an inextricable part of geography's ontological project', and in particular his call for a consideration of the geographies of environmental ethics.

Lynn's (1998a, 1998b) deployment of what he terms 'geoethics' should be seen as a significant attempt to address the issues that Proctor raises. Lynn (1998a: 280) uses the concept of geographic context to build a spatially sensitised 'argument on the moral status of animals'. This is needed because '[a]ll human activity, including moral conflict, occurs at sites embedded in situations, making geographic context a constitutive element of all ethical problems' (Lynn 1998a: 283). The significance of this is that the rather bland universalised 'ideal typologies' of biocentrism and ecocentrism are 'too rigid to adequately understand our earth' (Lynn 1998b: 231). The approach that I advocate here largely complements Lynn's geoethics, but there are two additional elements: first, I aim to raise issues to do with individual and collective relations, body-spaces and 'other' spaces which I feel merit consideration within geographies of human–animal relations; and second, I feel that developing an 'ethics of encounter', where all (always situated) encounters are considered to have some form of ethical 'freight', departs from, or perhaps extends, the geoethics position. This difference revolves around the Levinasian approach to ethics which is set out below. In this view the 'ethical ground' of lived encounters precedes and eludes formalised statements of ethics, even if they are 'geoethics' which are attempting to 'drop the commitment to uniform transgeographic principles [and instead to] approach moral problems with greater contextual sensitivity, allowing for a greater play and reciprocity between principles and particulars' (Lynn 1998b: 237). The difference is subtle, but may be significant in certain ways. Geoethics represents a powerful and cogent call for the spatial/contextual sensitisation of moral and ethical thought, but the problem still remains of what lies outside of that focus, however broadened. By adapting (or trying to adapt) Levinas's approach, all practices or encounters can be seen as ethically charged, and it is beholden on us to tune into this irreducible, rich ontology of ethical resonance, and from within that to build moral practices, rather than trying to extend the 'circle of light' while never being sure of what lies beyond it.

Heeding Whatmore (1997), this chapter is not intended as a straightforward backing of those calling for the extension of individualist rights from the human to the non-human realm. Rather I hope to show that the complex and often hidden geographies of (un)ethical human–non-human relations need to be addressed as part of a wider programme of envisioning how we might move 'forward', or at least to somewhere (else). As Whitford (1991: 14) points out, to envisage 'new social or ethical forms' is to confine the future within the conceptualisations of the present. The idea of the search for 'progress', or even moving towards utopia, is a case of moving away from dystopia. 'What we want is to get away from here. Where we hope to land ... is a "there" which we thought of little and knew of even less' (Bauman 1993: 224). 'Where we are' now needs the closest scrutiny, and the grimmest and brightest parts of it need clear narration.

Ethics as encounter

Proctor (1998) suggests that approaches to ethics can be divisible as descriptive, normative and meta-ethics. In this chapter the focus moves between the former two of these, but with the insertion of a further and fundamental view of ethics. This is the notion derived from Levinas that ethics is first a matter of being, that the ethical is the encounter between the Same/self and the Other, and that all interactions are ethical or have an ethical dimension.

At the outset it has to be acknowledged that there is a considerable degree of difficulty in developing a full-blown Levinasian account of human–animal ethical relations. When asked directly about human–non-human ethical relations (in Wright *et al.* 1988: 169–172), Levinas replied at one point that 'it is clear ... the ethical extends to all living things.' Yet he also made it plain that was he keen to distinguish the human, and also the human 'break ... with pure being', as central to his project, stating that '[t]he aim of being is being itself. However, with the appearance of the human – and this is my entire philosophy – there is something more important than my life, and that is the life of the other' (in Wright *et al.* 1988: 172). Thus, for Levinas the notions both of 'the face' – a tricky concept which seems to revolve around the presence of the other, and yet hints at a sort of ungainable 'wormhole' into the ultimate alterity of Otherness – and of the 'encounter' seem irrevocably set in the human realm.

Taking a pragmatist approach to these concepts (see Emel 1991), however, it seems that they can still be usefully deployed in considerations of human–animal relations. The importance of adopting an ethics of encounter is that in conventional ethical approaches – such as the 'descriptive', which 'characterise[s] existing moral schemes', or the 'normative', which constructs a 'suitable moral basis for informing human conduct' (Proctor 1998: 9) – the scope of what falls into ethical focus is formed in a complex dialectic between *what is* and what some assert *should be* of moral significance. The problem here is who (or what) is left outside of this scope, and plenty are. Those outside are rendered faceless and voiceless, and this has generally been the fate of animals in normative ethics. Only if someone from *within* argues for the inclusion of some previously ignored outsiders do they achieve even the possibility of an uncertain foothold in the process. As Lynn (1998a: 286) puts it, 'animals cannot organise and challenge the practice for themselves, they require human interlocutors to speak and act in their interests.' Levinas's approach to ethics, on the other hand, initially at least, does not deal in these forms of normative or even descriptive ethics. For Levinas, the encounter is *ethical* in a way prior to any conceptualisation or rationalisation of ethics. 'The term ethics always signifies the fact of the encounter, of the relation of myself with the Other' (Levinas, in Critchley 1992: 17). Thus, as Davis (1996: 48) points out, for Levinas, '[t]he ethical is the broader domain, where ethical experiences and relationships occur before the foundation of ethics in the sense of philosophically established principles, rules or codes.'

So the initial proposition here is that all encounters are ethical, or have an ethical resonance. This may vary in magnitude and may be deemed ethical or otherwise in more conventional terms, but, if 'everything in the universe is encounters, happy or unhappy encounters' (Deleuze, in Thrift 1996: 29), or if the world is ultimately just a 'scene of transactions' (Dewey, in Goodman 1995: 2), then this view of ethics may be a device with which to build away from the chronically partial and anthropocentric established constructions of ethics (and, it should be added, of 'agency' too). It may enable us to build accounts of ethical practice not from transcendental universalist positions but from within the practice of encounters. This may also offer a way out of the 'contractarian' foundation of some normative ethical frameworks (see Regan 1998), as Davis (1996: 51) describes:

> The decoupling of responsibility from reciprocity has been described as the decisive act which distinguishes Levinas's ethical theory from virtually all others. It would be a mistake for me to respect the Other because I expect anything in return: my obligation and responsibility are not mirrored by the Other's reciprocal responsibility towards me.

Levinas's concern for the irreducibility of the other, for 'an account of alterity which does not reduce the Other to the Same', seems to offer some potential in addressing the otherness of non-humans, for he is concerned 'to elaborate a philosophy of self and Other in which both are preserved as independent and self-sufficient, but in some sense in relation with one another' (Davis 1996: 51).

There is a need to open up unquestioned yet always situated encounters, as well as opening up the unevenness of encounters in terms of the weak and patchy normative ethics which do pronounce on the treatment of animals. This inevitably leads to a confrontation with the immense diversity of the world in terms of material and imagined spaces and places, and the practices therein, and with the vastly divergent ethics of encounter embedded within these. The treatment of animals in one form of situation or space would be deemed unethical in another form of space, say between intensive factory farm units and zoos. But, beyond this, many human–non-human interactions are ethically invisible in the normative sense. The partial codes which do exist do not reach to these places with any clarity or force, and this is particularly so of 'other spaces' and the spaces of *individual* non-human others. Thus, the prediscursive ethics of these encounters remain unexamined; yet, if they were to be so, they might be acknowledged as deeply unethical by existing normative codes and therefore a matter of great concern.

The ethical invisibility of non-humans in normative ethics

Normative ethics can usually be considered as being *a*geographical, in that they are dealing in the realm of the generalised and the universal. This is particularly so for the bulk of ethical deliberations which have focused on human and intra-human relationships. Those ethical conceptualisations which take the individual as a basic ethical unit, and which try to construct normative frameworks that are *universal,* have had no need of geography, for there is no need or possibility for any spatially based exceptions or variations to these ideas. Proctor (1998) stresses this apparent incompatibility of geographical and ethical theorisation, pointing out that on those occasions where geography has come at normative ethics and even meta-ethics from a 'grounded, contextualised, and often concrete perspective', this has been problematic because such approaches do not fit in with the style of most philosophical literatures. Within abstract non-space, it is the individual human subject which has often been the basis for normative ethical principles, and for the social structures which have sprung from this such as democratic principles, many religious and legal codes, and global initiatives such as the UN Human Rights pronouncements. It is painfully clear that there are all too many cataclysmic examples of the failure or disregard of this principle within intra-human relations, and this to some extent at least has been the focus of attention for much critically oriented geography (Proctor 1998).

As soon as these universalised and individualised aspects are taken from ethical consideration, as they have been in human–non-human ethical relations, unevenness, including spatial unevenness, is inevitable. There is also the possibility of a chronic dissipation of tension between practice and any norms which do exist, because of the evasions, omissions and confusions which can occur within these fragmented arrangements. This had led to a situation where the ethical dimension of many human–non-human relations, and the complex geographies of those relations, have remained unexamined. Often the ethics of commonplace interactions between human and non-humans are conceptually rooted in relationships between and within populations or groupings of some kind, and this in turn adds to a deeply uneven spatialisation of both normative and lived ethical relations.

Various considerations of the position of nature within modernity, or more narrowly of animals, show how nature and animals have generally been 'eclipsed in orthodox ethical discourse' (Whatmore 1997: 37). Serpell (1996: 218), in his account of the shift from hunter-gatherer to agriculturalist-based societies, shows how the rendering ethically invisible of the non-human other was a key manoeuvre within processes of modernisation:

> [T]he entire system [agriculture] depends on the subjugation of nature, and the domination and manipulation of living creatures. It is likely that this new relationship with animals and nature, with its conflicting combination of intimacy and enslavement, generated intense feelings of guilt. ... Faced with this conflict, new ideologies were required, ideologies that absolved farming people from blame and enabled them to continue their remorseless programme of expansion and subjugation with a clear conscience. Different societies adopted different ideologies, some more expedient than others. But, if the record of civilisation is anything to go by, it was the most ruthless cultures – the ones with the most effective distancing devices – who prospered most of all.

Ingold (1988: 15), in considering Tapper's account of the shift from a hunter economy to pastoralism and further into modernity in the form of 'modern factory farming', feels that within this process 'the animal moves from being a strange person to a familiar thing'. This positioning of non-human others in general becomes reflected in terms of normative ethics, where, as Clark (1977: 5) states, some orthodox ethical endeavours expressly declare that 'animals are effectively things, owed no more duties of justice, charity or religion "as neither are sticks and stones"'. Such views can be seen not as rare exceptions, but rather as reflecting the normal position of non-humans. Singer (1993: 79) indicates that in the highly influential contractarian ethics of Rawls and Gauthier, animals are excluded from the ethical sphere because it is a key element of these theories that 'the universalising process should stop at the boundaries of our [human] community'.

Of course, such views of animals, other non-human others and 'nature' more generally are now seen as central arenas of dispute within certain strands of environmental philosophy and ethics which have arisen in response to the ecological and ethical crises unfolding around us. In particular, attention has focused on the rationalising and hardening of such views of nature within 'the scientific revolution in Europe', which transformed views of nature and in so doing 'removed all ethical and cognitive constraints against its violation and exploitation' (Shiva 1988: xv). The role of key figures such as Descartes and Bacon in this separating off of nature, and the development of reductionist, mechanistic systems of knowledge with a view to mastering 'her', is charted by Pepper (1986, 1996), Plumwood (1993), Merchant (1992) and many others.

As powerful as that story has become, and as marked as its consequences are, it cannot be seen to be entirely monolithic. Midgley (1983: 46) points out that the rationalist tradition 'does not speak with a single voice' in relation to views of nature. She shows how other thinkers, particularly Leibniz, tried to occupy the 'yawning gap' left by Decartes, and concludes that 'even seventeenth-century rationalism … does not furnish us with a clear and unanimous licence to poison all the pigeons in the park' (Midgley 1983: 47). Attfield (1983) also warns against over-simplistic readings of Descartes and Bacon themselves in this respect. So, although animals and nature have been generally excluded from orthodox ethics, this does not result in a complete negation of moral considerations relating to animals. Clark (1977: 5) points out that there has been at least 'nominal genuflection in the direction of "avoiding unnecessary suffering"', and Midgley (1983) argues that in the work of Kant, Schopenhauer and others genuine calls for compassion can be found. In the end, though, she adds that the exclusion of animals from the central contractual forms of morality means that they, and others outside of this centre, can '*glide unnoticed into the shadows and be forgotten*' (Midgley 1983: 64, emphasis added). It is the tension between this general exclusion of non-human others and vestiges of concern stemming from other sources which has led to the ethical fragmentation of human–animal relations within these shadowy spaces, and has led Clark (1977: 13, emphasis added) to ask:

> If some animals have some rights, even of this negative sort, to be spared wanton ill-treatment, on what grounds do we deny them other rights, to life and happiness within their kind? *On what grounds can other animals be fair game?*

It has to be kept in mind that human–animal relations have been and remain driven by a heady mix of emotions which span from reverence to revulsion and from fetish to fear. Although nature gets excluded from orthodox ethics, and this certainly has had a political fall-out, some groupings of nature, in some places and in some times, are included in some form of, perhaps *de facto*, ethical considerations in other ways. These may be based on religious, spiritual, emotional or even aesthetic grounds. For example, Sheail (1976: 4) states that 'nature preservation in its earliest day was almost entirely an emotional issue'. Initiatives such as the 1869 Sea Birds Preservation Acts, and the forming of the Royal Society for the Protection of Cruelty to Animals (RSPCA) and the Royal Society for the Protection of Birds (RSPB), were responding to highly visible and wanton exhibitions of nature destruction such as shooting parties which sailed to sea bird colonies and shot thousands of birds to leave them floating dead on the water. Such concerns had parallels and links to the Romantic and transcendental movements, which were expressly orientated counter to Enlightenment/modernity (Wheeler 1993) in part due to their concern for and celebration of nature. Attitudes that they generated have fed into the development of the nature conservation movement, and eventually into modern environmentalism (Bunce 1994;

Cronon 1996; Pepper 1996). Their ramifications can also be found in continuing contemporary popular affections for nature (Bunce 1994), and particularly for certain landscapes (Macnaghten and Urry 1998) and certain animals. So, although there has been a general exclusion of animals and nature from the realm of ethical consideration, this has been distorted, partially reversed or even emphasised by other types of relationships with nature and animals, thereby adding to the deeply uneven and in some cases painfully contradictory relations between certain non-human others and human society.

Individuals and collectives

The construction of the modern world has been based on the uneven but deeply cutting relegation of the individual non-human other from the ethical field. This has significant implications for the ethics of human–animal interaction. As Serpell (1995: 825) states:

> [T]reating animals as groups of organisms (populations, species, ecosystems, and so on) creates ethical problems when it encourages people to ignore or devalue the well-being of the individual animals comprising those groups. An extreme example is provided by modern intensive farming. ... In the field of nature conservation, it is common to ignore or subordinate the interests of individual animals for the perceived good of their own or other species.

It is this process which in part shows such a marked geography of ethical relations. We relate to and act upon groups of animals in particular spaces. Here I deploy some brief examples to illustrate this point. Magpies in the UK can be killed as pests under the general licence of the Wildlife and Countryside Act of 1981. When an individual magpie is shot, or trapped, the person carrying this out is, in a very real sense, not in a relationship with the individual bird, but rather in a relationship with magpies as a population within a legislative and cultural space. The individual bird is seen purely as 'a magpie', with no individual visibility in terms of personality, consciousness, family relationships, and so on. In other cultural/legislative spaces, such as Norway and parts of Asia, magpies are treated more benignly (Taylor 1990). In the ethics of encounters between magpies and humans within these spaces, the individual is lost within a larger cultural construction of the overall magpie population. If magpies became a rare species within the UK and protected as an icon of conservation, as certain raptors are, individuals would be protected. Yet, although finding themselves within a different ethics of encounter, their treatment would still be driven by their membership of a given population in a given space rather than as an individual. This illustration is not meant to suggest that there is a straightforward correlation between our treatment of (wild) animals and their population density, but rather to show that it is relations with *collectives or populations* which often drive how we treat individual non-human others. It is the spaces and the territories in which these populations are conceptualised, moreover, that render what occurs here a distinctly geographical process.

There are clearly instances where growing scarcity within a particular space has not generated protection or a revaluing of a given species or population. For example, the extinction of large land predators such as the wolf in the UK shows that, in this case, a decline of a population in a territory did not produce a reversal of attitudes from one of pest to one of rare species to be protected. But here again, the relationship between any one individual wolf and the hunter killing it was driven by an understanding of wolves imagined as a threat, the wolf as enemy, the wolf as a collective construction. Emel (1995: 708–709) describes how the wolf

> was virtually eradicated in the continental United States (with one or two exceptions) and over most of Western Europe. In England, Germany, and France the wolf was particularly hated, and immigrants from those places to North America came with a very negative approach to the wolf.

In contrast to this, Emel (1995: 709) relates how wolves fared better 'in the Balkans and the eastern Alps, not only because these regions were less densely populated than other parts of Europe, but also because they lacked or never formed institutionalised systems of incentives for extermination'. In other words, there was and remains a distinct geography of institutionally and culturally inscribed human–wolf relations, and within this there is a geography of (un)ethical relations between humans and wolves which, in the end, is acted out in terms of specific encounters between individuals. The deeply upsetting descriptions of the treatment of wolves given in Emel (1995) is a reminder of the extent to which any individual non-human other, and the suffering that he/she might endure, is subsumed within wider constructions of populations or types. The same can be said of course for many, many other animals.

Although scarcity cannot be seen as a guarantee of some form of ethical consideration for any given species, abundance within populations of non-humans does often seem to have a significant effect on how that population, and thus individuals within it, are treated. This can be seen in Thorne's (1998: 1) work on kangaroos, where, through their extensive collective presence, the (un)ethical treatment of individuals is not only sanctioned but largely unquestioned. Here, 'animals are effectively "non-issues" in the sense that trading systems created around them are assumed defensible at a basic level ... because the animals are deemed abundant.' As Thorne points out, only if they were to decrease dramatically in numbers would kangaroos begin to enter the 'ethical' realm of the Convention on International Trade in Endangered Species of Wild Fauna and Flora (CITES) agreement and other endangered species campaigning. In CITES and other conservation initiatives, scarcity and population numbers are always key in determining what falls within or without consideration. The line between protecting or hunting or culling of, say, various whale species or elephants, and many other species, is one which is constructed around calculations of population levels in certain spaces of one sort or another. Although the resulting treatment of any individual in groups either side of this divide may well be markedly different, in *neither* case would the individual be the ethical focus.

The ethical invisibility of the individual non-human other has been and remains extremely useful and probably essential to modern societies. This has generally enabled humans to manipulate, exploit, displace, consume, waste and torture non-human individuals with impunity. Many examples could be taken to demonstrate this usefulness, and it is interesting to imagine transmitting the ethical invisibility of these others into a human context. For example, when it was proposed by Shell that the Brent Spa oil platform was to be disposed of by dumping at sea, the supporting scientific discourse argued that there would be only limited local effects on the environment and that it would not have any meaningful impacts on the wider environment. In other words, only individual organisms in the immediate vicinity of the dumping would suffer damage or die, while collectively the overall ecosystem would remain unaffected and populations would not be affected in any meaningful (statistical) way. Thus, the individual organisms who would suffer and die were indeed rendered ethically and statistically invisible. Individual non-human others are often ethically and politically invisible, then, and they become lost in the crowds of their own and other kinds. Such a positioning of non-human others has emptied the encounter of normative ethical consideration, and of more practical everyday moral or even emotional consideration. As Bauman (1993: 115) says, echoing Levinas, 'When the Other dissolves in the Many, the first thing to dissolve is the Face. The Other(s) is (are) now faceless.'

There are myriad spatial/topological formulations of ethical relations between humans and animals. Territories in which populations are counted and acted upon are just one prominent example. The interconnecting spaces of differing animal production/consumption systems, and the legislations which apply to them, are another. The topologies which connect these spaces, and the imaginary spaces which fold into these, add yet further dimensions. For example, farm animals *en masse* will bear certain overarching human constructions, and wild animals the same, which contribute to our relationships with, and treatment of, them. Within these broad groupings, differing types of animals will be conceptualised as sub-groups. Even different species and families of animals such as birds, mammals and reptiles will bear different ethical constructions, and other categories of grouping will cross-cut these such as native or alien, predator or prey, and so on. In other words, our constructions of individual animals depend on a complex overlaying of meanings and interactions, which in turn are dominated largely by constructions of groups, types, populations, and so on. The life spaces that these animals occupy hence map on to distinct geographies of ethics.

The (re)emergence of the ethically visible individual non-human other

I have already quoted Serpell's (1995) concern that nature conservationists – who are seemingly concerned with the ethics of human–animal ethical relations – tend to consider animals as populations or collectives, rather than as individuals. This issue is in fact subject to a good deal of debate and tension within and between groups relating to animal welfare and conservation (see Agar 1995). Some challenge the conservationist orthodoxy that the welfare of individual animals is straightforwardly subordinate to wider concerns of species and population welfare and conservation, as in the zoo captive breeding programmes. Some now insist that it is the individual animal which should be of ethical primacy. This then raises important issues about the ethical prioritising of human-generated abstract notions of species and populations above the corporeal beings which make up those populations. Here my aim is to point out that these and other questions are deeply entangled in the spatial materiality of the world. If such (un)ethical spatiality is inevitable, given the number and diversity of human–animal transactions taking place in differing spatial formations, it is still worth thinking about the ways in which animals can emerge into individual ethical focus.

Although there has to be some caution concerning the romanticisation of premodern nature–society relations, it does appear to be the case that some societies have possessed a substantially different base relationship with non-human others. This is a position which is often attributed to American Indian culture. Callicott (1994), although expressing caution about a too simplistic rendering of such ideas, does reveal that there may be some genuine grounds for these assumptions, and that within some instances of these cultures there was a place for the *individual* non-human other. Callicott (1994: 129) refers to Martin's work on the culture of the Ojibwa tribe, in which

> [n]ature, as conceived by the traditional Ojibwa, was a congeries of societies. Every animal, fish, and plant ... functioned in a society that was parallel in all respects to man's [*sic*]. Wildlife and plant life had homes and families just as man did.

Such views of non-human others are also described by Jay, in his account of the relationship between salmon and American Northwest Coast Indians:

> [S]almon were people who lived in houses far away under the sea. [They were] great generous beings whose gifts gave life. ... The Indians understood the salmon's gift involved them in an ethical system that resounded in every corner of their locale. The aboriginal landscape was a democracy of spirit where everyone listened.
>
> (In Cheney 1989: 123)

The moving away from these 'face-to-face' positionings of non-humans to making them 'faceless' things must contribute to the cruelty many face today, because '[h]umanity turns into cruelty because of the temptation to close the openness, to recoil from stretching out towards the Other' (Bauman 1993: 89). Any possible switch from relating to non-human others as collectives to relating to them instead as individuals has profound implications for how we live on this planet, and may be a significant narrative for the future. However, there are glimpses of some (re-)emergence, or perhaps persistence, of this latter kind of relation in modern society, and these clearly reflect the deep but often latent tensions within human–non-human relationships when compared with the fate of the unimagined other.

There has been a large output of theory addressing the aim of extending our ethical structures to take in some parts or all of nature. A few of these are based on the recognition of the non-human *individual* as an ethical entity (for commentaries, see Agar 1995; Attfield 1983; Whatmore 1997). Most obviously this is the drive behind the work of animal rights theorists such as Clark, Regan and Singer. A huge amount of thought and debate has gone into pursuing the holy grail of a clear, indisputable 'proof' that nature, or certain sections of nature (particularly animals), must be taken seriously in ethical terms, the assumption being that such a shift in ethical understandings will drag the practices of society toward more just and ecologically sustainable (Naess 1997) forms. But all of these endeavours meet with at least two sets of difficulties. First, the pushing back of the ethical horizon away from its anthropocentric epicentre runs into formidable obstacles, chiefly because established human ethical conceptualisation tends to focus not only on the human individual as an ethical unit, but also on the specific attributes which humans seemingly hold as the grounds of judging and acting within ethical terms (see Rolston 1999). Second, there is the problem of the high degree of complexity and contradiction which is the starting-point: that is, the messiness, as already noted, of present circumstances. There is a need to open up these spaces, and to attend to the fates that non-human others meet in them, fates that will vary wildly. This shows that these are far from being spaces where nothing of concern is happening, but rather are spaces where the ethics of the encounter are not being told.

There are plenty of instances where non-human others are lifted out of the collective and are to a degree at least recognised as individually ethically visible. This has been the case with both dolphins and certain primates. Dolphins have emerged as a distinctive conservation concern because of the remarkable characteristics that they appear to have. It is recognised that they are in some ways intelligent, compassionate and perhaps possess some form of language. These are the things which allegedly make humans into persons, and assertions that dolphins are persons on such grounds, and thus worthy of individual rights on a par to a human person, have even been tested in US courts (Midgley 1996). Similarly, considerations of primate rights, which would recognise them as individuals, have been debated on television in the UK, again because these beings seem to have, to a meaningful degree, some of the attributes that we see as making ourselves into significant individuals. Midgley (1996: 116) suggests that it is emotional fellowship which make creatures our fellow-beings and deserving of consideration in terms of being persons, and that dolphins and apes have the 'kind of social and emotional complexity' which makes such an emotional fellowship possible. Such gradual and (for some) grudging extensions of ethical consideration away from its narrow anthropocentric individualist ground is admitted not because we are truly trying to extend the ethical community, but rather because we recognise that some of 'us', or those like us, have been mistakenly excluded. Further, more inclusive extensions of the ethical community would have to deal with the notion of the other, those quite unlike us, and all of the ways in which humans and non-humans interact.

In stories such as *Black Beauty*, *Free Willy* or *Babe*, individualised animals do come into focus, albeit sometimes through anthropomorphic means, and such stories do create ripples in the exploited spaces of the collectives from which these animals are lifted. Popular UK television programmes such as *Pet Rescue* and *Animal Hospital* function by highlighting individual animal stories and the moving levels of love, concern and care given to particular animals, levels which begin to approach parity with the care of sick and injured humans.

Animals can hence emerge into individual (ethical) focus in all manner of ways. For one, it is striking that, when animals do emerge into individual (ethical) focus in media reports of 'animal incidents', this commonly results from some sort of spatial 'disruption' of usually unarticulated (un)ethical geographies. For example, consider whales being washed up on a beach, or trapped in ice, where incredible efforts are made to save them and to return them to the aquatic space where their fate again becomes unarticulated; or consider the 'Tamworth Two', two pigs whose escape from slaughter and eight days evading capture in the English countryside attracted 150 media crews and secured them eventual residence in an animal sanctuary (Ford 1998). Certain animals can also emerge, or be pushed, into individual focus through the imaginative lifting of them from their common kind. This is particularly so with 'celebrity' animals such as famous racehorses like Desert Orchid, and, most markedly, with the 'charismatic' animals which are so central to zoo economics. Here notable animals are given individual names, histories and so forth, and their lives, particularly in terms of 'love', procreation and even illness, are constructed and deployed like soap operas, as in the breeding history of ChiChi the panda in London Zoo. A host of household pets too are given names, and, perhaps, become loved as individuals with specific habits and characteristics which become visible within the intimate proximity of shared lives.

Problems with the individualist rights approach

Whatmore (1997) sees the extension of ethical consideration in the form of rights to individual non-human others, based on current liberalist notions of the individual, as problematic. In her assessment of the potential for building a form of *relational ethics* which extends beyond the human realm, she feels that such a manoeuvre might

> breach the impasse of individual ethics at a number of key points ... it releases 'nature' and non-human beings from their relegation to the status of objects with no ethical standing in the pursuit of individual self-interest, *without* resorting to the extension of this liberal conception of ethical agency to other animals.
>
> (Whatmore 1997: 47, emphasis in original)

Whatmore's description of where non-humans are in ethical terms now ('no ethical standing') is generally the same as what I have set out. But her concern is that to call for the extension of rights to non-humans through the extension of liberal individualised ethics is to call upon a model which has itself drawn considerable criticism from certain feminist and ecofeminist perspectives, and also which comes under pressure from notions of hybridity. Although 'the general' position of animals in terms of ethical relations is much as Whatmore states, in actuality there is a whole range of variation within that position, linked to the strong spatial character with which I am concerned. The implications of this warrant consideration for a number of reasons: in particular, it can expose the inconstancy of such relations and the unethical practice which is one side of that inconsistency.

Geographies of (un)ethical encounters

Be it the goldfish bowl, the rabbit hutch, the bull-fight ring, the animal hospital, the abattoir, the vivisection lab, the house with a cat flap, the factory farm, the 'humane' farm, these are among the spaces where (un)ethical encounters between humans and animals occur, and they are the spaces which determine to a significant degree the nature of the encounter. There is a need to start unpacking these often hidden geographies of encounter if we are to grasp the lived complexities of human–non-human relations. The work of Arluke and Sanders (1996) shows how close scrutiny of human–animal relations has to build from such a spatial grounding. They state that 'inconsistent behaviour towards animals is omnipresent in Western society' (Arluke and Sanders 1996: 5), and this realisation, and the view that 'the commanding presence of non-human animals in our society is largely taken for granted' (Arluke and Sanders 1996: 1), drives their aim of addressing human–animal relations from a sociological perspective. In their approach, which entails some ethnographic research, they adopt an overtly spatial orientation, investigating how human–animal relations differ within and between the differing spaces of an animal shelter and two laboratories which conduct research on primates. They also hint at a historical geography of human–animal relations by considering the case of animals in Nazi Germany. The relationships which some animals experience as pets in the domestic space of the human home provide a poignant counterpoint to the fate of millions of animals who never emerge from the collective of which they are one, or from the unhappy space within which they may spend their life. Although there are many reasons to be concerned over 'the pet trade', by examining the treatment of pets, and in fact the concept of 'pet' itself, it seems clear that some pets, often (but not exclusively) dogs and cats, become individually visible in emotional and even ethical terms. In the intimate sharing of domestic space where animals and humans are in close, even touching, contact, the space may well become ethically 'illuminated' (although, to repeat caution, in many cases this is not so).

The collectives of animals which are in (un)ethical relations to humans can be seen as formed in many differing circumstances, not least, as argued here, through highly spatial processes. The task of turning these spaces inside out, revealing the nature of what unfolds in them, is substantial, and cannot be summarised easily. Here I briefly offer three takes on the spatialities of ethics which warrant urgent consideration. The first entails the implications of animals being constructed as 'in place' and 'out of place' in a manner which complexly cross-cuts with the spatiality of human–animal relations; the second entails the implications of ethical invisibility for the spaces of animal bodies; and the third entails the extreme ethical remoteness of some 'other spaces' and 'other elements', taking water as a key example.

In place, out of place

Within the complex spatiality of human–animal relations, there are folded constructions of appropriate and inappropriate presences of animals in particular places. This is particularly marked in the reactions to animals as pets, animals in cities, and animals bound up in the constructions of places as pure or impure (Sibley 1995) and wild or tame. For example, in modern agricultural practices animals rarely 'get out of place'. Their existence from birth to death is spatially marshalled. But there are many historical and contemporary instances where farm animals have slipped out of place, or become regarded as 'out of place', as in earlier forms of agricultural trading networks where animals mixed with people in urban settings (Philo 1995). Other examples of complex contested constructions of animals being 'out of place' include cougars in California suburbs (Gullo *et al.* 1998), or feral cats in the Australian landscape, or once domesticated animals such as dogs and cats which have become free-living in urban settings (Griffiths, Poulter and Sibley, this volume). Such constructions are at once about animals, humans, spaces and places, and about how meanings and practices form in complex arrangements of tension and compression between these elements. The spatially articulated ethics of encounter with which I am concerned here are in part driven by these imaginative placements and displacements of animals within different spaces.

Other bodies

Many animals and other non-human others are in the unfortunate position of being ethically invisible as individuals, while being only too visible as bodies comprising economic resources of some kind or another. This is a particularly pressing issue because, as Emel and Wolch (1998: 3) point out, '[g]lobalization has augmented dramatically the circulation of animal bodies (whole and in parts).' This has ethical fall-out, particularly when contrasted with an individualised human ethics where the body, as the vehicle for the 'Cartesian individual' (Whatmore 1997: 38), effectively forms the ethical territory. Thus bodies are the medium to which ethical consideration either sticks or falls away, and which ethically charge, or otherwise, the spaces that the bodies concerned are occupying. Where the individual, the person, is the ethical unit, this ethics becomes spatially articulated: the space that is the body of the person is defined, and the spaces within which that body operates also become ethically defined. Not only in theory is the body itself an ethical territory, but the space around it also becomes so. Torture is therefore deemed unethical, as are more prosaic abuses of individuals through the environment in which they operate. But, if the individual body is *not* the ethical unit coming under ethical consideration but rather some collective of abstracted bodies (such as factory-farmed pigs), then the ethical impulse is not articulated in association with the spaces *of* the body and the spaces *around* the body. Consequently, the ethics here remain in the realm of the abstract, if they survive at all, and leave these spaces as ethical blanks. Without the presence of the ethically visible body to ground the ethical practices within these spaces, whatever ethical consideration there may be becomes generalised and dissipated via convention, markets, legislation, discourse and practice, often generating a tendency of downgrading towards lowest common denominators. Thus, the bodies of laboratory animals, Thorne's (1998) kangaroos, Emel's (1995) wolves and factory-farmed pigs are all opened up to unethical practice. Taking such an approach may enable a connection with emerging geographies of 'the body' which, although (re)considering the corporeal, do so almost exclusively in terms of the human body (see Pile and Thrift 1995). If 'space and [human] embodiment is an interdisciplinary problematic attracting the attention of geographers as well as many others' (Longhurst 1997: 496), then it needs little imagination to realise – in the light of the encounters which non-human bodies undergo – the expanded scope of the geographies of ethics, spaces and non-human bodies which might now be written. What of the implications of differing types of body, the non-mammalian bodies of reptiles for instance? What are the implications of this otherness of embodiment, and how does body scale cross-cut this? The enormity of the bodies of whales and the smallness of insects seem to have to set up body-space challenges to our ethical sensibilities.

Other space(s)

My additional concern is that (otherly embodied) lives may be positioned yet further towards the margins, or beyond, of our ethical imagination due to the very spaces that they occupy being profoundly other to the one(s) that humans generally occupy. These are lives which may be lived in spaces such as water, soil, sky, (perpetual) darkness, forest canopy, and so forth, and (again) those lived on differing scales. I will not develop any form of typology of these other spaces, but will proceed by considering some of the encounters which take place in the profoundly other spaces of water. Water constructs forms of life spaces which are markedly alien to the 'airy' spaces that we humans inhabit. Water has a number of unique and peculiar qualities (see Farber 1994: 13–14), and to live in it must present utterly differing ways of embodied being. Space, scale, dimension, direction, light, sound, pressure, movement/resistance, gravity, body/element interactions are totally other to those of the air. Bodies and senses are (mostly) exclusively adapted to one or the other. My suggestion is that our (human) ethical imaginations also find water a hostile, impenetrable space, and that as a result many of the lives lived there are ethically invisible to us. The obvious exceptions to this, notably the concern for dolphins and some whales, may be due to their being in some ways like us (live young-bearing, air-breathing, language-using mammals), thus (partially) overcoming the otherness of body- and life-space.

It is striking that in all of the debate, controversy and ongoing disputes within the European Union about fishing quotas, relatively little attention is paid to the ethical dimensions of these issues. The over-fishing and fish quota debates are conducted almost entirely in terms of economics, employment, territorial rights, stock depletion and rescuing sustainable-use equilibrium. The fact that these are (perhaps some of the last) wild ecosystems, now being strip-mined, hardly arouses any debate at all. Visser (1998: 22) gives a graphic account of this process, a startling glimpse of human intrusion into this other world:

> Factory ships have grown to a 4000-ton capacity or more, dragging huge nets and hauling them up every four hours, 24 hours a day. Sonar devices and spotter aircraft hunt down the prey. 'Rockhoppers' see to it that rocks present no obstacle to shaving the ocean floor. 'Tickler chains' create dust and noise in order to flush out every lurking fish. Such a ship leaves a desert in its track. No matter how large the holes of a net's mesh, dragging means that once a number of fish have been caught all the holes fill up; the tiniest fish is trapped as surely as the largest. Gill nets, made of almost invisible monofilament, are not dragged, but can save the cost of bait by trapping fish by the head and strangling them; a lost and drifting gill net can strangle fish for up to five years.

Kurlansky's (1997) account of the fishing out of the almost unimaginably abundant cod grounds of the north-west Atlantic, a thousand-year story climaxed by the types of technology highlighted above, shows the immense scale of these encounters which have passed by and continue to pass by largely unconsidered in ethical terms.

When our gaze can penetrate these spaces, what we see is equally startling and disturbing. In one of the many natural history films made by the BBC and fronted by David Attenborough, an underwater camera followed an ocean-going fishing trawler as the catch that had been made was sorted on deck. As the boat proceeded, the camera recorded a steady rain of differing dead creatures – those of no economic value thrown back overboard – as they sank down past the camera to the seabed below. Here, maybe, they would be eaten by sharks, which had been caught, had their fins cut off and then thrown back into the sea, but which survive by squirming across the sea bed. The bi-kill of fishing is recorded at 50 per cent of the total catch, since in the pursuit of some prey, others are indiscriminately slaughtered. Big-game fishing, shark-hunting and fishing more generally are still devoid of any widespread ethical consideration. This, when compared to the concern for *some* land-based human–animal relations, shows just how distant from our ethical vision are these other beings living in this profoundly other form of space. As Smith (1998: 23) proposes, factors of distance and otherness have long been seen as critical in the dissipation of ethical concern, citing

Ginzburg's (1994: 108) use of Aristotle's assertion that 'the nearness of the terrible makes men [*sic*] pity. ... Men also pity those who resemble themselves.'

Another example of the failure of our ethical gaze to penetrate watery spaces is the factory farming of salmon. Apart from a number of environmental management problems which have attracted some attention, there would appear to be significant ethical issues here which apparently go completely unconsidered. Salmon have long been held up as icons of the 'wonder of nature' because of their remarkable life-cycle. Ironically, it is the very spatiality of this, the distances, the navigation feats and the returning to the same river to spawn, which has made it so. This has been reflected in the building of salmon steps to allow them to pass river engineering features. But now millions of salmon are denied this life-cycle, having instead the dimensions of a fish-farm cage as their life territory. The consequences of this inability of salmon to respond to whatever drives their natural cycles no doubt cannot be easily assessed, but, suffering or not, it is surely an issue which needs some sort of ethical assessment. Yet this is apparently lacking, again perhaps due to the very otherness of these beings and the spaces in which their stories unfold.

Conclusions

The suggestion is that there are distinct and hidden geographies of ethical relations between humans and non-humans. This geography of (un)ethical relations involves the spaces and patterns by which we classify and act upon differing groups of animal and other non-human others. Although individuals are generally ethically invisible, the collective to which they are allocated will likely be in a distinct (un)ethical relationship with human society, and thus the individual's ethical position will differ from that of an individual in another group. Understandings of current human–nature relations need to confront this issue. Not only does it show a complex rather than simple terrain upon which we are trying to build, or to move away from, it also demonstrates the deeply inconsistent and hypocritical relations that we currently have with most non-human others.

There is a need to investigate these ignored geographies of the non-human world and the fragmented unethical practices which abound there. It is only, as Lynn (1998a) suggests, when these geographies are opened up that new ethical developments can even be considered. Whatmore (1997: 43) feels that constructing new forms of ethical community will involve 'displacing the fixed and bounded contours of ethical community', recognising instead the specific embodiment and practice of ethical communities. This builds upon the recognition of 'an embodied and practically engaged self ... from what human beings do in the world ... so as to rediscover the totality of [his/her] practical bonds with others' (Kruks, in Whatmore 1997: 43). Thus, there is embodiment and spatiality here which reflects the mistrust of narrow, universal, abstractly conceived and applied objectivistic doctrines. This spatiality may be the ground of neo-Aristotelian ethics (Thrift 1996: 36), which is 'an active and practical form of ethics founded in an evaluative sensibility arising from the concrete experience of specific situations'. Here the ethics of the encounter are foregrounded, rather than lost. Understanding the spatiality of such encounters and the unethical nature of their ethical content at present, the best of them and the most horrible, has to play a part in this process. This returns us to an adapted Levinasian ethics demanding 'some account of how, without universalisation, the encounter with the Other can be at the foundation of a moral society' (Davis 1996: 52).

Acknowledgements

I am very grateful to Sarah Whatmore and Lorraine Thorne for comments on earlier drafts of this chapter, and for the extremely detailed, thoughtful and encouraging work of the two editors which has gone into the later drafts. All remaining inadequacies rest with the author alone.

References

Agar, N. (1995) 'Valuing species and valuing individuals', *Environmental Ethics* 17: 397–415.

Anderson, K. (1995) 'Culture and nature at the Adelaide Zoo: at the frontiers of "human" geography', *Transactions of the Institute of British Geographers* 20: 275–294.

Anderson, K. (1997) 'A walk on the wild side: a critical geography of domestication', *Progress in Human Geography* 21: 463–485.

Arluke, A. and Sanders, C.R. (1996) *Regarding Animals*, Philadelphia: Temple University Press.

Attfield, R. (1983) *The Ethics of Environmental Concern*, Oxford: Blackwell.

Bauman, Z. (1993) *Postmodern Ethics*, Oxford: Blackwell.

Bunce, M. (1994) *The Countryside Ideal: Anglo-American Images of Landscape*, London: Routledge.

Callicott, J.B. (1994) *Earth's Insights: A Survey of Ecological Ethics from the Mediterranean Basin to the Australian Outback*, Berkeley: University of California Press.

Casey, E.S. (1998) *The Fate of Place: A Philosophical History*, Berkeley: University of California Press.

Cheney, J. (1989) 'Postmodern environmental ethics: ethics as bioregional narrative', *Environmental Ethics* 11: 117–134.

Clark, S.R.L. (1977) *The Moral Status of Animals*, Oxford: Clarendon Press.

Critchley, S. (1992) *The Ethics of Deconstruction: Derrida and Levinas*, Oxford: Blackwell.

Cronon, W. (1996) 'The trouble with wilderness, or, getting back to the wrong of nature', in W. Cronon (ed.) *Uncommon Ground: Rethinking the Human Place in Nature*, New York: W.W. Norton and Co.

Davis, C. (1996) *Levinas: An Introduction*, Cambridge: Polity.

Emel, J. (1991) 'Ecological crisis and provocative pragmatism', *Environment and Planning D: Society and Space* 9: 384–390.

Emel, J. (1995) 'Are you man enough, big and bad enough? Ecofeminism and wolf eradication in the USA', *Environment and Planning D: Society and Space* 13: 707–734.

Emel, J. and Wolch, J. (1998) 'Witnessing the animal moment', in J. Wolch and J. Emel (eds) *Animal Geographies: Place, Politics and Identity in the Nature–Culture Borderlands*, London: Verso.

Farber, T. (1994) *On Water*, Hopewell, NJ: Ecco Press.

Fitzsimmons, M. and Goodman, D. (1998) 'Incorporating nature: environmental narratives and the reproduction of food', in B. Braun and N. Castree (eds) *Remaking Reality: Nature at the Millennium*, London: Routledge.

Ford, C. (1998) 'A tail of two pigs', *The Spark* 13: 30.

Ginzburg, C. (1994) 'Killing a Chinese mandarin: the moral implications of distance', *New Left Review* 208: 107–120.

Goodman, R.B. (1995) *Pragmatism: A Contemporary Reader*, London: Routledge.

Gullo, A., Lassiter, U. and Wolch, J. (1998) 'The cougar's tale', in J. Wolch and J. Emel (eds) *Animal Geographies: Place, Politics and Identity in the Nature–Culture Borderlands*, London: Verso.

Ingold, T. (1988) 'Introduction', in T. Ingold (ed.), *What is an Animal?* , London: Unwin Hyman.

Kurlansky, M. (1997) *Cod: A Biography of the Fish that Changed the World*, New York: Walker and Co.

Longhurst, R. (1997) '(Dis)embodied geographies', *Progress in Human Geography* 21: 486–501.

Lynn, W.S. (1998a) 'Animal, ethics and geography', in J. Wolch and J. Emel (eds) *Animal Geographies: Place, Politics and Identity in the Nature–Culture Borderlands*, London: Verso.

Lynn, W.S. (1998b) 'Contested moralities: animals and moral value in the Dear/Symanski debate', *Ethics, Place and Environment* 1: 223–244.

Macnaghten, P. and Urry, J. (1998) *Contested Natures*, London: Sage.

Merchant, C. (1992) *Radical Ecology: The Search for a Liveable World*, London: Routledge.

Midgley, M. (1983) *Animals and Why They Matter*, Athens: Georgia University Press.

Midgley, M. (1996) *Utopias, Dolphins and Computers: Problems of Philosophical Plumbing*, London: Routledge.

Naess, A. (1997) 'Sustainable development and the deep ecology movement', in S. Baker, M. Kousis, D. Richardson and S. Young (eds) *The Politics of Sustainable Development: Theory, Policy and Practice within the European Union*, London: Routledge.

Pepper, D. (1984) *The Roots of Modern Environmentalism*, London: Routledge.

Pepper, D. (1996) *Modern Environmentalism: An Introduction*, London: Routledge.

Philo, C. (1995) 'Animals, geography and the city: notes on inclusions and exclusions', *Environment and Planning D: Society and Space* 13: 655–681.

Pile, S. and Thrift, N. (eds) (1995) *Mapping the Subject: Geographies of Cultural Transformation*, London: Routledge.

Plumwood, V. (1993) *Feminism and the Mastery of Nature*, London: Routledge.

Proctor, J.D. (1998) 'Ethics in geography: giving moral form to the geographical imagination', *Area* 30: 8–18.

Regan, T. (1998) 'The case for animal rights', in S.M. Cahn and P. Markie (eds) *Ethics: History, Theory and Contemporary Issues*, Oxford: Oxford University Press.

Rolston III, H. (1999) 'Ethics on the home planet', in A. Weston. (ed.) *An Invitation to Environmental Philosophy*, New York: Oxford University Press.

Serpell, J. (1995) 'A consideration of policy implications: a panel discussion', *Social Research: In the Company of Animals*, 62: 821–826.

Serpell, J. (1996) *In the Company of Animals: A Study of Human–Animal Relations*, Cambridge: Cambridge University Press.

Sheail, J. (1976) *Nature in Trust: The History of Conservation in Britain*, Glasgow: Blackie.

Shiva, V. (1988) *Staying Alive: Women, Ecology and Development*, London: Zed Books.

Sibley, D. (1995) *Geographies of Exclusion: Society and Difference in the West*, London: Routledge.

Singer, P. (1993) *Practical Ethics* (Second Edition), Cambridge: Cambridge University Press.

Smith, D. M. (1998) 'Geography and moral philosophy: some common ground', *Ethics, Place and Environment* 1: 7–34.

Taylor, K. (1990) 'One for sorrow, two for joy', *BBC Wildlife* 8(2): 80–88.

Thorne, L. (1998) 'Kangaroos: the non-issue', *Society and Animals* 6: 167–182.

Thrift, N. (1996) *Spatial Formations*, London: Sage.

Thrift, N. (1999) 'Steps to an ecology of place', in D. Massey, P. Sarre and J. Allen (eds) *Human Geography Today*, Oxford: Polity.

Visser, M. (1998) 'Hooked: review of Kurlansky's *Cod* (1997)', *London Review of Books* 20(8): 22–23.

Weston, A. (ed.) (1999) *An Invitation to Environmental Philosophy*, New York: Oxford University Press.

Whatmore, S. (1997) 'Dissecting the autonomous self: hybrid cartographies for a relational ethics', *Environment and Planning D: Society and Space* 15: 37–53.

Whatmore, S. and Thorne, L. (1998) 'Wild(er)ness: reconfiguring the geographies of wildlife', *Transactions of the Institute of British Geographers* 23: 435–454.

Wheeler, K.M. (1993) *Romanticism, Pragmatism and Deconstruction*, Oxford: Blackwell.

Whitford, M. (1991) *Luce Irigaray: Philosophy in the Feminine*, London: Routledge.

Wolch, J. and Emel, J. (1995a) 'Bringing the animals back in', *Environment and Planning D: Society and Space* 13: 632–636.

Wolch, J. and Emel, J. (eds) (1995b) Guest edited issue: 'Bringing the animals back in', *Environment and Planning D: Society and Space* 13: 631–730.

Wolch, J. and Emel, J (eds) (1998) *Animal Geographies: Place, Politics and Identity in the Nature–Culture Borderlands*, London: Verso.

Wright, T., Hughes, P. and Ainley, A. (1988) 'The paradox of morality: an interview with Emmanuel Levinas', in R. Bernasconi and D. Wood (eds) *The Provocation of Levinas: Rethinking the Other*, London: Routledge.

14 Afterword

Enclosure

Michael J. Watts

The animal scrutinises [Man] across a narrow abyss of non-comprehension. ... The man too is looking across a similar, but not identical, abyss of non-comprehension. And this is so wherever he looks. He is always looking across ignorance and fear.

(Berger 1980: 24)

In the zoo, as Berger (1980: 24) famously put it, animals constitute a living monument to their own demise: 'everywhere', he says, 'animals disappear'. In posing the question, 'why look at animals?', Berger sought to show that there is a central paradox residing in our experience of the zoo and its inhabitants. It is a source of popular appeal and considerable attraction – 'millions visit the zoos each year out of a curiosity which was both so large, so vague and so personal that it is hard to express' – and yet in the zoo 'the view is always wrong ... like an image out of focus' (Berger 1980: 20, 21). Inevitably zoos disappoint. The excitement of the wild is replaced by alienation, lethargy, isolation, incarceration and boredom. The spark which has historically linked human and animal has been extinguished. At the heart of the zoo's paradoxical status is a sort of twin alienation. On the one hand, the zoo is a prison – a space of confinement and a site of enforced marginalisation like the penitentiary or the concentration camp. And on the other, it cannot subvert the awful reality that the animals, from whatever vantage-point they are viewed, are 'rendered absolutely marginal' (Berger 1980: 22). The inescapable fact is that the zoo recapitulates the relations between humans and animals, between nature and modernity. It demonstrates, as Berger says, a basic ecological fact of loss and exclusion – the disappearance and extinction of animals – through an act of incarceration.

Berger was at pains to connect the zoo as a monument to loss, as a space of confinement, to a human crisis, specifically to the disposal, or perhaps more appropriately the enclosure, of the peasantry. In quick succession, the emergence of the zoo – that is to say, the marginalisation of the animal – is followed by the marginalisation of a class – the peasant – for whom familiarity with, and a wise understanding of, animals is a distinguishing trait. The second half of the twentieth century, as Hobsbawm (1994: 289) noted in his magisterial book *The Age of Extremes*, contained 'the most dramatic and far-reaching social change ... which cut us off for ever from the world of the past, [namely] the death of the peasantry'. If the world-renowned San Diego Zoo memorialises wildlife extinction and animal loss, perhaps the monument to the death of the peasantry is the folk museum replete with its artefacts and arcana of rural life. Both historical losses are, in Berger's vision, 'irredeemable for the culture of capitalism' (Berger 1980: 26).

I have begun with Berger because he offers a productive way of thinking about the geography of animals, and the panoply of issues – farm animals, electronic zoos, hunting debates, feral cats, zoos and colonialism – raised by the various contributors to this volume. To put the matter starkly, one might say that the relation between animals and modernity can be construed as a gigantic act of *enclosure* – necessitating, of course, loss and displacement – which contains a double-movement. The first, of which the zoo is the exemplary modern instance, is to accomplish what Greenblatt (1991: 3) in *Marvelous Possessions* calls 'the assimilation of the other'. By putting Marx to the service of the assimilation process, Greenblatt (1991: 6) suggests that the zoo, for example, is part of the 'reproduction and circulation of mimetic capital'. A zoo is what he calls a cultural storehouse, part of the proliferation and circulation of representations which become, for Greenblatt (1991: 6), 'a set of images and image-making devices that are accumulated, "banked" as it were, in books, archives, collections ... until such time as new representations are called upon to generate new representations'. This mimetic quality of capitalism suggests that the representations – of animals, of ecosystems, of nature – are social relations of production. The representation both is the product of the social relations of capitalism and is a social relation itself 'linked to the group understandings, status hierarchies, resistances and conflicts that exist in other spheres of the culture in which it circulates' (Greenblatt 1991: 6). The zoo is, then, product and producer (Malamud 1995: 12). The zoo places animals in captivity ('preservation', 'salvage ecology') as a response to – the product of – the devastating ecological consequences of modernisation. Equally, the zoo culture also serves as an institutionalised means, a scientific means no less, to represent animals in quite specific ways and generate culturally mimetic portrayals of itself.

The zoo 'matters' to the extent that its images and representations of animals achieve a reproductive power – that is to say, by speaking to our humanity, or the philanthropy of capitalists, or enhancing environmental sustainability, or perpetrating the distinctions between the wild chaos of nature and the order of a rational capitalist world. The fact that zoos disappoint, to return to Berger, is simply to assert that the hegemony of representations is never complete or fully secured, and that the zoo cannot possibly cover its tracks. The zoo is unequivocally about loss and captivity, and the very antithesis of the fecundity and freedom which nature purportedly signifies. Not simply a product, the zoo is an unwitting producer 'capable of decisively altering the very forces that brought it into being' (Greenblatt 1991: 6). In this sense, then, the zoo is a sort of metaphor for thinking about animals and capitalism; it offers an experience of nature 'that presents itself as mimetic of a larger animal macrocosm' (Malamud 1995: 12) within the great cosmos of capitalist commodities. Many of the chapters in this volume approach the question of nature and of animals as marvellous possessions, as part of the complexities of the reproduction and circulation of mimetic capital.

There is, however, a second vectoring in the double-movement of animals and modernity. Berger gestures to it in his invocation of the historic demise of the peasantry. Properly speaking, the historical reference-point here is the agrarian question of the mid- and late nineteenth century in Europe (Kautsky 1988). What, said Kautsky, in his classic book *The Agrarian Question*, is happening to the peasantry? In his still relevant account of the German agricultural sector in the last quarter of the nineteenth century, the agrarian question encompassed accumulation (how were surpluses extracted from the peasant-dominated agricultural sector?), science (how were the forces of production being revolutionised?) and politics (what were the political implications for the German Social Democratic Workers' Party of the disintegration and differentiation of the peasantry?). Of course, Kautsky was at pains to point out that the 'disappearance' of the peasantry was a more complex and uneven process than the orthodox Marxist predictions might suggest. Nevertheless, his book demonstrated clearly that Berger's concern with the historic loss of the peasantry was, at the same time, the process by which land-based and livestock activities were *industrialised*:

> The transformation of agricultural production into industrial production is still in its infancy. [But] bold prophets, namely those chemists gifted with an imagination, already are dreaming of the day when bread will be made from stones and when all requirements of the human diet will be assembled in chemical factories. ... But one thing is certain. Agricultural production has already been transformed into industrial production in a large number of fields. ... This does not mean that ... one can reasonably speak of the demise of agriculture ... but [it] is now caught up in the constant revolution which is the hallmark of the capitalist mode of production.
>
> (Kautsky 1988: 297)

In the current parlance, peasant agriculture was subject to the twin processes of 'appropriation' and 'substitution' (Goodman *et al.* 1987). The former spoke to the means by which more of production (including input provision) was provided by off-farm industry, while the latter identified the increasing trend to produce food and fibres in factories as a fully industrial process comparable to non-agricultural forms of manufacture. On one side stood drip irrigation and genetically modified crops, and on the other the feedlot or the industrial manufacture of artificial sugars.

To return to Berger again, the demise of the peasantry (which marks the disappearance of animals) is very much about the industrial and scientific assault on animals within the food provisioning system. What Kautsky saw in its infancy was the process by which science was harnessed to the problem of the industrialisation of livestock and crop production. Genetics lay the groundwork for the direct application of new technology – and the drive for mechanisation – to the plant and the animal. To quote Geidion's (1948: 6) magisterial book *Mechanization Takes Command*, 'what happens when mechanization encounters organic substance?' He properly addressed the question of the mechanisation of agriculture from the eighteenth century, and the means by which bread and meat were mechanised, displacing the artisan crafts of butchery and bread-making, and of course transforming popular taste. Even in 1948 when the book was published, Geidion (1948: 7) could observe: 'Still of unmeasured significance is mechanization's intervention in the procreation of plants and animals.' Geidion was of course charting the process by which the high-yielding corn variety or the milk cow laced with BST had been converted into *sites of accumulation*. In so doing the animals and the plants had been, as it were, transformed from *within*. The zoo represents the transformation from *without*: the extinction of species which attended the ecological devastation wrought by industrialisation. One might say, then, that the alienated, lethargic elephant in Berger's description of the zoo and the genetically modified sheep are both monuments to the historic loss which he invokes in contemporary capitalism. One memorialises the ecological (biodiversity) crisis of modernity; the other the scientific and industrial reconstitution of nature (Haraway 1991). I would simply wish to make the point that both elements of this double-movement can be understood as forms of enclosure – and by definition confinement, incarceration, discipline and subjection. Both necessitate some sort of assimilation of the other, but unlike the zoo, in which the reproduction and circulation of mimetic capital predominates, the industrial milk cow or battery hen *is* capital, although livestock too in the name of all commodity production has, to return to Greenblatt, its own stock of images, representations and fetishistic qualities.

It is to this second aspect – the industrialisation and subjection of animals and livestock in food provisioning systems – that I wish to turn briefly. Running through my remarks, however, is an overarching concern with violence and suffering which links the zoo to the feedlot. As Malamud (1995: 179) says, the zoo is a locus of pain. Its cruelty reveals itself in the obvious torment and alienation experienced by the animals, but also in more subtle, invisible ways. A colleague at

Stanford explained to me not long ago that one of her students – a newly minted PhD in neuro-biology – is in great demand by zoos as a consultant with the specific purpose of determining the intensity and duration of petting that sheep and goats can withstand in child 'petting zoos' before permanent cranial damage is inflicted. But this is no less the case for industrial agriculture, and it is hardly surprising that the key works on animal rights – one thinks of Singer's *Animal Liberation* (1975) – often start with accounts of the suffering endured by battery hens or pigs destined for the meat-processing plant.

I want to speak about the lowly and humble chicken. According to the 1995 *Poultry Yearbook* published by the United States Department of Agriculture, 7 billion chickens were sold in the United States in 1994 (roughly thirty per person). In 1991, chicken consumption per capita exceeded beef, for the first time, in a country which historically has had a veritable obsession with red meat. The fact that each American man, woman and child currently consumes roughly 1.5 pounds of chicken each week reflects a complex vectoring of social forces in post-war America. First, there has been a change in taste driven by a heightened sensitivity to health matters and especially the heart-related illnesses associated with red meat consumption. Second, there is the fantastically low cost of chicken meat, which has in real terms *fallen* since the 1930s (a century ago Americans would eat steak and lobster when they could not afford chicken). And not least there is the growing extent to which chicken is consumed in a panoply of forms (Chicken McNuggets, say) which did not exist twenty years ago and which are now delivered to us by the massive fast-food industry – a fact which itself points to the reality that Americans eat more and more food outside of the home. Food consumption 'away from home' is, by dollar value, 40 per cent of the average household food budget.

Broilers are overwhelmingly produced by family farmers in the US, but this turns out to be a deceptive statistic. They are grown by farmers under contract to enormous transnational enterprises – referred to as 'integrators' in the chicken business – who provide the chicks and feed. The growers, who are not organised into unions and who have almost no bargaining power, must borrow heavily in order to build the broiler houses and the infrastructure necessary to meet contractual requirements. Growers are not independent farmers at all. They are little more than underpaid workers – what we might call 'propertied labourers' – of the corporate producers who also dominate the processing industry. Working in the poultry-processing industry, in which the broilers are slaughtered and dressed and packaged into literally hundreds of different products, is one of the most underpaid and dangerous jobs in the country (a new US government report published in 1998 – *New York Times*, 9 February 1998: 12 – cites almost two-thirds of all poultry-processing plants as in violation of overtime payment procedures!). Immigrant labour – Vietnamese, Laotian, Hispanic – now represents a substantial proportion of workers (almost wholly non-union) in the industry. The largest ten companies account for almost two-thirds of broiler production in the US. Tyson Foods, Inc., the largest broiler producer, accounts for 124 million pounds of chicken meat per week, and it controls 21 per cent of the US market with sales of over $5 billion, two-thirds of which go to the fast-food industry. According to Don Tyson, the Chief Executive Officer of Tyson's, his aim is to 'control the centre of the plate for the American people'. The vast majority of chickens sold and consumed are broilers (young chickens), which, it turns out, are rather extraordinary creatures. In the 1880s there were only 100 million chickens. In spite of the rise of commercial hatcheries early in the twentieth century, the industry remained a sideline business run by farmers' wives until the 1920s. Since the first commercial sales (by a Mrs Wilmer Steele in 1923 in Maryland, who sold 357 in one batch at prices five times higher than today), the industry has been transformed both by the feed companies, which began to promote integration and the careful genetic control and reproduction of bird flocks, and by the impact of big science, often with government support. The heart of the US chicken industry is in the ex-slave-holding and cotton-growing South. Until the Second World War, the chicken industry was located primarily in the Delmarva peninsula in the mid-Atlantic states (near Washington, DC). During the 1940s and 1950s, the industry moved south – restructuring as it did so – from which emerged the large integrated broiler complexes, what in current parlance is now called flexible or post-Fordist capitalist organisation. The largest producing region is Arkansas – the home state of President Clinton – and the chicken industry has been heavily involved in presidential political finance and lobbying, including a recent case in which the Secretary of Agriculture, Dan Espy, was compelled to resign in 1995. The lowly chicken, in other words, reaches into the White House.

The post-war history of the national and global chicken or 'broiler' industry is one of mindboggling dynamism and growth. Advances in breeding, disease control, nutrition, housing and processing have conferred upon US integrators some of the lowest production costs in the world. Since 1940, the industry's feed conversion rate has declined precipitously from three pounds of feed per pound of liveweight to under two pounds. Over the same period the average broiler liveweight has increased from 2.89 lb to 4.63 lb, and the maturation period – the time required for a bird to reach market weight – has plummeted from over seventy days to less than fifty. The result is what was called in the 1940s the 'perfect broiler'. Avian science has now permitted amazing rates at which the birds add weight (almost five pounds in as many weeks!). The average live bird-weight has almost *doubled* in the last fifty years; over the same period the labour input in broiler production has fallen by 80 per cent! The broiler is the product of a truly massive R&D campaign; disease control and regulation of physiological development have fully industrialised the broiler to the point where it is really a cyborg: part-nature, part-machine (Boyd forthcoming). Our understanding of chicken nutrition now exceeds that of any other animal, including humans. High technology and industrial production method has also been the key to the egg industry. A state of the art hen-house holds 100,000 birds in minuscule cages stretching the length of two football fields; it resembles a twenty-first-century high-technology torture chamber. The birds are fed by robot in carefully controlled amounts every two hours around the clock. In order to reduce stress, anxiety and aggression (which increases markedly with confinement and increased egg-laying), the birds are housed in low-intensity light or wear red contact lenses, which, for reasons that are not clear, reduces feed consumption and increases egg production!

All of this has been driven by a small number of large integrators who dominate the chicken commodity system, what is technically referred to as the chicken *filière*. Tyson Foods – the largest broiler integrator – currently accounts for almost a quarter of the US ready-to-cook market, and is representative of a class of global corporations who supply some of the most aggressively transnational companies, namely the fast-food industry, who are typically the leading edge of capitalist expansion as markets open up under the pressure of neo-liberal reforms. The US is the largest producer and exporter of broilers, with a sizeable market share in Hong Kong, Russia and Japan. Facing intense competition from Brazil, China and Thailand, the global chicken industry is driven by the lure of the massive Chinese market, and by the newly emerging and unprotected markets of Eastern Europe and the post-Soviet states. As any consumer might surmise, the world chicken market is highly segmented: Americans prefer breast meat (and most production of such is for the home market), while US exporters take advantage of foreign preference for leg quarters, feet and wings to fulfil the large demand from Asia. The chicken is a thoroughly global creature – in its own way not unlike the global car or global finance (Dixon and Burgess 1997).

I wish to focus here on the role of big science and the productivity revolution (see Boyd forthcoming; Boyd and Watts 1996). Two forces were central to the radical increases in biological and labour productivity, and in the *reconstitution* of the broiler itself. First, the move to year-round confinement, made possible by the discovery of Vitamin D in 1926, facilitated the shift to industrial broiler production. Second, the federal government stepped in during the Depression years to lay the foundations for the post-war productivity increases and the application of what one might call 'big science' to the industry. Although chicken-breeding experiments had been established by the US Department of Agriculture as early as 1912, it was not until 1933, with the establishment of the National Poultry Improvement Plan (NPIP), that the government emerged as a major force in facilitating the development of new techniques in disease control, breeding and husbandry. Gradually, mortality rates were reduced – approximately

30 per cent by the end of the 1930s – which permitted the move to more intensive confinement operations.

What proved to be decisive during the war and immediate post-war years,
however, was the revolution in primary breeding (Boyd forthcoming; Bugos
1992). Until the 1940s, farmers had relied on pure breeds that had been developed for egg production, with little concern for meat qualities. With increased
demand for chicken meat during the war, breeders began to focus their attention
on the development of specialised breeds for meat production. In 1944, the
search for the broad-breasted chicken began in earnest as the A&P food chain
launched a series of national breeding contests. Striking illustrations of early
retailer power in product design, the two 'Chicken of Tomorrow Contests' in
1946 and 1951, which greatly accelerated the development of the modern broiler
industry by establishing cross-breeds or hybrids as the standard throughout the
industry, ushered in the age of the 'designer chicken'. Employing principles
pioneered in the hybridisation of corn and other crops, primary breeders developed standard pedigrees of male and female lines that combined to produce a
superior bird. By the 1950s, such cross-breeds provided the genetic basis of the
modern broiler industry, and breeders focused on fine-tuning their pedigrees to
meet the ever more exacting demands for genetic uniformity and quality assur-
ance. On the eve of integration in the broiler industry, moreover, the 'biological
lock' of hybridisation strengthened the boundary between the primary breeders
and the rest of the industry – a boundary that has persisted to this day.
Complementing these dramatic advances in breeding were substantial public
investments in nutrition, disease control and confinement technologies during the
immediate post-war period, much of which was sponsored by the land grant
universities and the federal government. The development of high-performance
rations combined with the use of Vitamin B-12 and antibiotics in feed dramatically
reduced mortality and increased feed conversion efficiency. At the same time, the
completion of rural electrification facilitated substantial improvements in environmental control and labour productivity within confinement operations.
Whereas in 1940, an average 250 'man-hours' were required to raise 1000 birds
to maturity, by 1955 the required time had dropped to 48 hours.

Boyd's (fothcoming) path-breaking research has shown that in terms of biological productivity, mechanisation and industrialisation, the chicken has no peer in
contemporary food provisioning. The prosaic chicken has been made over in such
a way that it can only be thought of as bionic. In 1935 the average weight of the
broiler was 2.8 lb, it took 112 days to reach market weight and had a feed conversion ratio (lb feed/lb broiler) of 4.4; fifty years later the figures are respectively
4.6 lb, 43 days and 1.9. It is the leading edge of just-in-time flexible production,
and a foundational example of the design of nature to meet the industrial labour
process. The chicken, moreover, is quickly becoming *the* global meat. It was the

combination of confinement, nutrition and growth research, and genetic improvement that produced by the 1950s a chicken which was '*made to order to meet the needs of the meat industry*' (Warren 1958: 16, emphasis added). By the 1980s the new biotechnologies entered the broiler industry, specifically the life sciences companies intent on genetic engineering to develop breeds keyed to the sale of proprietary health products. Furthermore, with the famous Chakrabarty ruling in 1980, substantial interest was aroused in the possibility of transgenic chickens subject to patent protection. The possibility of a full genetic map of the chicken – assisted by the fact that DNA can be isolated from nucleated red bloodcells, unlike in most mammalian species – is well underway, which would 'allow selective improvement to proceed on the basis of genotype rather than phenotype, representing a very significant expansion in "breeding power"' (Boyd forthcoming).

The chicken-genome project, then, marks a twenty-first-century act of Frankensteinian enclosure. First, *confinement* marks both the shift from open range to broiler houses, but also a process of industrial integration within a highly oppressive broiler complex, and which has as its counterpoint the enclosure of the chicken contractor (Boyd and Watts 1996). Second, the *designed animal*, the 'designer chicken', establishes the extent to which nutritional and genetic sciences have produced a 'man-made' broiler to fit the needs of the industry. It is, then, an archetypical (and, as we shall see, in some senses a limiting case) cyborg (Haraway 1991). There is something grotesque about the creation of a creature which is a turbo-charged growth machine, steroidally enhanced, producing in unprecedentedly short periods of time enormous quantities of flesh and muscle around a sort of Frankenstein-like skeleton. Third, the chicken is transformed into a *site of accumulation*; this is reflected in its curious physiognomy and anatomy ('all breast and no wings'). Harvey's (1998) important account of the body as an accumulation strategy focuses on three moments: productive consumption of the working body; exchange; and individual consumption. What is striking about the chicken – indeed the meat industry more generally – is the extent to which the 'working body' has not simply been 'Taylorised' in some way, but actually constructed physically to meet the needs of the industrial labour process. The poultry industry in this sense combines the worst of productive consumption of the human body (the appalling working conditions and health deficits associated with working on the line) with the most horrifying forms of reconstituted nature. Nineteenth-century work conditions meet up with twenty-first-century science.

Harvey (1998) spends less time on the dialectics of nature itself, however, namely the fact that the effort to convert the body into a site of accumulation runs up against the ethical, social and biological limits of the commodification of the biological realm (see Polanyi 1944). This is nowhere clearer than in the broiler industry. It is perfectly obvious that the chicken industry is on the verge of a major biological, and specifically health, crisis. On the one hand, the dangers of disease – the Hong Kong-derived avian flu in 1998 is a case in point – are enormous, driven in part by the susceptibility of fragile and vulnerable chickens to new pathogens which are able to destroy flocks very quickly. One part of this crisis is driven by resistance to antibiotics – the R-plasmid problem – and another part by emergent diseases. Furthermore, efforts to increase breast-meat yields have created a high propensity for musculoskeletal problems, metabolic disease, immuno-deficiency and male infertility (Boyd forthcoming) – which is to say, high bird mortalities. At the social level, the health problems of the industry are legion. In the US, for example, where the salmonella problem and chicken are virtually synonymous, the human health effects from resistant bacteria of animal origin, and from antibiotics used as feed additives, has remained largely unregulated. In June of 1999 (*Le Monde*, 13 June, 1999: 210), frightening news of dioxin-contaminated chickens in Belgium exposed once again the trade-off between cost-driven growth promotion and long-term vulnerability associated with increased antibiotic-resistant pathogens and feeding additives. The popular concern with labelling and moving back to free-range chickens sharpens the contradictions of the industrial and productivist model of agriculture. And finally there are the *ethical questions* raised by confinement, battery-hens and by the suffering imposed by mass production.[1] Singer, who once sat in a cage in the middle of Melbourne to publicise the plight of battery hens, specifically argues that the 'simple' chicken deserves to be protected from unnecessary pain (see *NewYorker*, 6 September 1999). Animal welfare questions are, in this regard, no longer local or parochial concerns; indeed, they have reached the bargaining table of the World Trade Organisation. For example, the European Union has been concerned, on the one hand, to pressure trade partners to meet its own animal welfare standards, and, on the other, to limit animal rights considerations against the 'competitive position of EU producers in a market liberalised under WTO agreements' (Fisher 1999: 13). On 15 June 1999, in a significant breakthrough, the EU Agriculture Council finally decided to prohibit the use of battery cages by 2012.

If my prosaic chicken-story reveals anything, it is that the monuments to animals which Berger invokes are on shaky foundations, and in their different ways cracked and pitted with the contradictions of the modernist impulse to commoditise and to industrialise nature. One response to the crisis of nature modernised has, of course, been articulated in terms of animal rights and social justice, and it is to this matter that I shall briefly turn in my conclusion (see Coetzee 1999; Wolch and Emel 1998). In the United States, at least, it is clear that the animal rights movement has considerable popular and academic credibility and legitimacy. To cite two cases, courses on Animal Liberation are offered at the Harvard Law School, and the *New York Times* recently ran a leader on the Great Ape Legal Project (part of the Animal Legal Defense Fund), which is now widely understood to have compelling grounds for bringing a suit to trial that is 'the inevitable culmination of challenges [which the legal profession is] …

starting to make to the principle that animals are property without legal rights' (*NewYork Times*, 22 August 1999: 2). Typically the philosophies of animal rights are expansions of largely individualist, human-centred moral discourses and accordingly put pressures on questions of moral status, duties, rights and the like (see Benton 1993). As Benton (1993: 3) says, such positions focus on 'how individual deliberating moral agents should treat the individuals of other species which they encounter'. Indeed, some of the most compelling and articulate of animal rights philosophers see any preference of the ecosystem (i.e. the community) over the individual animal as evidence of fascist leanings (Regan and Singer 1976).

This is not the place to evaluate the robustness of the animal rights position as such – although I am convinced that the critiques of Benton and others are compelling – but instead I shall take seriously the claim of fascism, and push it in a rather different direction. Quite specifically, I want to dwell on the oft-noted fact that some of the strongest articulations of zoophilia were associated, as has often been noted, with Nazi ecology. Indeed, new work by Proctor (1999a, 1999b) extends the argument further, showing how Nazi research on environmental cancer, on the use of tobacco, on organic approaches to agriculture, and on many aspects of preventative health were both original and 'vigorous'. The point here is not of course to suggest that animal rights, ecological thinking and the anti-smoking movement are *ipso facto* fascist; or conversely, as Proctor notes, that one can rescue honour from the horrors of the Nazi era. But in reading, for example, Ferry's (1995) account of Nazi ecology, what is striking is the depth of commitment to nature's rights. The monumental law of 1933 (*Tierschutzgesetz*) which secured this protection expressed both a radical critique of anthropocentrism and a call for the rights of nature, rights which existed *sui generis* not with respect to people (Corni 1990; Ferry 1995: 99). Cruelty against wild or domesticated animals warranted severe punishment because cruelty was not associated with the notion that human sensibility must be protected. As Ferry points out, what runs through the laws is an authentic recognition of difference. Rights in animals and in nature defend the idea of an original human state distinguished by the unity of local nature and culture. The National Socialist conception of ecology (and its attendant rights) encompasses 'the notion that the *Naturvolker*, the "natural peoples", achieve a perfect harmony between their surroundings and their customs' (Ferry 1995: 105). Unlike the rootlessness of liberalism, invocation of nature's rights assures that culture prolongs nature and is co-eval with an animal way of life. Rights discourse in this way is put to the service of reductionism, of a primitive state in which animals and human coexist, and which is the basis of cultural (i.e. human) difference (versus, for example, assimilation). Of course this can and must be read as a sort of Nazification of science, but it is a complex process in which, to cite Proctor (1999a: 23), 'we don't want to forget Mengele's crimes but we should also not forget that Dachau prisoners produced organic honey'.

My point in raising these issues is twofold, each of which speaks to the complexity of invoking both science and liberal rights in any account of animals and modernity. Like Proctor (1999a), I do not think that there is a necessary totalitarian tendency within modern science, but rather that routine science is no more incompatible with violence and cruelty than is fascism with the promotion of public health. And second, in invoking animal rights, the fascist experience draws attention to the process by which a rights discourse can be rendered reductive and essentialist, and in so doing open up the possibility for a unity of nature and culture to find a true essence – an original people or nation. Rights, in this expression, can become part of the mythos, the mythic core, as Griffin (1991: 27) calls it, of fascism. It is here that a strong individualist account of animal rights carries its own dangers. To conflate human and non-human animal rights as identical is to forfeit the advantages of what Benton (1993: 210) calls a 'naturalistic (but non-reductionistic) view of human nature and the human/animal continuism that goes with it'. As Benton (1993: 210) himself notes, three features of moral rights (if the liberal-individualist model of animal rights is adopted) travel badly across the species boundary: reciprocity in mutual recognition of rights; the links between natural rights and citizenship; and rights as autonomously made moral claims. To suggest otherwise, it seems to me, is to come close to the sort of organicism that is the hallmark of various sorts of political fundamentalism.

Acknowledgements

This chapter draws upon some joint work on the poultry industry conducted with William Boyd of the Energy and Resources Group at the University of California, Berkeley. His forthcoming dissertation on the US poultry industry, and his forthcoming work, represents a major and original analysis of the industry and its socio-biological dynamics.

Note

1 See, for example, the activities of United Poultry Concerns and the work of Davis (1995).

References

Benton, T. (1993) *Natural Relations: Ecology, Animal Rights and Social Justice*, London: Verso.

Berger, J. (1980) 'Why look at animals?', in *About Looking*, New York: Pantheon.

Boyd, W. (forthcoming) 'The real subsumption of nature', *Technology and Culture*.

Boyd, W. and Watts, M. (1996) 'Agro-industrial just-in-time', in D. Goodman and M. Watts (eds) *Globalizing Food*, London: Routledge.

Bugos, G. (1992) 'Intellectual property protection in the American chicken-breeding industry', *Business History Review* 66: 127–168.

Coetzee, J. (1999) *The Lives of Animals*, Princeton: Princeton University Press.

Corni, G. (1990) *Hitler and the Peasants*, New York: Berg.

Davis, K. (1995) 'Thinking like a chicken', in C. Adams and J. Donovan (eds) *Animals and Women*, Durham, NC: Duke University Press.

Dixon, J. and Burgess, J. (1997) *When Local Elites Meet the WTO: Chicken as the Meat in the Sandwich*, Research Paper 3238, The University of Newcastle, Australia.

Ferry, L. (1995) *The New Ecological Order*, Chicago: University of Chicago Press.

Fisher, C. (1999) 'Animal welfare likely to be on the WTO menu', *Bridges* 3(6): 13.

Giedion, S. (1948) *Mechanization Takes Command*, New York: Oxford University Press.

Goodman, D., Sorj, B. and Wilkinson, J. (1987) *From Farming to Biotechnology*, Oxford: Blackwell.

Greenblatt, S. (1991) *Marvelous Possessions*, Chicago: University of Chicago Press.

Griffin, R. (1991) *The Nature of Fascism*, London: Routledge.

Haraway, D. (1991) *Simians, Cyborgs and Women*, London: Routledge.

Harvey, D. (1998) 'The body as an accumulation strategy', *Environment and Planning D: Society and Space* 16: 401–421.

Hobsbawm, E. (1994) *The Age of Extremes*, New York: Pantheon

Kautsky, K. (1988) *The Agrarian Question* , London: Unwin Hyman.

Malamud, R. (1995) *Reading Zoos*, New York: New York University Press.

Polanyi, K. (1944) *The Great Transformation*, Boston: Beacon.

Proctor, R. (1999a) 'Why did the Nazis have the world's most aggressive anti-cancer policy?', Paper presented to the Environmental Politics Seminar, University of California, Berkeley. 12 November.

Proctor, R. (1999b) *The Nazi War on Cancer*, Princeton: Princeton University Press.

Regan, T. and Singer, P. (eds) (1976) *Animal Rights and Human Obligations*, Englewood Cliffs, NJ: Prentice-Hall.

Singer, P. (1975) *Animal Liberation*, New York: Pantheon.

Warren, D. (1958) 'A half-century of advances in the genetics and breeding improvement of poultry', *Poultry Science* 37(1): 4–20.

Wolch, J. and Emel, J. (eds) (1998) *Animal Geographies: Place, Politics and Identity in the Nature–Culture Borderlands*, London: Verso.

Index

Note: Page numbers in **bold** type refer to figures, page numbers in *italic* type refer to tables and page numbers followed by 'n' refer to notes.